BRE Building Elements series

ROOFS AND ROOFING

Performance, diagnosis, maintenance, repair
and the avoidance of defects

Third edition

WITHDRAWN

H W Harrison, P M Trotman and G K Saunders

 bre press

ii

BRE is the UK's leading centre of expertise on the built environment, construction, energy use in buildings, fire prevention and control, and risk management. BRE is a part of the BRE Group, a world leading research, consultancy, training, testing and certification organisation, delivering sustainability and innovation across the built environment and beyond. The BRE Group is wholly owned by the BRE Trust, a registered charity aiming to advance knowledge, innovation and communication in all matters concerning the built environment for the benefit of all. All BRE Group profits are passed to the BRE Trust to promote its charitable objectives.

BRE is committed to providing impartial and authoritative information on all aspects of the built environment. We make every effort to ensure the accuracy and quality of information and guidance when it is published. However, we can take no responsibility for the subsequent use of this information, nor for any errors or omissions it may contain.

BRE, Garston, Watford WD25 9XX
Tel: 01923 664000
enquiries@bre.co.uk
www.bre.co.uk

BRE publications are available from
www.brebookshop.com
or
IHS BRE Press
Willoughby Road
Bracknell RG12 8FB
Tel: 01344 328038
Fax: 01344 328005
Email: brepress@ihs.com

Requests to copy any part of this publication should be made to the publisher:
IHS BRE Press
Garston, Watford WD25 9XX
Tel: 01923 664761
Email: brepress@ihs.com

Front cover photographs:
Left: Lighthouse™, Kingspan Offsite,
               on the BRE Innovation Park
Top right: St Pancras Station, London
Middle right: Slate mansard roof on housing
Bottom right: T5, Heathrow Airport, under construction
               Courtesy of BAA

BR 504
© Copyright BRE 1996, 2000, 2009
First published 1996
Second edition published 2000
Third edition published 2009
ISBN 978-1-84806-092-0

CONTENTS

PREFACE TO THE FIRST EDITION

This book is about roofs and roofing: the materials and products, methods and criteria which are used in the construction of these elements of buildings. A more specific purpose is to draw the attention of readers to features of roofs and roofing which ensure good performance, and the corollary to this, the things that can and do go wrong so that they can be avoided. It is in no sense a text book of construction practice; simply drawing from BRE experience to provide the building professional with an illustrated checklist of points of concern with roofing, and where to obtain further information and advice.

Faults which formerly occurred on a substantial scale (for example those relating to bracing and binding of trussed rafter roofs) are still occurring frequently. Over the years this has justified BRE continually bringing such faults to the notice of designers and contractors by means of leaflets. If some of the errors seem to be elementary, this is simply a reflection of what occurs in practice. This book can therefore be seen as a continuation of the practice of holding a mirror up to the industry so that its members can recognise what they are doing, or what they are not doing, and take corrective action before faults or defects occur rather than after they have occurred.

There are many individuals and firms with whom communication needs to be established. Since 1980, there has been a considerable growth in the number of specialist roofing firms: an estimated 7,000 contractors in 1988. In that year the value of work undertaken by the roofing industry was just over £262 million of which over half was on maintenance and repair work.

Readership

This publication is addressed primarily to building surveyors and other professionals performing similar functions, such as architects and builders, who maintain, repair, extend and renew the national building stock. It will also find application in the education field, where lecturers can find indication of where the construction syllabus needs attention.

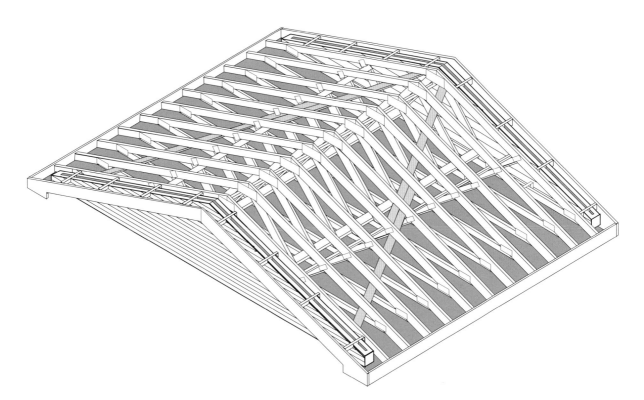

Bracing in a domestic trussed rafter roof. It is this bracing that helps to convert a collection of trused rafters into a stable structure

There is no shortage of guidance on roofing, as BRE has consistently said for many years. The problem is that people do not use the guidance that exists. In order to try to remedy that situation, the advice given in the later chapters of this book concentrates on practical details. However, there also needs to be sufficient discussion of principles to impart understanding of the reason for certain practices. Without that understanding, practitioners have not been sufficiently encouraged to follow correct procedures until they too make the same mistakes, or overlook the same precautions, as previous generations have done.

There is sufficient discussion of the underlying structure to enable an understanding of the behaviour of the whole roof, without going very far into engineering design principles. To give an example, included in the book is an examination of wind and how it affects different profiles of roofs, so that an appreciation can be imparted of the need to fasten down adequately the various kinds of solutions against wind action. But there is no intention to provide the reader with sufficient information to undertake the structural design of a roof.

Scope of the book

The text concentrates on those aspects of construction which, in the experience of BRE, lead to the greatest numbers of problems or greatest potential expense, if carried out unsatisfactorily. It follows that these problems will be picked up most frequently by maintenance surveyors and others carrying out remedial work on roofs. This is the reason why the material dealing with pitched tiled and slated roofs is far greater in quantity than that dealing with, for example, shingled roofs or long-span roofs.

Although most of the information relates to older buildings, surveyors may be called upon to inspect buildings built in relatively recent years. It is therefore appropriate to include much material concerning observations by BRE of new buildings under construction in the period from 1985 to 1995.

Many of the difficulties which are referred to BRE for advice stem from too hasty an assumption about the causes of a particular defect. Very often the symptom is treated and not the cause, and the defect recurs. It is to be hoped that this book will encourage a systematic approach to the diagnosis of roof and roofing defects.

The case studies provided in some of the chapters are selected from the files of the BRE Advisory Service and the former Housing Defects Prevention Unit, and represent the most frequent kinds of problems on which BRE is consulted. They are not meant to be comprehensive in scope since the factors affecting individual sites are many and varied.

As has already been said, this book is not a textbook on building construction. Hence, the drawings are not working drawings but merely show either those aspects to which the particular attention of readers needs to be drawn or simply provide typical details to support text.

Since there would otherwise be repetition of basic information, the book works cumulatively; that is to say,

the basic construction of a roof type, and an elementary statement of the principles which underlie it, are given in the early chapters dealing with the more simple roofs. These examples and illustrations are progressively supplemented by more complex examples appropriate to other roof types as they are described in the later chapters. Duplication is thereby reduced to a minimum. Furthermore, the standard headings within the chapters are repeated only where there is a need to refer the reader back to earlier statements or where there is something relevant to add to what has gone before. I have assumed that readers will know many of the standard abbreviations used in the industry (dpc, PVC, PTFE, etc.) and I have declined to spell them out; but others, some arguably, are more obscure and I have spelt them out (at some risk of creating strings of adjectives and offending against rules of good English championed by H W Fowler!).

Excluded from the scope of the book, primarily since it is not possible to offer concise advice, is any detailed consideration of those forms of roofs and roofing which demand very specialised understanding and treatment, such as timber roof structures for heritage buildings, especially medieval timber roofs of all kinds; stone built vaults and spires; the elegant structures of Victorian railway stations; heavy roofs such as those used for car parking or helicopter landing areas; and the more complex kinds of reinforced concrete shells and suspension structures. Advice on these kinds of roofs can be obtained from BRE Technical Consultancy or from a variety of organisations or consultants specialising in such buildings and structures.

In the UK, there are three different sets of building regulations:
- the Building Regulations 1991 which apply to England and Wales,
- the Building Standards (Scotland) Regulations 1990, and
- the Building Regulations (Northern Ireland) 1994.

There are many common provisions between the three sets, but there are also major differences. Although the book has been written against the background of the building regulations for England and Wales, this is simply because it is in England and Wales that most BRE site inspections have been carried out. The fact that the majority of references to building regulations are to those for England and Wales should not make the book inapplicable to Scotland and Northern Ireland.

Some important definitions

For the purposes of this book, a roof is defined as the upper structure for and covering of a building, and roofing as the materials which form that covering. It is a matter for individual interpretation of the circumstances when a wall becomes a roof and vice versa, but for present purposes a roof is taken as any inclination of more than 5° from the vertical. This covers mansard slopes but not vertical tile hanging. The definition of a flat roof is given in Chapter 2.11.

Since the book is mainly about the problems that can arise in roofs, two words, 'fault' and 'defect', need precise definition. Fault describes a departure from good practice in design or execution of design; it is used for any departure from requirements specified in building regulations, British Standards and codes of practice, and the published recommendations of authoritative organisations. A defect (a shortfall in performance) is the product of a fault, but while such a consequence cannot always be predicted with certainty, all faults have the potential for leading to defects. The word failure has occasionally been used to signify the more serious defects and catastrophes.

Where the term 'investigator' has been used, it covers a variety of roles including a member of BRE's Advisory Service, a BRE researcher or a consultant working under contract to BRE.

Because the term 'separating wall' has been used in the construction industry from the earliest days, and is still in current use, I prefer to use it in this book as a generic term despite the comparable term 'compartment wall' which is found in the national building regulations.

I have accepted, with considerable reluctance, the international standard spelling of sulfur.

ACKNOWLEDGEMENTS

Photographs which do not bear an attribution have been provided from my own collection or from the BRE Photographic Archive, a unique collection dating from the early 1920s.

To the following colleagues, and former colleagues, who have suggested material for this publication or commented on drafts, or both, I offer my thanks:

M G Atkins, C Bain, J C Beech, R W Berry, Dr P A Blackmore, Dr R N Butlin, A H Cockram, Dr N J Cook, Dr J P Cornish, R N Cox, Diana M Currie, R J Currie, Maggie Davidson, Dr J M W Dinwoodie, Dr V Enjily, A R Fewell, Dr L C Fothergill, Dr D Gardiner, G J Griffin, J H Hunt, C Hunter, E J Keeble, Dr A J Lewry, Dr P J Littlefair, A P Mayo, Dr D B Moore, Penny Morgan, B S H Muckley, A J Newman, R E H Read, J F Reid, D G Rennie, M R Richardson, A D Russell, C H Sanders, G K Saunders, C M Stirling, J R Southern, J Thomson, P M Trotman, C H C Turner, P L Walton, A Weller, J M West and Dr T J S Yates, all of the Building Research Establishment.

J May of the British Board of Agrément commented on the sections on flat roofs, and J A King of the Rural Development Commission, Adela Wright of Society for the Protection of Ancient Buildings, and Dr-Bruce-Walker of Historic Scotland commented on the section on thatch. In addition, acknowledgement is given to the original authors of Principles of modern building, Volume 2, from which several passages have been adapted and updated.

I am also indebted to F Nowak who gave me considerable help with the drawings, with the many different aspects of construction technology, and especially with the practicability of work on site.

HWH
June 1996

PREFACE TO THE SECOND EDITION

Since first writing this book, there have been few changes to the technical contents of British Standards and building regulations. The nature of roofs and their constituent materials is that they will remain fairly constant except insofar that they will need to reflect developing European Standards and trade.

While there have been new editions of the national house condition surveys, the housing stock has changed relatively slowly: between 1991 and 1996 in England, for example, only 1% of dwellings were lost and 4% were added. For the numbers of dwellings having roofs of different materials, and the proportion which suffer from faults, the new data are unlikely to differ by more than one or two percentage points from the 1991 analyses. A spot check was carried out from the 1996 English survey on roofs covered or partly covered with natural slate and stone, which revealed 3.4 million dwellings barely

different from 1991. The data for all surveys, therefore, have not been revised. Even so, of all faults in the external envelope of dwellings, those in roof coverings recorded in the 1996 English House Condition Survey form nearly one-quarter of the total for the external envelope, the others being spread between external walls, dpcs, chimneys, doors and windows. These data cannot be compared with Figure 1.3, which includes all faults.

BRE research has been reflected in changes to the chapters on lightning protection (2.6) and thatched roofs (3.7).

The list of references has been updated to reflect replacement or revised publications, mainly British Standards; additional publications have been added as 'Further reading'.

HWH
March 2000

PREFACE TO THE THIRD EDITION

The third edition of this book is being published at a time when the UK construction industry is facing a significant reduction in its work load, and nearly a decade after the second edition was prepared. That decade has seen massive changes in public awareness of the need for sustainability in construction, and the introduction of the Code for Sustainable Homes in November 2006, which since May 2008 has formed a basis for assessment of the acceptability of the design of new housing in England and Wales.

But similar needs exist for the whole of the UK's future building programme, new-build hospitals, factories, educational buildings and other long-life buildings which will provide challenges to designers in meeting the conditions brought about by anticipated climate changes and the need to be carbon-neutral and to conserve our dwindling natural resources. There have also been significant changes in British Standards which increasingly reflect those taking place in Europe.

However, the UK cannot afford year-on-year to renew more than a very small percentage of the stock of existing buildings, and the need for intelligent conservation and upgrading of the old stock is arguably of equal if not more importance.

It is against this background that this third edition of *Roofs and roofing* has been prepared. In addition to thorough revision of the chapters on the more traditional forms of construction, such as tiling and slating, completely new chapters have been prepared on:

- extensive lightweight green roofs,
- modern methods of construction,
- roof-mounted photovoltaic systems,
- thermal insulation in lofts,
- loft conversions,
- single-layer membranes.

New sections have been introduced as appropriate into existing chapters, including:
- new forms of metal roofing,
- siphonic roof drainage,
- new materials technologies,
- improved protective finishes for timber, metals and concrete.

Where appropriate, each chapter now contains a section dealing with provisions that may become necessary to accommodate climate change (eg increased rainfall, stronger winds and higher temperatures).

There have been considerable changes too in the standards covering roof drainage which have been reflected in the revised text.

Approximately one-quarter of the photographs are new to this edition.

HWH
PMT
GKS
June 2009

ABOUT THE AUTHORS

H W (Harry) Harrison is an architect. After short periods at a large UK building contractor and a small architectural practice, he spent 34 years at BRE, finishing as Head of BRE's Construction Practice Division. At BRE he carried out research into many aspects of the design and specification of buildings and building components, especially of weathertightness, accuracy and jointing. He has served on British and International Standards Committees responsible for putting into practice the results of BRE and other research, and for several years was the secretary of Commission W60, the Performance Concept in Building, of the International Council for Building Research, Studies and Documentation. He was a founder member of the International Modular Group. In later years, he became a specialist in building defects, and was responsible for the Housing Defects Prevention Unit, and the BRE Advisory Service. In the 1990 Queen's Birthday Honours List he was appointed a Companion of the Imperial Service Order for services to building research.

P M (Peter) Trotman served a student apprenticeship with Bristol Aero-Engines Ltd (now Rolls Royce Ltd), qualifying as a mechanical and structural engineer. Peter joined BRE in 1967, initially in the Public Health Engineering Section carrying out research into water services and drainage. Since 1975 he has been involved in the BRE Advisory Service with a particular interest in all aspects of dampness. He has made site inspections of many hundreds of properties in the UK, Europe and South America, reporting on their condition and recommending remedial measures. Appointed Head of the BRE Advisory Service in 1990, he has managed programmes of site investigations, lectured to construction professionals and prepared many BRE technical publications. He has served on several BSI committees and was a member of the British Wood Preserving and Damp Proofing Association (BWPDA) property care technical committee. He was for many years Coordinator for the International Council for Research and Innovation in Building and Construction (CIB) Commission W86 — Building Pathology, and Chairman of the Trustees for Upkeep — the independent charity for training in property maintenance.

G K (Gerry) Saunders

Gerry Saunders is a mechanical engineer by qualification having obtained his ONC and HNC in 1970. He started work in the Scientific Civil Service in September 1966 for the Forest Products Research Laboratory at Princes Risborough. In the Timber Mechanics Section he carried out strength tests on timber and timber joints and this work led to the introduction of short finger joints and the truss plates that are commonly used in roofs today.

Later, the Laboratory became part of the Building Research Establishment, and Gerry's work was on failures of metal skin sandwich panels, measurement of the moisture content in walls of timber-framed houses and pre-normative work on the impact resistance of lightweight walls and claddings.

In 1980 Gerry transferred to the Flat Roofs and Sealants Section where he began his long association with the roofing industry for which he is best known. He carried out research to study the movement of foamed plastic insulants used in roofs and also helped to develop the understanding of the heat flow in inverted roofs. He developed an expertise in the behaviour of bitumens, plastics and rubbers used in roofing which was recognised when he became Chairman of B546/1, the British Standards committee responsible for bitumen sheets and BS 747. In 2004, after supporting many European Committees as the UK Expert, he was elected to the Chair of CEN TC254 which deals with flexible sheets for waterproofing.

Since the privatisation of BRE in 1997, Gerry has become known for his work as a roofing expert and consultant helping to resolve problems with roofs.

1   INTRODUCTION

The main aim of this edition remains the same as for previous editions — it is to remind readers of the quality of construction of roofs and roofing which is necessary to ensure good performance of the fabric for the defined life of the building, and the corollary to this, to remind them also of the things that can and do go wrong, so that they can be avoided and to achieve all this against the background of continually rising standards.

The intended readership remains the same: building surveyors and other professionals performing similar functions, such as architects and builders, who maintain, repair, extend and renew the national building stock. But it is also important that those persons undergoing training for the construction industry are made aware of the sometimes rather elementary mistakes that occur in the design office and on site, and this places an additional reponsibility on those in charge of the construction syllabus.

The book is not a textbook of construction practice; it simply draws from BRE experience to provide the building professional with an illustrated checklist of points of concern, and where to obtain further information and advice.

Figure 1.1: The present rapid pace of change in construction techniques is well illustrated in this view of low-rise buildings on the BRE Innovation Park which includes prototype dwellings built to the requirements of the Code for Sustainable Homes[1]. On the left is a fairly traditional-shaped pitched gabled roof, while other examples feature monopitch construction which facilitates the installation of roof-mounted PV systems, flat roofs using sedum mats, and at the top right is an experimental installation of small wind turbines. Also featured in the examples are other instances of the involvement of the roof in rainwater harvesting, and construction using structural insulated panels (SIPs)

1.1 BACKGROUND

The majority of roofs, of whatever kind of structure or covering, perform well. However, there is evidence from BRE sources that avoidable defects in roofing occur too often (Figure 1.2). These sources include:

- surveys of housing, of both new construction and rehabilitation,
- evidence from the UK house condition surveys (undertaken every five years), and also,
- particularly for building types other than housing, the past commissions of the BRE Advisory Service and current BRE consultancy projects.

The National House Building Council reported in 1993[2] that pitched roof problems were growing in number, with one of the main problem areas being the traditional 'cut' roof structure; that is to say, one built on site, not prefabricated.

Figure 1.2: A flat roof showing signs of water penetration. Ponding can quickly expose defective jointing of sheets

RECORDS OF FAILURES AND FAULTS IN BUILDINGS

BRE Advisory Service records

A BRE Advisory Service survey dating from 1970 put roof and roofing failures at about one-quarter of all those recorded, with some building types showing proportionately more failures (Figure 1.3). There were only slightly greater numbers of cases of rain penetration through walls than through roofs. Ten years later, the figures had changed somewhat, and roof and roofing failures then formed around one-third to one-half of all failures investigated, depending on building type. These failures represented a huge waste of the nation's resources of materials and manpower for remedial work.

In the years up to 1985, approximately equal numbers of housing and non-housing cases were referred to BRE for advice, though there were slight differences in the relative numbers of cases of flat and pitched roofs. Flat roof investigations continue to outnumber pitched roof by about 2.5 to 1 in housing, and by about 3.5 to 1 in non-housing (1992 figures, which are the latest available).

Investigations of roof finishes during a sample three-year period, 1985–87, were slightly more numerous than those involving the structure, although the numbers of cases were approximately equal for factories and flats. With swimming pool roofs, however, there were five cases involving the structure, and only one involving finishes.

In a 1987–89 survey report[3], a more detailed picture of the situation can be found (Figure 1.4).

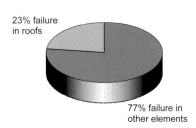

23% failure in roofs

77% failure in other elements

Roofs in all kinds of buildings 1970

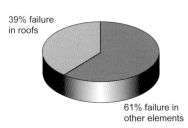

39% failure in roofs

61% failure in other elements

School roofs 1970

46% failure in roofs

54% failure in other elements

Swimming pool roofs 1970

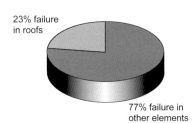

23% failure in roofs

77% failure in other elements

School roofs 1980

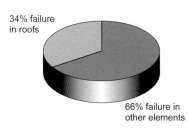

34% failure in roofs

66% failure in other elements

Factory roofs 1980

Figure 1.3: Failures in roofs. Data from BRE Advisory Service

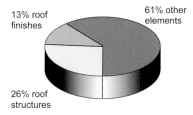

Design faults, non-housing

13% roof finishes
61% other elements
26% roof structures

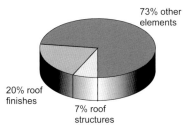

Execution faults, non-housing

73% other elements
20% roof finishes
7% roof structures

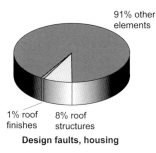

Design faults, housing

91% other elements
1% roof finishes
8% roof structures

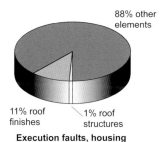

Execution faults, housing

88% other elements
11% roof finishes
1% roof structures

Figure 1.4: Faults in roofs, 1987–89 Data from BRE Advisory Service

Non-housing design and execution faults

Design faults in the roof structure accounted for 1 in 4 of the sample, and roof finishes 1 in 8, a total of about two-fifths of all faults in all elements investigated.

For execution, the corresponding figures were 1 in 15 for structure and 1 in 5 for finishes, a total of just over one-quarter of all faults.

A small detailed sample of investigations on roofing carried out by BRE in the four years, 1988–91, showed that more commercial and industrial buildings than dwellings now feature in the investigations carried out by BRE.

The BRE Advisory Service was reduced following privatisation of BRE in 1997. BRE no longer offers free advice and operates on a consultancy basis. This consultancy work has been growing and it is now possible to review this and highlight problem areas. The following are examples of problems investigated.

Traditional pitched roofs
The main issues of concern are:
• condensation and
• dimensional stability of some artificial slates.

Condensation
The condensation problem is associated with the 'cold sealed roof' where the only ventilation is through the vapour permeable underlay. This roof construction has been observed to have suffered from condensation occurring on roof timbers shortly after the new-build process. Once the initial drying period is complete the problem of condensation is much reduced and satisfactory performance achieved. This form of construction may be transitory as the preference for insulating at rafter level increases.

Dimensional stability of artificial slates
Some artificial slates have been observed to 'curl' so they no longer lie flat. This curl can be sufficient to pull out the tail rivet. There is as yet no defined method of assessment in terms of amount of dimensional change. It has been suggested in judgement on a particular case that those products which do not move sufficiently to lift out the tail rivet can be deemed as being fit-for-purpose and acceptability becomes a matter of aesthetics.

Flat roofs
The main issues of concern for flat and low-sloped roofs are:
• the waterproofing membrane, despite the availability of high-performance bitumen systems and single-ply plastics and rubbers and
• condensation in insulated metal roofs over high moisture-producing areas.

Waterproofing membrane
The use of the plastics and rubbers has increased dramatically over the last 10 years and consequently a high proportion of problem roofs investigated have single-ply membranes. The most problematic construction involves the inverted roof construction and it appears the membranes can be damaged during the construction process by following trades. In particular, damage can occur to the single-ply membrane during the building phase where plant machinery is added to the roof later.

Condensation in insulated metal roofs
There appears to be a problem with condensation forming when insulated metal roofs are used over high moisture-producing areas such as swimming pools. The fixings used often penetrate the vapour control layer, allowing moisture to pass into the roof construction.

Housing design and execution faults
Design faults in the roof structure accounted for 1 in 12 of the sample. The faults found in roof finishes were negligible.

For execution, the corresponding figures were negligible for roof structure and 1 in 9 for roof finishes. These represent about 1 in 5 of all faults investigated.

BRE Defects Database records
There is further detailed information from a very much larger sample for housing in BRE's Quality in Housing Database which is summarised in BRE's Information Paper IP 3/93[4]. Although the survey was conducted some time ago, current investigations show that the data most probably have not changed much over the intervening years, showing that the industry sometimes does not learn from experience. This database recorded actual inspections by BRE researchers or external consultants working under BRE supervision. It is possible to see from these records exactly where mistakes were being made (Figure 1.4) and by whom; and who, ostensibly, needed further

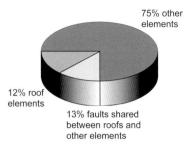

75% other
elements

12% roof
elements

13% faults shared
between roofs and
other elements

All faults

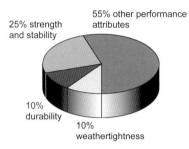

55% other performance
attributes

25% strength
and stability

10%
durability

10%
weathertightness

All roofing faults, performance affected

Figure 1.5: Proportions of faults in roofs, 1993. Data from BRE Housing Defects Database

guidance and what form that guidance should take.

Types of failure include:
• weatherightness,
• condensation,
• cracking,
• strength and stability,
• durability and
• sound.

The database showed that in 1994, of the 1241 types of fault concerning a single element of roofing (about one-quarter of the overall total of faults of all kinds), just over half occurred in the roof element itself and the remainder at the junctions with other elements (eg separating and external walls). In about one-quarter of these faults, for example rain penetration or cracking, defects had already occurred, while the remainder all had incipient defects or potential for failures built in. Just over one-quarter of all faults related to the single attribute of strength and stability, 1 in 10 to weathertightness, about the same number to durability, and the remainder to multiple performance attributes (eg sound and fire performance together) (Figure 1.5).

The proportion which roofing faults formed of the total number of faults observed depended on

the kind of construction (eg timber frame or traditional masonry).

For faults with their origin in execution, it was estimated that around twice as many were due to lack of care as to lack of knowledge. One conclusion which might follow from this difference is that there is a stronger case for guidance on roofing faults than for other elements, at least as far as work on site is concerned. This situation is probably equally true today.

In about one-quarter of all faults, an estimate was made of the cost of correction of each one: the average was around £100 (at 1990 prices) equating to about £200 at 2008 prices. When this is multiplied by the number of faults repeated on each dwelling, the total sums become far from negligible.

In the site observations carried out by BRE there has normally been no weighting for the degree of importance of the consequences of faults. It can be observed though that many failures in roofs could potentially be of a serious nature, with the consequences of faults far outweighing the costs of their correction or avoidance. In a sample of approximately 10% of all roofing faults recorded in the 1990 studies, about one-quarter were assessed to be potentially hazardous, about one-third costly, and the remainder merely inconvenient.

The majority of faults in roofs and roofing of housing originated with site practices current before the 1939–45 war. Only about one-sixth of the total related to practices introduced since the war. Having said that, however, evidence from the new-build survey of highly insulated dwellings contained within the Quality in Housing Database published in 1993[4] indicated a misunderstanding of the seriousness of the consequences of inadequate ventilation and the need for elimination of thermal bridges in spite of the provisions in standards and building regulations (Figure 1.6). Approximately one-sixth of all infringements of such requirements for newly constructed roofs, as with other elements, have proved to be (in the view of BRE investigators) infringements of the building regulations in force at the time. In later surveys which showed that the provisions of Approved Documents to the England and Wales building regulations were not followed, a high proportion of the chosen solutions were considered (in the view of BRE investigators) not to comply with the regulations. Summaries of the actual faults found are given in the appropriate sections of the chapters which follow.

No detailed analysis has been done of the proportions of each of the faults relating to roofing

Figure 1.6: Mould growth caused by condensation in a domestic roof, in turn caused by lack of ventilation. Blocking the eaves gap with insulating material, so preventing air movement between the outside atmosphere and the roof space, is a common cause of condensation in roofs

which are the responsibilities of the various parties to the building process, whether it be roofing for new-build or for rehabilitation. The split, though, for all elements taken together is about one-tenth to materials and components with the remaining 90% equally divided between design and site responsibilities. There is no reason to suppose that the distribution for roofing differs greatly from these overall figures.

House Condition Surveys

The first edition of this book drew information on the numbers and condition of domestic roofs from the auxiliary tables accompanying the English, Welsh, Scottish and Northern Ireland House Condition Surveys conducted in 1991–1993[5–8]. Tables giving this precise information are no longer published, so it has not been possible to update some of the data to be found in the earlier edition. However, some of the data is still gathered, and a limited analysis of the data for England and Northern Ireland has been carried out specifically for this book.

An important qualification must be made about these data. The faults identified by the surveyors ranged quite widely in importance. Estimates were made of the proportions of each roof affected by such faults (in tenths of total areas) and of the relative urgency of the repair necessary. For the purposes of this book it is considered sufficient to know how many roofs are affected.

The latest versions of the UK House Condition Surveys are:
- English House Condition Survey 2005[9],
- Scottish House Condition Survey 2004/5[10],
- Welsh House Condition Survey 2004[11],
- Northern Ireland House Condition Survey 2005[12].

In 1996 there were 20,321,000 dwellings in England, and in the 10 years from then until 2006 the number had increased to 21,781,000. 17,977,000 (82%) were houses and bungalows and 3,804,000 (17%) were flats of

Table 1.1: Data from Stock Profile, English House Condition Survey[5]

Date of construction	No. of dwellings	% of total
Before 1919	4,731,000	21
1919–1944	3,808,000	17
1945–1964	4,279,000	19
1965–1982	4,928,000	22
Post 1982	4,035,000	18

Table 1.2: Data from Figure 2, Scottish House Condition Survey[6]

Date of construction	No. of dwellings	% of total
Before 1919	414,180	18
1919–1944	345,150	15
1945–1964	552,240	24
1965–1982	575,250	25
Post 1982	414,180	18

various kinds. Table 1.1 gives data on numbers built by date of construction.

Similarly, in 1996 there were 2,123,000 dwellings in Scotland, increasing to 2,301,000 by 2005. 65% were detached, semi-detached or terraced houses, and the remainder (35%) were flats of various kinds. Table 1.2 gives data on numbers built by date of construction.

Up-to-date figures for Wales and Northern Ireland were not available at the time of writing.

BRE PUBLICATIONS ON ROOFS AND ROOFING

Principles of modern building
Principles of modern building, Volume 2: Floors and roofs[13], was first published in 1961. In many respects it proved to be a milestone in the application of science to building construction. The chapters on roofs are now showing their age, not only because the examples date from the 1950s, and all the dimensions in the text are imperial, but also because such performance requirements as thermal insulation are minimal in today's terms. However, since the book deals with

principles, in some respects it is less out of date than might be imagined.

The general principles stated in *Principles of modern building* which have stood the test of time form the basis of the introductory sentences of some of the chapters in this roofs book. Descriptions of agents affecting the roof, such as the weather, are also taken in the main from BRE texts which have been widely used for other publications.

BRE publications
The BRE Digests, Information Papers, Good Building Guides and Reports listed in the references and further reading sections offer a fairly comprehensive coverage of roofs and roofing, although, since they have been published over a span of many years, they can have received very little cross-referencing between them all. These publications have been drawn upon to a considerable extent in this book which therefore provides a useful key to BRE material that is relevant to roofs.

BRE Reports
A number of BRE reports contain sections relevant to roofs and roofing which may not be apparent from their titles. Those relevant to roofs of housing include: *Assessing traditional housing for rehabilitation*[14], *Surveyor's check list for rehabilitation of traditional housing*[15], and *Quality in traditional housing*, Volumes 2 and 3[16].

Information on the roofs of non-traditional housing may be available in BRE Reports on particular building systems and in *Non-traditional houses*[17].

CHANGES IN CONSTRUCTION PRACTICE OVER THE YEARS

Pitched roofs
Swings in the fashions for roofing design occur just as with other aspects of architectural design. Traditional shallow pitched, lead covered roofs (or 'leads' as Samuel Pepys used to call them) have been and still are common in heritage buildings, mainly churches and cathedrals.

Pitched roofs of various kinds of structure and covering have

Figure 1.7: A fine old arched tie-beam roof

Figure 1.9: A glazed roof over an atrium in a department store

continued to form the majority of roofs since medieval times (Figure 1.7). It is thought that a fair majority of the nation's stock of buildings still have pitched roofs of one kind or another. The maintenance of the coverings of such roofs for the most part is not particularly onerous and, in normal circumstances, has proved to be well within the capabilities of most surveyors, architects and roofing contractors. However, in the case of the underlying structure, for example the cast and wrought iron spans of Victorian railway stations, there can be problems which need the advice of specialists (Figure 1.8).

On the other hand, shell concrete roofs, popular at the time of the Festival of Britain in 1951 and discussed at some length in

Principles of modern building[13], have largely gone out of fashion, though some buildings roofed with them will remain in use. Now it is the large-span glazed or semi-glazed atrium roof, often in vaulted shapes, which is in vogue (Figure 1.9). There have also been considerable developments, for example in the bending of large steel sections, opening up a whole new range of shapes for designers to exploit.

Rather more common these days is the sheeted low pitched or flat roof of the typical DIY store or supermarket. These roofs often have steeply pitched tiled perimeters; and they are not entirely without their problems to judge from the number of failures. Amongst them there is an element of innovatory

design. It seems that designers do not always fully foresee all the intricate geometry of some of their designs for 'high-tech' solutions, so condensation, rain penetration, workmanship and supervision then become particular pinch points.

In the domestic short-span field there has been an almost complete swing away from the strutted purlin roofs of interwar and early post-war years towards trussed rafter roofs. These came into widespread use in 1964: economical to construct but of limited utility to the occupier in terms of storage space. Future years may see a gradual reintroduction of roofs which allow use of the space within; for example by means of attic trusses. Unpublished information from Housing Association Property Mutual indicated that during the three years, 1991–93, most new houses in a large sample employed trussed rafter roofs (Figure 1.10) or composite roofs (composite roofs are part-trussed rafter and part-cut). About 1 in 10 of these houses had cut roofs. During this timespan the numbers of new houses having flat roofs was negligible. Since the mid-1980s, a further noticeable swing in fashion has been from hips to plain gables and back to hips.

One tendency which has become very clear is that of increasing complexity of the geometry of buildings, and more especially of roofs. It is also very clear that defects increase in direct proportion to increases in complexity of geometry of the surfaces of buildings; in other words at the intersections of different planes. This is not an argument for a return to simple shapes so much as an argument for taking extra care in both design and construction when complex shapes are to be used.

Flat roofs

A swing in architectural fashion away from the pitched roof in favour of the flat roof occurred with the so-called modern movement led by the Bauhaus in the 1920s. This theme was continued throughout in the world – the UK was no exception (Figure 1.11) – and taken up with enthusiasm by UK architects with

Figure 1.8: Newly restored Victorian train shed roof at St Pancras Station, London

Figure 1.10: Attic truss providing clear headroom in the centre part of the roof

Figure 1.12: An early post 1939–45 war school. Very few such schools had pitched roofs

the building systems for housing and schools of the late interwar and post-war periods (Figure 1.12). The flat roof was also the only conceivable solution for some of the convoluted plan forms adopted in those times. It is the poor performance of some of these, particularly system-built roofs, coupled with lack of tolerance to thermal and moisture movement of structures both immediately adjacent to the roofs and of the roofs themselves, that gave flat roofing a bad name. Fortunately,

newer designs and materials mean that there has been a progressive reduction in the incidence of defects over the whole of the flat-roofed building stock, even though there will still be a residue of defective flat roofs to attend to.

There rarely seemed to be serious problems with mastic asphalt, provided it was laid in accordance with standard practice and laid on a stable substrate. Problems have been encountered when mastic asphalt was laid over a substantial thickness

of efficient insulation and suffered from cracking of asphalt during very low winter temperatures. The edge details are also required to withstand much higher movements when laid over highly insulated lightweight decks and can fail. No such problems occur when mastic asphalt is used with concrete decks.

The real difficulty came with built-up bitumen felt flat roofs which failed in large numbers in the 1950s and 1960s. Particularly when exacerbated by poor maintenance strategies, owners of large building stocks with these roofs, such as county councils and the Property Services Agency, found themselves responsible for substantial costs for their repair. The roofs which failed were largely those using organic felts fully bonded to the deck immediately below them. Consequent thermal movement of the substrate from exposure to solar radiation caused breakdown of the waterproof membrane after a relatively short life. The problem of failure due to thermal movement was further exacerbated by the introduction of thermal insulation. The first type chosen was extruded polystyrene which has a particularly large dimensional change with temperature compared with other insulants. This was waterproofed using the organic felts and could fail in as little as 3 years.

BRE Advisory Service undertook a major investigation in 1989 on the performance of recently introduced kinds of flat-roofing membranes. In a large number of site investigations, no leaks were found through the membrane which could not be explained by faulty detailing. Figure

Figure 1.11: A 1930s flat-roofed house. Lack of adequate projections, as might be provided by eaves, has led to staining from rainwater run-off

Figure 1.13: Checking the adhesion of a BS 747 class 5 built-up roof sheet overlap

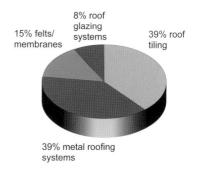

Figure 1.14: UK roofing products market by value, 2006. Data from Roofing Market Report[20]

1.13 shows one of the roofs being inspected by a BRE investigator.

For roofs which have suffered damage or failure, there needs to be a reappraisal of the original design, as it would not do to replace a roof which has proved to be inadequate with a replica.

With the introduction of newer materials (polyester reinforcing fleece) and a better observance of good practice, there has been a progressive reduction of the incidence of defects in newly built flat roofs.

Materials

Since *Principles of modern building*[13] was written, the main development has been an increased awareness of the effects of construction on the environment and the need to provide for sustainability.

Environmental and sustainability issues

Environmental issues, such as, for example, climate change, fossil fuel depletion, embodied energy, minerals extraction and waste disposal, need to be taken into account when specifying roofs and roofing in sustainable new construction, and with refurbishment of older buildings. These issues relate in the main to roof configuration. The roof's contribution to environmental issues ranges between 2% and 12% of the total for the building, depending in the main on the number of stories and on the plan form of the building and assuming that minimum standards given in building regulations are met.

In principle, low mass and minimal industrial processing promote best conditions for the environment. Timber-framed pitched roofs covered in tile or natural slate yield the smallest impacts on the environment because of the lightweight form of the construction, with light-steel systems coming next. Man-made slates have a greater impact because of their greater consumption of energy and resource-intensive materials.

Flat roofs are a different matter, since the form of construction has a major effect on its impact on the environment. Timber decks are most beneficial, with steel decks coming next. Poorest performers are concrete flat roofs, stemming in the main from the high mass and quantities of the materials used rather than in embodied energy consumed in their manufacture.

So far as thermal insulation is concerned, BREEAM[18] and *The Green Guide to Specification*[19] ratings are not sensitive to the type of insulation used, provided that the material is not ozone-depleting.

The durability of the weatherproofing membrane plays a major role in the environmental impact of flat roofs, with long life and reduced maintenance tending to off-set resources consumed in materials manufacture.

Other issues

Other significant changes have included the following.
- There has been an increase in the use of rigid sheeted roofs, sometimes of double

construction, following increased thermal insulation standards,
- On-the-slope insulation has come into more general use, for example of the prefabricated kind that rebates over rafters, demanding a higher standard of accuracy in construction than when such insulation could be cut to fit. The industry still has much to learn about these new forms of insulation.
- Major changes in the use of materials include the progressive discontinuation of asbestos cement in its various forms, such as corrugated sheet and preformed deck, and the virtual elimination of compressed strawboard as a deck material. BRE investigators encounter fewer roof decks of woodwool and chipboard than used to be the case, but many more of orientated strand board and plywood.

The past few years have seen a parity developing in the market between roof tiling and metal roofing systems (Figure 1.14). Tiling and metal roofing systems now dominate the market, with metal roofing increasing in popularity following increases in industrial and commercial developments and a concomitant reduction in the popularity of pitched roofs which has affected the tiling sector[20].

Safety

One aspect of the construction of roofs that does not seem to change is the distressing frequency of accidents. Falls associated with roofwork account for more deaths and injuries in the construction industry than any other activity. The Health and Safety Executive (HSE) identifies roofwork as a particular hazard. Indeed, some 28 workers on average are killed each year, and several hundred are seriously injured as a result of working on roofs. In 1992, HSE ran a campaign to secure improvements in the safety record of the industry and has published a booklet, *Safety in roofwork*[21]), which advises on safety procedures. A duty to provide training to ensure the health and safety of their employees is placed upon employers by Section 2(1) of the Health and Safety at Work, etc. Act 1974[22] and the most recent CDM Regulations[23].

Main causes of accidents identified from HSE data

The data from HSE are restricted to falls from height and so do not identify all accidents associated with roofing. Falls from height are due to:
- falls through fragile roofs,
- falls through fragile rooflights,
- falls from unprotected edges.

Falls on fragile roofs were mainly through old fibre/asbestos cement sheets. Some are associated with the installation of the liner sheet to built-up metal roofs. Design options are limited. An improved fibre cement sheet with inner reinforcement is available for new buildings and should be used. In terms of durability, fibre cement sheet has been used successfully in farm buildings. Other roofing systems could be specified but their significantly higher cost reduces the likelihood of their being used, despite the fact that they could lead to the reduction of accidents in the future because of their non-fragile nature.

Falls through rooflights are related to the fragility of the plastics glazing material used. An Industry/HSE Advisory Committee for Roofwork (ACR) has developed a fragility test for rooflight assemblies based on a falling weight[24]. However, materials which are non-fragile when they are installed may become fragile over time. Significant risks to roof workers are associated with rooflights which

are installed in the same plane as the roofing sheet. These are used mainly on profiled metal or fibre cement roofs where the rooflight has a similar profile. The design decision is again one of material selection. Because of the uncertainty when a non-fragile rooflight might become fragile over time, a roof design could include easy access points for installing nets below the rooflights when the roof is being accessed for repair or maintenance.

Falls from the edges of roofs are not related to any particular type of roof. There were also falls from unprotected openings, included as edges, although they were not from the perimeter of the roof. A fall from a ladder while cleaning a gutter could be avoided on a flat roof by placing the gutter away from the perimeter of the roof. Although not generally considered good practice from a waterproofing point of view, it would be safer for the installer and maintenance personnel. One fall from the edge of a roof occurred whilst resetting a tiled gable end. The design decision might be to avoid the use of tiled gable ends but this is unlikely in practice even though it would remove the risk. In this particular case, the accident could have been avoided. Most could have been obviated by using a properly assembled access tower[25, 26].

Construction Design & Management (CDM) Regulations

The CDM Regulations[23] were first introduced in 1994 with a view to reducing the accidents associated with construction work. Since roofs are assembled at height the construction has been a major cause of accidents, some of which have been fatal. The CDM Regulations were revised and reissued in 2007.

The CDM principles are that the construction and maintenance must be considered at the design stage and a health and safety plan drawn up. However, it is considered by most designers that providing points to attach safety harnesses is sufficient. This is only a low-level strategy which, although an improvement on no provision, still requires more attention to be given

to making the roof a safe place to work.

CDM also covers the construction phase and the maintenance phase, eg access to service items such as header tanks must be given proper consideration[27].

Other relevant legislation or guidance is contained in:
- Construction (Working Places) Regulations[28],
- Construction (Lifting Operations) Regulations[29],
- Construction (General provisions) Regulations[30],
- Workplace (Health, Safety and Welfare) Regulations 1992 Approved Code of Practice and Guidance[31].

CLIMATE CHANGE

The UK Climate Impacts Programme (UKCIP)[32] has provided climate change models enabling potential risks to the UK to be identified. The latest models available for discussion were produced in 2002 and are known as the UKCIP02 scenarios.

There are four scenarios based on predictions of the amount of emissions into the earth's atmosphere that may occur due to human activities. These are low, medium–low, medium–high and high. The UKCIP02 climate models assumed that the scenario based on the effects of medium to high emissions of CO_2 would be the most likely. In short, the predictions are for hotter and drier summers and wetter but milder winters.

This means that roofs could be exposed to:
- higher temperatures: more often and for longer,
- more UV light,
- greater intensity of rainfall,
- changes in wind behaviour. (It is not certain how wind behaviour may change. The frequency of high wind events may increase and greater wind speeds may occur.)

Although the scenarios are based on the effect of greenhouse gas emissions, they also take account of a slowing down of the Gulf Stream. The predictions assume that changes in weather patterns due to change in the flow of the Gulf Stream occur relatively slowly compared with those predicted for the emissions models. The predicted changes are made for the next 100 years, which covers the intended lifetime of most buildings being designed currently.

Further information is available on the UKCIP website[32]; see also Hunt (2007)[33].

Good practice guidance

Good practice guidance is found in a number of publications:
- national building regulations,
- British Standard codes of practice,
- industry guides,
- manufacturer's literature and
- independently produced documents (eg BRE publications are sources of good practice guidance for roofing).

A number of British and European product standards also provide useful additional information. Most of these documents refer to British or European standards relating to specific design guidance on topics such as wind loads or drainage.

Building regulations

An important requirement in any roof design is to take account of the current Building Regulations.

England and Wales

Guidance is given in a set of Approved Documents (ADs)[34]. The most relevant regulations to roofing design are:
- AD A: Structure,
- AD B: Fire safety,
- AD C: Site preparation and resistance to contaminates and moisture. (The 2004 document includes references to condensation previously included in A to D, and F),
- AD F: Ventilation,
- AD L: Conservation of fuel and power.

Scotland

The most relevant documents are Sections 1, 2, 3, 4 and 6 of the Technical Handbooks, Domestic and Non-domestic[35].

Northern Ireland

The most relevant documents are Technical Booklets D, E, F and K[36].

British Standards

The main British Standards for roofing are:
- BS 5427-1: 1996[37],
- BS 5534: 2003[38],
- BS 6229: 2003[39].

Product standards or codes of practice referred to in these publications are too numerous to list here.

Industry guides

Industry guidance is also produced and typical examples are:
- Single Ply Roofing Association[40]. Industry best practice in the design, installation and maintenance of single ply roofing systems.
- Metal Cladding and Roofing Manufacturers Association. Technical guidance available on their website[41].
- British Flat Roofing Council/ CIRIA[42].

BRE publications

The following BRE publications are also relevant:
- Roofs and roofing pack[43] This pack brings together over 20 published titles from BRE providing guidance and advice.
- Digest 419[44]. This Digest helps the designer, specifier and flat roofing contractor to specify durable bituminous sheets.
- Digest 486[45]. This Digest summarises the views of experts in the roofing industry who have put forward recommendations as to how roof design may mitigate the effects of climate change.

Structural aspects

The roof has to be able to support itself and to transmit other loads through the supporting structure to the ground. Loads on roofs are referred to as dead, imposed and live loads.

Guidance can be found in BS 6399, Parts 1-3[46].

Proposed modifications to good practice guidance

The modifications to good practice guidance are grouped by design parameters. Two approaches can be

taken into account for the effects of climate change on roof design:
- modify key factors in all the calculations to establish design parameters (eg increase the basic wind speed by a factor),
- apply a modification that takes account of such factors but may be simpler to apply (eg increase the number of fixings per unit area).

By adopting these modifications the roof may be over-designed. However, although this may incur modest extra construction costs, it could save much more in the future as repairs are avoided and periods of maintenance extended.

Changes in working practice

The installation of roofs may be influenced by climate change but this may not significantly affect the current design of the roof though it may affect workmanship issues such as air tightness and thermal bridging.

It should be recognised that working practices will evolve as weather patterns change. Stronger winds will increase the number of days when work cannot safely be carried out on roofs. Wind loads on partially completed roofs can cause failure of the components already installed because they are not designed to withstand such wind forces, whereas the fully installed system could. This is often ignored and may be a problem in the future. The solution is to ensure all wind loads are considered for each element during the construction phase. Extra fixings may be required.

Conclusions

The predicted changes in climate over the coming decades may well have a significant effect on several aspects of roof design. However, there are many relatively straightforward ways in which current practice can be amended to cope with the expected conditions. Some aspects are not easy to deal with and changes in design, technology and detailing will be needed, particularly in relation to enhanced standards of insulation.

1.2 REFERENCES

[1] **Communities and Local Government (CLG).** The Code for Sustainable Homes. Available from www.communities.gov.uk/thecode or from www.planningportal.gov.uk

[2] **National House Building Council.** Keeping the roof on claims. Standards Extra No 4. Amersham, NHBC, 1993

[3] **Trotman PM.** An examination of the BRE Advisory database compiled from property inspections. Proceedings of the International Symposium 'Dealing with Defects in Building', Varenna, Italy. September 1994 (copies available from the author at BRE)

[4] **Harrison HW.** Quality in new-build housing. Information Paper IP 3/93. Bracknell, IHS BRE Press, 1993

[5] **Department of the Environment.** English House Condition Survey: 1991. London, The Stationery Office, 1993 (Analyses of so far unpublished data were carried out especially for *Roofs and roofing*. Further information may be obtained from BRE Technical Consultancy)

[6] **Scottish Homes.** Scottish House Condition Survey 1991. Survey report. Edinburgh, Scottish Homes, 1993

[7] **Welsh Office.** Welsh House Condition Survey 1993. Cardiff, Welsh Office, 1993

[8] **Northern Ireland Housing Executive.** Northern Ireland House Condition Survey 1991. First report of survey. Belfast, Northern Ireland Housing Executive, 1993

[9] **Communities and Local Government.** English House Condition Survey 2005. London, CLG, 2007

[10] **Scottish Executive Development Department.** Scottish House Condition Survey 2004/5. Edinburgh, Scottish Executive, 2006

[11] **Welsh Assembly Goverment.** Living in Wales 2004: Welsh Housing Quality Standard. Cardiff, Welsh Assembly Government, 2006

[12] **Northern Ireland Housing Executive.** Northern Ireland House Condition Survey 2006. Belfast, The Northern Ireland Housing Executive, 2008. Available as a pdf from www.nihe.gov.uk

[13] **Building Research Station.** Principles of modern building, Volume 2: Floors and roofs. London, The Stationery Office, 1961

[14] **BRE.** Assessing traditional housing for rehabilitation. BR 167. Bracknell, IHS BRE Press, 1990

[15] **BRE.** Surveyor's checklist for rehabilitation of traditional housing. BR 168. Bracknell, IHS BRE Press, 1990

[16] **BRE.** Quality in traditional housing, Volume 2: An aid to design. Volume 3: An aid to site inspection. BRE Report. London, The Stationery Office, 1982

[17] **Harrison H, Mullin S, Reeves B & Stevens A.** Non-traditional houses: Identifying non-traditional houses in the UK 1918–75. BR 469. Bracknell, IHS BRE Press, 2004

[18] **BRE.** BREEAM. See www.breeam.org.uk

[19] **Anderson J, Shiers D & Steele K.** The Green Guide to Specification. An environmental profiling system for building materials and components. 4th edition. BR 501. Bracknell, IHS BRE Press, and Oxford, Wiley-Blackwell, 2009

[20] **AMA Research.** Roofing Market Report UK 2007. Cheltenham, AMA Research, 2007. Available from www.amaresearch.co.uk

[21] **Health and Safety Executive (HSE).** Safety in roofwork. Guidance Booklet HS(G)33. London, HSE, 1987. (To be revised)

[22] **HMSO.** Health and Safety at Work, etc. Act 1974. Chapter 37. London, The Stationery Office, 1974

[23] The Construction (Design and Management) Regulations 2007. London, The Stationery Office, 2007

[24] **The Advisory Committee for Roofwork (ACR).** The Red Book – Test for non-fragility of roofing assemblies ACR[M]001-2005 (third edition). ACR Materials Standard. London, ACR

[25] **Saunders GK.** Designing roofs with safety in mind. Information Paper IP 7/04. Bracknell, IHS BRE Press, 2004

[26] **The Advisory Committee for Roofwork (ACR).** Practical methods of providing edge protection for working on roofs. London, ACR, 2009

[27] **Joyce R.** CDM regulations 2007 explained. London, Thomas Telford, 2007

[28] **HMSO.** Construction (Working Places) Regulations. London, The Stationery Office, 1966

[29] **HMSO.** Construction (Lifting Operations) Regulations. London, The Stationery Office, 1961

[30] **HMSO.** Construction (General Provisions) Regulations. London, The Stationery Office, 1961

[31] **Health and Safety Commission.** Workplace (Health, Safety and Welfare) Regulations 1992. Approved Code of Practice and Guidance L24. London, The Stationery Office, 1992

[32] **UK Climate Impacts Programme (UKCIP).** Climate change scenarios for the United Kingdom. The UKCIP briefing Report. Norwich, The Tyndall Centre for Climate Research, 2002. See www.tyndall.ac.uk

[33] **Hunt JCR.** Climate change and civil engineering challenges. Proc Inst Civil Engineers: 2007 (November)

[34] **Communities and Local Government (CLG).** The Building Regulations 2000.
Approved Documents:
  A: Structure, 2004
  B: Fire safety, 2006
  C: Site preparation and resistance to contaminates and moisture, 2004
  F: Ventilation, 2006
  L: Conservation of fuel and power, 2006
London, The Stationery Office. Available from www.planningportal.gov.uk and www.thenbs.com/buildingregs

[35] **Scottish Building Standards Agency (SBSA).** Technical standards for compliance with the Building (Scotland) Regulations 2009
Technical Handbooks, Domestic and Non-domestic:
  Section 1: Structure
  Section 2: Fire
  Section 3: Environment
  Section 4: Safety
  Section 6: Energy
Edinburgh, SBSA. Available from www.sbsa.gov.uk

[36] **Northern Ireland Office.** Building Regulations (Northern Ireland) 2000.
Technical Booklets:
  D: Structure, 1994
  E: Fire safety, 2005
  F: Conservation of fuel and power.
  F1: Dwellings, F2: Buildings other than dwellings. 1998
  K: Ventilation, 1998
London, The Stationery Office. Available from www.tsoshop.co.uk

[37] **British Standards Institution.** BS 5427-1: 1996. Code of practice for the use of profiled sheet for roof and wall cladding on buildings. Design. London, BSI, 1996

[38] **British Standards Institution.** BS 5534: 2003. Code of practice for slating and tiling (including shingles). London, BSI, 2003

[39] **British Standards Institution.** BS 6229: 2003. Flat roofs with continuously supported coverings: Code of practice. London, BSI, 2003

[40] **Single Ply Roofing Association.** Design guide for single ply roofing. Available from www.spra.co.uk.

[41] **Metal Cladding and Roofing Manufacturers Association.** Technical guidance available from www.mcrma.co.uk

[42] **British Flat Roofing Council/CIRIA.** Flat roofing. Design and good practice: Design guide. 1993.

[43] **BRE.** Roofs and roofing pack. AP 207. Bracknell, IHS BRE Press, 2005.

[44] **BRE.** Flat roof design: bituminous waterproof membranes. Digest 419. Bracknell, IHS BRE Press, 1996

[45] **Saunders GK.** Reducing the effects of climate change by roof design. Digest 486. Bracknell, IHS BRE Press, 2004

[46] **British Standards Institution.** BS 6399: Loading for buildings. London, BSI
  Part 1: 1996 Code of practice for dead and imposed loads (AMD 13669)
  Part 2: 1997 Code of practice for wind loads (AMD 13392) (AMD corrigendum 14009).
  Part 3: 1988 Code of practice for imposed roof loads (AMD 6033) (AMD 9187) (AMD 9452)

FURTHER READING

British Standards Institution. BS EN 12056-3:2000. Gravity drainage systems inside buildings. Roof drainage, layout and calculation.

British Standards Institution. BS 5250:2002 Code of Practice for contol of condensation in buildings

British Standards Institution. BS 1925:1988. Specification for mastic asphalt for building and civil engineering (limestone aggregate).

British Standards Institution. BS 476-3:2004. Fire tests on building materials and structures. Classification and method of test for external fire exposure to roofs.

CIRIA and British Flat Roofing Council. Flat roofing: design and good practice. Book 15. London, CIRIA, 1993

Health and Safety Commission. Management of health and safety at work. Management of Health and Safety at Work Regulations 1992. Approved Code of Practice. London, The Stationery Office, 1992

Health and Safety Executive. Construction (Head Protection) Regulations 1989. Guidance on Regulations. London, HSE, 1990

Ruberoid Building Products. Ruberoid Blue Book: Flat roofing: A guide to good practice. 2003.

Technical committees of the Concrete Tile Manufacturers Association, Clay Roof Tile Council and the National Federation of Roofing Contractors. Roof tile fixing specification. The zonal method user's guide. March 2006.

2   THE BASIC FUNCTIONS OF ALL ROOFS

Functions to be taken into account in the design of roofs, apart from protecting the building from wind and rain, include:
- the maintenance of appropriate internal environments by control of heat loss,
- prevention of condensation,
- control of transmission of external noise,
- prevention of fire spread from external sources,
- provision of daylighting where appropriate, and
- many other important considerations vital to the performance of the building as an entity.

These functions will be considered in turn in this chapter as they apply to all buildings and therefore to all roofs. In the remaining chapters of this book, they will be considered in turn for each of the basic roof configurations chosen for examination.

For the future, the possible effects of climate change need to be taken into account. These will be concerned mainly with changes in temperature, wind and rainfall and will be dealt with in the appropriate chapter. The United Kingdom Climate Impacts Programme (UKCIP) has produced a series of scenarios which can be examined at www.ukcip.org.uk/[1]. The predictions made can be generalised as longer hotter summers and milder and moister (increased number and severity of rainfall events) winters. BRE Digest 499[2] gives recommendations for designing roofs for climate change in the form of modifications to good practice. These recommendations are included in each chapter as appropriate.

Figure 2.1 A complex roof will need good maintenance to keep the weather out

2.1 STRENGTH AND STABILITY

As *Principles of modern building*[3] pointed out, there are three primary structural systems that can be adopted for spanning across space: the chain, the arch and the beam. The characteristic features are:

- *the chain:* all its parts are in tension,
- *the arch:* all its parts are in compression,
- *the beam:* the top material is in compression and the bottom material is in tension.

Each system has its advantages and disadvantages.

The chain
The chain, catenary, hanging chain, or cable or suspended stay provides the lightest form of structure and is used in the longest spans. However, this form of structure can be inherently flexible and provision therefore has to be made for sometimes substantial movements.

The arch
The arch provides a suitable form for spanning large spaces only if the materials or components from which it is constructed are suitable to resist the large compressive forces which may be experienced. Horizontal thrust can also be large on shallow arches, and the simplest way of limiting that thrust is to provide a tie between the supports.

The beam
The beam provides the most versatile solutions for shorter spans. Considerable attention has been given to the development of the most efficient sections of beams, often using a combination of different profiles, sections and reinforcement techniques to exploit the characteristics of the chosen materials.

Individual spans using all three of the above techniques can be one-way or two-way. Roofs can be made up of different combinations of all three systems:

- suspended cable stayed or strutted main spans carrying intermediate spans of beams,
- two-way spanning vaults,
- one-way spanning arches carrying secondary one-way spanning arches, beams, beams on beams, or space frames.

The combinations are almost infinite, but the examination of the efficiencies and economics of each combination is totally beyond the scope of this book. All that can be done is to provide information relating to the most common variations with, perhaps, a brief mention of a few of the less frequently found forms.

However, every roof needs to be sufficiently strong to carry the self-weight of the structure together with intermittent loads — for example those due to environmental effects [eg snow (Figure 2.2) or wind] or maintenance — and it must do this without undue distortion or damage to the building, whether perceptible or imperceptible to its occupants.

These requirements, as far as the structure of roofs is concerned, are embodied in the various national building regulations:

- the Building Regulations 2000[4], which apply to England and Wales;
- the Building Standards (Scotland) Regulations 2004[5]; and
- the Building Regulations (Northern Ireland) 2000[6].

Figure 2.2: Snow redistributed by wind

As an example, requirements of the building regulations for England and Wales in relation to structural issues are:

'(1) The building shall be constructed so that the combined dead, imposed and wind loads are sustained and transmitted by it to the ground

   (a) safely; and

   (b) without causing such deflection or deformation of any part of the building, or such movement of the ground, as will impair the stability of any part of another building

(2) In assessing whether a building complies with sub paragraph (1) regard shall be had to the imposed and wind loads to which it is likely to be subjected in the ordinary course of its use for the purpose for which it is intended.'

STRUCTURAL REQUIREMENTS
The roof needs to be built on a supporting structure that will adequately transmit any loads through to the walls or other vertical supports. The loading on a roof is in the form of dead loads, imposed loads and live loads. Live loads include wind loading.

The requirements are given in the supporting documents of the national building regulations:

- England and Wales: AD A[4],
- Scotland: Section 1 of the Technical Handbooks, Domestic and Non-domestic[5],
- Northern Ireland: Technical Booklet D[6].

There is more guidance in BS 6399: Parts 1–3[7].

Changes have recently been introduced to the methods of ensuring compliance with the building regulations relating to structural design. As an example, consider the situation in Scotland. After a period when self-certification of structural design was permitted and alleged to be found wanting, a new certification scheme has been introduced involving independent assessment of the design. The approved certifier takes responsibility for the whole of the structural design – not separately for each individual building element, which must be seen as a significant improvement over the previous situation[8]. The forthcoming implementation in the UK of Eurocodes by 2010 will also see significant changes to practice.

Some evidence of how well the roofs of existing houses are performing in relation to dead and live loads is given in the various national house condition surveys. Some 15 years ago around 1 in 25 dwellings in England showed some problem with the roof structure. The oldest houses, as might be expected, showed structural problems more frequently than later ones (Table 2.1). Occurrences post 1965 show a significant increase in the period 1945–64. It is interesting to speculate on the reasons for this, coinciding as it did with the increase in the use of trussed rafters and concomitant reduction in skills in pitching traditional roofs.

Figure 2.3 illustrates the problems of sagging, humping and spreading; data for their incidence for all dwellings some 15 years ago and in 2005/6 is given in Tables 2.2 and 2.3.

In 2005/6 for all dwellings (including houses, bungalows and flats) structural faults were identified in the roofs of 20.3% of all dwellings (approximately 1 in 5). These figures represent a significant reduction in

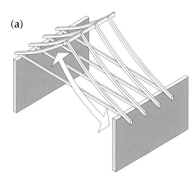

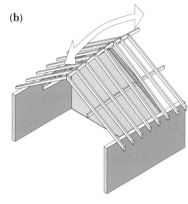

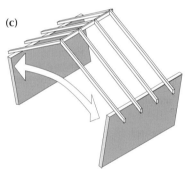

Figure 2.3: (a) Sagging, (b) humping, (c) spreading

recorded faults over the intervening 15 years.

In 1991, out of the 431,000 dwellings deemed unfit for occupation in England, the condition of the roof structure and coverings played a major role in 284,000 cases. By 2005/6 the criteria had changed, but 352,376 dwellings (1.6% of all dwellings) failed to meet the decent homes standard by reason of faults in their roofs.

In 1991, Northern Ireland dwellings showed 1 in 20 having faults in the roof structure, with more than three-quarters of them occurring in pitched roofs. In 2005/6 for all dwellings (including houses, bungalows and flats) faults were identified in the roofs of 8.3% of all dwellings (approximately 1 in 12). Table 2.4 gives data for the incidence of sagging, humping

Table 2.1: Reported problems with roof structure. Data from English House Condition Survey circa 1990[9]

Date of construction	Proportion of dwellings (approx.)	% of total
Before 1850	1:9	11.0
1945–1964	1:70	1.4
Post 1965	1:35	2.8
All dates	1:15	6.6

Table 2.2: Reported problems with roof structure. Data from English House Condition Survey circa 1990[9]

Type of problem	Proportion of dwellings (approx.)	% of total
All problems	1:15	6.6
Sagging	1:90	1.1
Humping	1:50	2.0
Spreading	1:100	1.0
Pre 1899	1:50	2.0
1918–1945	1:200	0.5
Post 1945	Negligible	

Table 2.3: Reported problems with roof structure. Data from English House Condition Survey 2005/6[10]

Type of problem	Proportion of dwellings (approx.)	% of total
All problems	1:5	20.3
Sagging	1:32	3.1
Humping	1:166	0.6
Spreading	1:250	0.4

Table 2.4: Reported problems with roof structure. Data from Northern Ireland House Condition Survey 2005/6[11]

Type of problem	Proportion of dwellings (approx.)	% of total
All problems	1:12	8.3
Sagging	1:142	0.7
Humping	1:1000	0.1
Spreading	1:500	0.2

and spreading for all dwellings in Northern Ireland in 2005/6. 5,528 dwellings (0.8% of all dwellings) failed to meet the decent homes standard by reason of faults in their roofs.

There is no current comparable information available from Scotland and Wales and enquiries should be addressed to their respective departments.

To design for every destructive eventuality or natural hazard would be prohibitively expensive, however, so it is customary in design to establish limits for acceptable risks of failure; for example, in codes of practice, or, in the case of loading, to design for loads which have an estimated probability of exceedance. The main points in relation to the requirements of the building regulations for each category of roof are dealt with in the appropriate chapters.

WIND

Gales in the UK cause extensive damage to buildings each year; most of this damage is to roofs. In the severe gales of early 1990, around 1.1 million houses were damaged and gales since then have continued to take their toll.

Two-thirds of the costs of the damage were from roofs. Following this, BRE published a leaflet[12] advising owners of pitched lightweight roofs to have their buildings examined and to have remedial work carried out where necessary.

Pitched lightweight roofs have always been a problem, as Figure 2.4 indicates.

Changes have been made to the rules for design since the 1939–45 war. In particular, the 1952 wind loading code did not take account fully of the effects of building shape and wind directions, and gave, as a result, lower wind loads than would be obtained today using BS 6399-2[7].

Although much of the damage in the 1990 gales was to old roofs, many newer and well maintained houses suffered too. Roof failures account for, on average, one death every two years, and about 12 or so injuries per year. Flying slates and tiles alone account for about two injuries per year in those accidents which are recorded. It is estimated that many more go unreported and unrecorded. These casualty rates must be kept in perspective and it should be remembered that they are very small in comparison with some other causes. It is a matter for the collective judgement of society, operating through building regulations and British Standards, whether such rates are acceptable, as it would be expensive to uprate all standards to provide for better protection. Nevertheless, it is the responsibility of all concerned with roofing to understand the effects of wind on the structures for which they are responsible, to assess the risk of failure in any one case, and the consequential effects on their professional indemnity insurance. Assistance and advice may be

available from manufacturers and suppliers of roofing materials in the form of calculations for determining fixing requirements.

Two approaches can be taken into account for the effects of climate change on roof design.
- Modify key factors in all the calculations to establish design parameters (eg increase the basic wind speed by a factor).
- Apply a modification that takes account of such factors but may be simpler to apply (eg increase the number of fixings per unit area).

By adopting these modifications the roof may be over-designed. However, it could save much more in the future as repairs are avoided and periods of maintenance are extended.

The live load on a roof is due mainly to wind. There is a higher degree of certainty that winter storms will increase, but extreme events, such as the storm of 1987, may also occur more frequently. All the codes of practice for roofing refer to BS 6399[7] for guidance on calculating wind loads. The UKCIP02 scenarios are predictions for future weather patterns based on computer simulations[1]. Some of the weather predictions have a high degree of certainty of becoming reality (eg hotter, drier summers). Some individual features of the weather, such as wind behaviour, are harder to predict and consequently the certainty of the prediction actually occurring is less. The uncertainty of these scenarios means that recommendations for increasing the design wind speed are not justified, based on the scenarios alone. However, there is the risk that wind speeds may increase. In this case, a designer may decide to make modifications with a view to reducing future costs or liability. Design wind speed is specified in BS 6399-2[7]. This could be increased by a factor of 10% with only minor increases in construction costs.

Wind speed

Wind speed varies depending on the geographical location of the site, the altitude of the site above

Figure 2.4: Gale damage to aluminium-sheeted pitch roofs in the 1950s

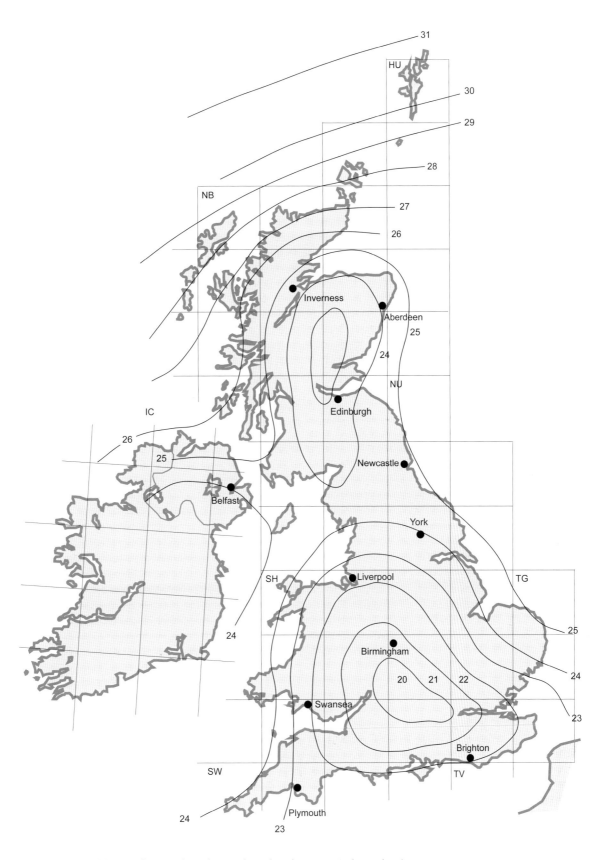

Figure 2.5: A map of the UK showing the reference basic hourly mean wind speed (m/s)

sea level, the direction of the wind, and the season of the year. More local effects such as height above ground level, topography and terrain also affect the speed and direction of the wind at a particular site. In the immediate vicinity of a building or a group of buildings, the wind changes speed and direction rapidly, depending on the form and scale of the buildings. Wind speeds in coastal regions are generally greater than in inland areas; for instance, the speeds near the coasts of southern England are some 10–25% greater than at the same altitudes in places in the centre of southern England. The highest wind speeds occur in the north of the British Isles, the highest basic gust wind speeds used in design according to BS 6399-2[7] being 56 m/s in the far north west of Scotland.

The contour lines (isopleths) in Figure 2.5 give the maximum wind speeds which have 2% probability of being exceeded in any year. This map is compiled from records of the speed of steady wind and gusts, prepared continuously at meteorological stations at various locations in the UK, and then averaged to give mean wind speeds for each hour. The records are then adjusted statistically to give the basic wind speed for design purposes. This map shows basic wind speed at 10 m above ground level in 'standard' country terrain at sea level. To find the wind speed at a particular site, the basic wind speed is then corrected for altitude and all the other factors described below.

Altitude
Speed over reasonably level ground increases by 10% with each additional 100 m of altitude above sea level. About half of the UK is below 200 m with wind speeds at most 20% greater than those at sea level.

This effect due to altitude accounts only for changes in wind speed caused by slowly changing topography (Figure 2.6). More rapid topographic changes (hills, cliffs, escarpments and ridges) cause local changes in wind speed (for more details see BRE Digest 346, Part 5[13]).

Wind direction
In the UK the strongest winds are the 'prevailing' winds. Throughout the country, these prevailing winds come from between the west and south west; the direction does not vary significantly with location (Figure 2.7). The wind blows from the prevailing direction nearly one-and-a-half times as often as from the opposite point of the compass. The frequency of winds from other directions is between these extremes. BS 6399-2[7], however, calls for designs taking into account wind from any direction.

Another form of diagram, a 'wind rose', is often used to show the frequency of wind (and often wind

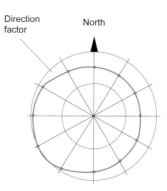

Figure 2.7: Directional variation of extreme winds in the UK

speeds) blowing from each direction. The length of each arm of most wind roses is proportional to the frequency of wind blowing from that direction, each direction covering a 30° angle. This diagram can be related visually to wind-sensitive and wind-protective features on site plans.

Data on wind direction are also available in tabular form by application to the Meteorological Office.

Height above ground level
The speed of the wind is reduced, due to surface friction, when it passes over the ground; the more so if the ground is rough.

There are two basic categories of terrain to be provided for in design for wind action: country and town. However, there are modifying factors for the distance from the sea, and from the edge of town. Over the sea, wind speed is greatest and turbulence is least. Over towns, wind speed is less and turbulence is greatest, irrespective of altitude. For country, the values of wind speed are somewhat lower than those for town.

Wind speed increases with height above ground level, being slowed down to zero at ground level by surface friction. Figure 2.8 shows how the wind speed varies with height in an area of smooth terrain such as the sea or a very large lake, and in a rough area such as a town. In the rough terrain the wind tends to skip over the buildings, leaving sheltered areas. As a result, the effective height of buildings for wind calculations can be reduced.

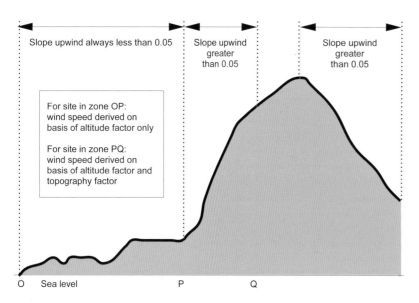

Slope upwind always less than 0.05

Slope upwind greater than 0.05

Slope upwind greater than 0.05

For site in zone OP: wind speed derived on basis of altitude factor only

For site in zone PQ: wind speed derived on basis of altitude factor and topography factor

O   Sea level   P   Q

Figure 2.6: The effect of altitude on wind speed (vertical scale is exaggerated)

(a)

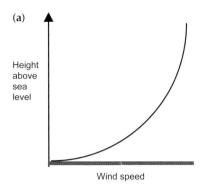

(b)

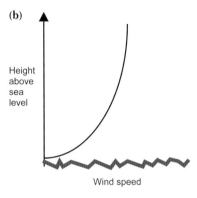

Figure 2.8: Wind speed varying with height: (a) smooth terrain, (b) rough terrain

This change in the speed and turbulence of wind does not take place instantaneously as the wind passes from a smooth surface, such as the sea, to a rough surface, such as a town: there is a gradual transition. So to predict the speed of the wind from a particular direction, at a particular site in the country, it is necessary to measure the distance the wind has travelled from the sea and make corrections. If the site is a town, a further similar correction is then made for the distance of the site from the edge of the town. Figure 2.9 illustrates this. This method will give different values of wind speed for winds coming from different directions, and is over and above the difference due to the climatic variation in wind speeds for different directions discussed earlier.

Gusts

Wind speeds vary greatly from place to place and from moment to moment due to turbulence caused by convection and friction of the air over the ground. The scale of the turbulence also varies greatly. Major eddies may extend over several kilometres and last several minutes. At the other extreme, small, though possibly severe, eddies caused by minor obstructions such as buildings may extend over only a few metres and last only a fraction of a second.

The critical factor for the design of a particular building or its roof is the speed of the gust which is just large enough to envelop the whole building or roof, or the whole component when designing a part such as a roofing panel, or in calculating the suction on an individual tile.

The increase in the wind speed due to gusts is greater for small buildings and greater still for small components such as roofing tiles. With large buildings, the smaller gusts will not act simultaneously, so they tend to cancel each other out. For a building 10 m high with a diagonal of 20 m, at a site 100 km inland where there are no special topographic effects, the gust speed for the design of the building as a whole will be rather more than 50% greater than the steady wind speed.

For a component on the same building with a diagonal less than about 5 m (eg a roof light) the gust speed will be rather more than 60% greater than the steady wind speed. For components as small as tiles and slates, it will be correspondingly greater still.

Topography

For predicting wind speeds, topography is classified into three shapes:

• valleys,
• hills and ridges, and
• escarpments and cliffs.

The behaviour of the wind depends on the upwards slope of the topographic feature. There are three distinct cases:

• gentle topography,
• shallow topography,
• steep topography.

Gentle topography
Gentle topography has a gradient of less than 5%. Its impact on wind speeds is included in the correction due to altitude which was described earlier.

Shallow topography
Shallow topography has gradients of 5–30%. It is the type found most often in the UK.

With shallow topography, the wind follows the shape of the slope (Figure 2.10). In the case of the shallow valley, the speed of the wind flowing across the valley drops as it enters the valley and picks up to its original speed again as it leaves the valley (Figure 2.10a); but no

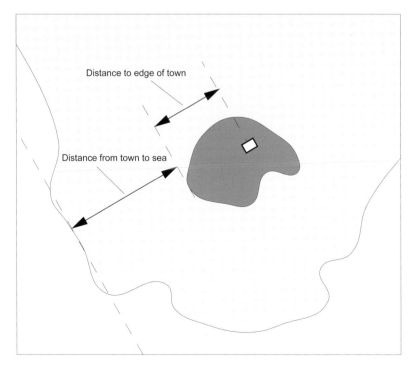

Figure 2.9: Correction for the distance of a building from the edge of a town

(a)

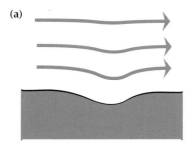

(b)

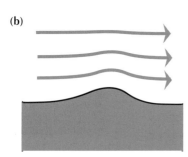

(c)

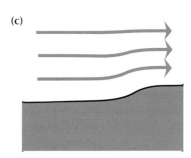

Figure 2.10: Wind speed: shallow topography (a) valley, (b) hill or ridge, (c) escarpment or cliff

reductions in design wind speed are recommended for this or any other direction. When the wind flows along the valley, it is not significantly changed.

Wind blowing over a shallow hill or ridge, or escarpment, accelerates up the upward slope to a maximum at the summit or crest (Figures 2.10b and c). The acceleration begins before the slope at a distance from the summit or coast of about 1.5 times the length of the slope. The effect on wind speed varies with height above ground level and this effect is greatest at ground level. For the steepest shallow slope, with a gradient of 30%, the basic wind speed will increase between ground level and the summit or crest by 60% due to this topographic effect. For lower slopes, the increase in mean wind speed at the summit or crest can conveniently be calculated as twice the gradient, up to a maximum of 60%: so for a 10% slope the increase is 20%.

In the case of a shallow hill or ridge, the wind then decelerates on the downwind slope, returning to its initial speed at a distance beyond the top of some 2.5 times the length of the upward slope (Figure 2.10b).

In the case of a shallow escarpment, the wind decelerates more slowly, eventually reaching a new speed appropriate to the change in altitude (Figure 2.10c). The effect on the winds flowing down escarpments is symmetrical to those flowing up them.

When the wind blows from a direction which is not at right angles to the slope of a topographic feature, such as an escarpment, the effect is to make the angle of the slope shallower, so the acceleration of the wind will be less.

Steep topography
Steep topography has a gradient of more than 30%. The reason for defining topography in this way is that for slopes greater than 30%, the wind behaves quite differently. Instead of hugging the crown, the wind breaks away from it, creating a separate flow in the space it bypasses (Figure 2.11).

In the case of a steep valley, the wind jumps across the valley, leaving a sheltered region in the valley itself (Figure 2.11a); but, for design purposes, no reduction of wind speed is recommended for any wind direction. When the wind blows along the valley, no shelter occurs.

In the case of a steep hill or ridge, or escarpment, the wind breaks away from the ground before the upwind slope at the point where a 30% slope would have begun (Figures 2.11b and c). The flow rejoins the ground at the summit or crest. So the main flow is, for practical purposes, the same as for a 30% slope. However steep the slope is, the acceleration of the wind at the summit or crest is no more than it would be for a 30% slope.

In the immediate vicinity of a building or group of buildings, wind changes speed and direction rapidly depending on the form and scale of the building. Severe eddies and vortices occur (eg at corners, eaves and ridges, and at any projections such as chimneys, dormer windows

(a)

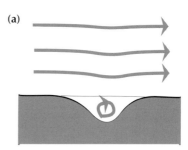

(b)

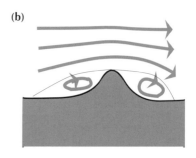

(c)

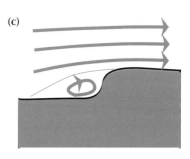

Figure 2.11: Wind speed: steep topography (a) valley, (b) hill or ridge, (c) escarpment or cliff

or tank rooms) thereby creating positive and negative pressures (eg suction on the roofing). The slope of the roof is also an important factor.

When designing for wind, these effects are not taken into account by computing wind speeds at all points on the building. Instead, empirically derived pressure coefficients are chosen at critical points for the form and scale of the building. These are then used to modify the basic wind pressure derived from the speed of the wind at the appropriate height of the site.

Seasonal factors
Seasonal factors can be relevant to construction work. The highest speed extreme winds in the UK are expected in December and January. In the summer months of June and July, extreme winds may be expected to be only about 65% of the winter extremes; for the 6-month summer period from April

to September, normal wind speeds are only expected to be 84% of those in the 6-month winter period, October to March. Tables of these seasonal factors are published for 1-month, 2-month and 4-month periods starting in each month of the year[14].

Wind loading for buildings with pitched and flat roofs is further discussed in Chapters 2.10 and 2.11, respectively.

For the assessment of new materials, the Method of Assessment and Test No 27[15] of the European Union of Agrément (UEAtc) specifies the use of a test regime which includes the use of various suction loads on specimens. The tests are of 500 cycles at each pressure change in 1 kPa steps up to 10 kPa. The specimens are then examined for various types of failure including, for bonded systems:
• delamination of the waterproofing system,
• failure of the waterproofing/adhesive interface,
• failure of the adhesive,
• failure of the adhesive/deck interface,
• failure of the deck,
and for mechanically fixed systems:
• delamination of the waterproofing system,
• tearing of the waterproofing material at the point of fixing,
• pull-out of the fixing from the deck.

Further tests under the UEAtc regime are specified for resistance to peeling.

A test method has been developed as a European Technical Approval Guideline (ETAG) for testing the fastener and membrane for mechanically-fixed single-ply membranes. This allows CE-marking of such a system.

BRE can provide realistic test regimes; for example, using computer-controlled simulation of actual wind-flow patterns (Chapter 5.2).

SNOW

Snowfall in the UK varies from year to year, and with geographical location and altitude. Snow loads tend to increase generally with altitude. Snow density can vary considerably (eg from 0.05 to 0.2 g/cm³), and density can increase by compaction if snow lies for long periods. Compaction will not, however, increase the actual loads on the roof unless rain saturates the unmelted snow.

Extremes of snow loads can arise in two ways:
• through heavy snowfall, and
• through drifting in windy conditions.

Ground snow measurements (which have been converted into loads in Figure 2.12 and corrected to a datum of 100 m above sea level to remove the influence of altitude) are carefully made to exclude the effects of drifting. It is rare in the UK for snow to lie sufficiently long for successive falls to accumulate, and extremes are therefore most likely to ensue from a single snowfall. Sheet ice can also form in certain circumstances, though rarely to thicknesses greater than 50 mm, giving maximum loads of about 0.45 kN/m².

Although basic values for ground snow load are available for the UK, there are no corresponding systematically collected meteorological data for snow load distributions on roofs. The natural accumulation of snow on a roof is influenced not only by the amount and condition of the snow but also by the wind speed, wind direction, and the shape, size and form of the building and those surrounding it.

Snow tends to accumulate at obstructions to form local drifts, and asymmetric snow loads can occur on some forms of pitched and curved roofs, particularly those with valley gutters.

Until the problem of drifting snow was identified as a particular matter for attention in design, roofs were designed for a uniform load allowance. So, before the publication of BS 6399-3[7], the designer was not required to consider the possible variety of snow load distributions on roofs in the UK. Roof slopes of less than 30°, with no access other than for cleaning and repair, were designed for a uniformly distributed imposed load of 0.75 kN/m² measured on plan and for a concentrated load of 0.9 kN. For roof slopes greater than 30°, the intensity of the uniformly distributed load was modified by the application of a linear reduction factor which varied between 1 for a slope of 30° and 0 for a slope of 75° or more. The earlier codes made no allowance for drifting and accumulation of snow on particular areas of a roof, and took no account of the regional variation in snowfall in the UK.

Now roofs have to be designed to take into account the different possible snow load distributions. There is more information in BRE Digest 439[16]. On some small roofs where significant drifting is unlikely to occur, the loading is taken to be a uniformly distributed load approach similar to, but not identical with, the previous method. Where a roof is altered, and an obstruction formed at which snow could accumulate in a local drift, the possibility of increased loads has to be taken into account.

The calculations for dead and imposed loads already take into account ponding of water on flat roofs. Since winters are predicted to be milder, no increase in snowfall is anticipated so loads calculated for snow remain the same as for the current climate. Dead loads are not expected to be affected by climate change other than by changes in roof design (eg using some designs of green roofs may impose higher dead loads on the roof structure).

Increased loads could be indirectly driven by climate change. The reduction of the impact of buildings on the environment together with the need to limit energy consumption has historically led to an increase in the thickness of thermal insulation incorporated into roof design. These measures are part of reducing the effects of human activities which may have led to climate change occurring. The thickness of thermal insulation has been determined by national building regulations which have been amended many times to reduce heat loss from buildings.

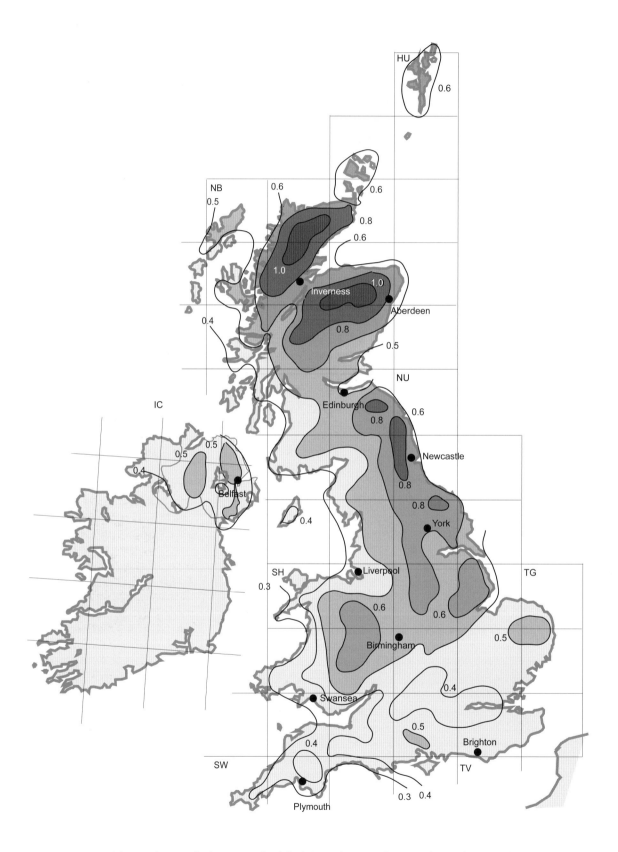

Figure 2.12: A map of the UK showing the basic snow load (kN/m²) on the ground (not on the roof) 100 m above sea level. These values must be corrected for altitude, and they do not allow for drifting; they have a probability of exceedance of 0.02

SMALL-SPAN ROOFS

There have been few problems due to snow loading on small-span domestic roofs of traditional construction. This gives confidence that roof members — sized, for example, according to Section 2B of Approved Document A of the Building Regulations 2000 (England and Wales)[4] — will be satisfactory under likely snow loads. However, where alternative designs of roofing are proposed which cannot follow these simple rules, calculations for snow loading have to be carried out.

BS 6399-3[7] permits a simplified set of load conditions for the design of some small buildings where all the following criteria are complied with:

- no access is provided to the roof except for cleaning and maintenance,
- the total roof area is not larger than 200 m², or the roof is pitched and not wider than 10 m with no parapets,
- no other buildings are within 1.5 m distance,
- there are no abrupt changes in height greater than 1 m (eg from dormers or chimneys) or, if there are such abrupt changes in height, the lower area is to be not greater than 35 m²,
- there are no other areas on the roof which could be subject to asymmetrical snow loading through, for example, manual or wind snow clearing.

If all the above conditions are satisfied, then the simplified procedure may be used. This procedure is based on consideration of a uniform snow load according to geographical location (with a specified minimum value). The restrictions applied ensure that any drifting which does occur on the roof will not be significant and will be covered by the uniform load case given (in addition to the concentrated load requirement).

LARGER-SPAN ROOFS

Since the 1970s there have been a number of cases where larger-span roofs have suffered serious structural damage due to the accumulation of large quantities of snow. The estimation of snow loads on larger non-domestic roofs is complex, and it is not possible to provide simple design rules.

Loads to be taken into account in design are covered in BS 6399-3[7]. The design procedures take into account regional variations in loading, and also the redistribution of snow by wind. A detailed exposition of the effects of snow on roofs can be found in the *Handbook of imposed roof loads*[17]. BRE Digest 439[16] also applies.

BS 6399-3[7] gives load shape coefficients for local drifting of snow, and uses the snow load statistical factor to produce estimates of loads with probabilities of exceedence different from 0.02. Digest 439[16] presents background information and design requirements for local snow drift loads on roofs to accord with the amended BS 6399-3. For the purposes of calculations of snow loading, roofs up to 30° pitch are assigned the same maximum uniform snow loading as a flat roof. For a pitched roof of 30° or more, the asymmetric snow load case is different.

For buildings not designed to the simplified requirements for small buildings, the following load conditions must be considered:

- the uniformly distributed snow load,
- asymmetric snow load due to redistribution by the wind,
- local drifting,
- partial loading due to snow removal.

These snow loads must be considered in addition to other loads arising from access and maintenance such as:

- uniform distributed load measured on plan dependent on degree of intended access to the roof,
- concentrated load dependent on degree of intended access to the roof.

These load conditions (listed above) are not additive, and for most cases are considered individually. It is assumed that:

- sheet ice maximum loading will not coincide with maximum snow loads,
- additional loads from people and their equipment will not coincide with maximum snow loads,
- uniformly distributed snow load and loads due to redistribution of snow do not coincide.

It is also assumed that drifts form only up to the height of an obstruction (but valleys are assumed to be filled).

Provided that significant rainfall does not follow extreme snowfall, the imposed loads due to the accumulation of meltwater on a flat roof will not exceed the maximum snow load on the roof. On roofs which are not flat but are capable of retaining water (eg a pitched roof with parapets), blocked drainage systems will cause large localised loads. Snow boards in the gutters will assist in keeping drainage paths clear (Chapter 7.1).

2.2 DIMENSIONAL STABILITY

Expansion and contraction of any part of the building fabric subjected to variations of moisture content and temperature will have the potential to cause problems if not accommodated in the design. Where elements are totally restrained, the stress developed is the product of the strain and Young's modulus, and will lead to:

- compressive forces for expansion, and
- tension forces for contraction.

As a general rule, all common building materials will be subject to thermal expansion and contraction. So far as materials used in roofing are concerned, it is the larger components that need most consideration. An example might be the use of a glass-fibre reinforced polyester (GRP) valley gutter; chopped strand GRP has a coefficient of linear thermal expansion of 20–35 per °C × 10⁻⁶, and components made of this or other plastics (especially in the range 50–150 per °C × 10⁻⁶) will tend to move more than some metals of comparable sizes. Steel, for example, has a coefficient of 12 per °C × 10⁻⁶ (BRE Digests 227–229[18–20]).

Moisture movement is mainly a property of porous materials; thus metal, glass and well-fired ceramics have zero movement, while concrete, calcium silicate, porous clay and stone have reversible movements in the range 0.02–0.2%. Concrete and calcium silicate products also have an irreversible drying and carbonation shrinkage in the range 0.02–0.1%, and fired clay has an irreversible initial expansion in the range 0.02–0.1%. In most real situations there is some restraint, though rarely complete restraint, afforded to materials undergoing these movements.

Where two differing materials are joined, differential movement can occur which usually exacerbates the problem. The best practice is to try to accommodate movements at the smallest scale, provided this does not prejudice the structural, thermal or rain exclusion functions. Therefore, provided roofing components are simply shaped, not too thick, not too large, have movement-tolerant fixings and have joints at their periphery which accommodate movement, the movement strains and the corresponding loads (stresses) are generally small and can be borne by the element or panel. With large structures, if one or more of these requirements are not met, stresses may accumulate over large areas and damage may result. In small structures such as detached domestic houses, it is sometimes possible to omit explicit movement design and to depend on restraint from other elements to accommodate movement loads.

Progressive movement occurs at the edges of roofs (eg at eaves) mainly as a result of ratcheting where movement resulting from expansion (usually thermal) is not fully compensated by contraction, and the material, eg in a concrete deck, cracks. The resulting crack may be filled by loose particles or debris, which keeps the ratcheting process going. There might also be differential movement to take into account, for example, in timber-frame construction. Design information on movements of the underlying structures is given in BRE Digests 227–229 [18–20], clause 20 of BS 5628-3[21], and section 7 of BS 8110-2[22].

ACCIDENTAL AND MAINTENANCE LOADS

During service, it can be assumed that the roofs of buildings will be subjected to irregular applications of accidental and maintenance loads of widely varying magnitudes. Loads may cause damage which ranges from no more than unsightly disfigurement of the roof surface to breaking and dislodgement of tiles with serious risk to the safety of both occupants entering and leaving, and also to passers-by.

One common occurrence noted in site inspections is that of inadequate support for wide valley gutters in pitched roofs, particularly those that have been prefabricated and which provide easy routes for maintenance traffic, but at considerable risk of fracturing or puncturing the surface of the gutter.

Impact damage will normally be restricted to that caused by ladders and mobile platforms used for maintenance access. Limited damage by missiles may also occur where vandalism is a problem. With respect to explosions and impacts from vehicles and missiles such as the parts of planes, in normal buildings it is not economic to design for zero damage. In special cases, though, such as nuclear power stations, roofs should be fully designed for these events. In normal roofs, the main principle adopted is to limit damage to as small a section of the roof as possible, enough that the remainder of the structure survives and provides an escape route for occupants.

2.3 EXCLUSION AND DISPOSAL OF RAIN AND SNOW

In addition to carrying structural loads, the roof must perform several other functions depending on the use to which the building is to be put.

There is an absolute requirement to prevent snow and water reaching the interior of a building, in contrast with some of the other functional requirements where some shortfall may be tolerable, at least for brief periods. Within the terms of building regulations, control of moisture penetration is a functional requirement, and the building needs to be designed and built adequately to resist such penetration.

Two kinds of rainfall intensity need to be considered in relation to roofs:

- rain falling approximately vertically,
- rain driven by wind.

Both categories contribute to the total quantities of rainwater needing disposal, but the second category particularly affects the weathertightness of lapped roofing, such as tiles and slates, and even the direction and extent of lap of larger sheets.

The capacity of any roof drainage system should be adequate to dispose of intense rainfall that usually occurs in summer thunderstorms. Although the upland areas of the north and west of the UK receive higher average rainfalls than eastern areas, it is south east England where very intense short duration rainfalls are more frequent.

The Building Regulation requirement for England and Wales H3 (1) states:

'Adequate provision shall be made for rainwater to be carried from the roof of the building'. [4]

Although keeping out the rain could be considered as one of the prime performance requirements of roofs in general, a surprisingly high number have proved to be defective in this respect. Evidence from the English House Condition Survey 1991[9] shows that nearly 1 in 12 of all houses had defects in their valley gutters or flashings, and 1 in 5 had defects in their rainwater disposal systems. These data are unlikely to have changed by more than 1 or 2 percentage points in the intervening years.

DESIGN RAINFALL INTENSITIES

For eaves gutters, the rainfall intensity is given in Figure 2.13. This shows rainfall intensities expressed in litres/sec/m^2 and range from 0.012 litres/sec/m^2 in the middle of Wales to 0.022 litres/sec/m^2 north of London and other parts of eastern England. This map shows the intensity for a 2-minute duration storm with a return period of 1 year. The 2 minutes refers only to the storm's peak, so the actual rainfall will last longer than 2 minutes. The data used to generate the maps are historical. No account of climate change has been taken, so it is recommended that a higher rainfall category should be selected in most circumstances. The next highest category may be appropriate, pending the accumulation of revised data.

The flow into an eaves gutter depends on the area of the surface being drained and whether the surface is flat or pitched and the

Table 2.5: Calculation of drained area. Data from Table 1, Approved Document H[4]

Type of surface		Effective design area
1	Flat roof	plan area of relevant portion
2	Pitched roof at 30°	plan area of relevant portion × 1.29
	Pitched roof at 45°	plan area of relevant portion × 1.50
	Pitched roof at 60°	plan area of relevant portion × 1.87
3	Pitched roof over 70° or any wall	elevational area × 0.5

angle of pitch. The effective area is calculated using Table 2.5.

Table 2.6 shows the maximum effective area which should be drained into the gutter sizes which are most often used. These sizes are for a·gutter which is laid level, half round in section with an outlet, having a sharp-edged (rather than rounded) profile, at only one end and where the distance from a stop end to the outlet is not more than 50 times the water depth. At greater distances, the capacity of the gutter should be reduced. Table 2.6 shows the smallest size of outlet which should be used with the gutter.

Where the outlet is not at the end, the gutter should be of a size appropriate to the larger of the areas draining into it. Where there are two end outlets they may be up to 100 times the depth of the flow apart.

The gutters should be laid with any fall towards the nearest outlet and laid so that any overflow in excess of design capacity will be discharged clear of the building,

Figure 2.13: Rainfall intensities for design of gutter and rainfall pipes (litres/sec/m²)
© Crown copyright. Reproduced from Approved Document H of the Building Regulations (England & Wales)[4]
under the terms of the Click-Use Licence)

Table 2.6: **Gutter sizes and outlet sizes. Data from Table 2, Approved Document H**[4]

Maximum effective roof area (m²)	Gutter size (mm diameter)	Outlet size (mm diameter)	Flow capacity (litres/sec)
6.0	—	—	—
18.0	75	50	0.38
37.0	100	63	0.78
53.0	115	63	1.11
65.0	125	75	1.37
103.0	150	89	2.16

Note: Refers to nominal half-round eaves gutters laid level with outlets at one end and having a sharp-edged profile. Outlets having a round-edged profile allow smaller downpipe sizes

reducing the risk of overspilling of rainwater into the building or of causing overloading of the structure. Previous generations of designers were of course familiar with these risks: the gargoyles on medieval churches are examples of solutions to the problem (Figure 2.14).

A simplified design procedure for rainwater disposal from domestic pitched and flat roofs is given in Chapter 3.1 and BRE Good Building Guide 38[23].

For gutters other than eaves, for example valley gutters, parapet gutters, and drainage systems from flat roofs, where overspill from these would have serious consequences if water enters the structure, then the rainwater system should be designed

Figure 2.14: A gargoyle over the hopper head on a church

following BS EN 12056-3[24]. Four categories of design rate are given together with relevant maps.

In certain areas of the UK subject to high levels of driving rain it may be worthwhile to provide verge gutters in addition to eaves gutters, to cope with rainfall driven across the slope of the roof which might otherwise be blown off the verge of the roof.

SIPHONIC ROOF DRAINAGE SYSTEMS

The use of siphonic storm-water drainage was in its infancy when the first edition of this book was prepared. Subsequently, there has been increasing interest in the use of the technique, since, for a larger building, it offers a number of potential advantages over hitherto universally specified gravity drainage systems. The technique has been used on both refurbished and new buildings. Advantages include:

- suction-induced full-bore pipe flows and hence a reduction in the number of down-pipes, particularly useful when duct space is limited. There may also be a concomitant reduction in below-ground drainage runs.
- removal of the necessity for sloping horizontal discharge pipes (flat discharge pipes can be more easily accommodated within floor voids).

Performance requirements are contained in BS EN 12056-3[24] and design guidance, together with testing, commissioning and maintenance is given in BS 8490[25].

The outlets each consist of a conventional grille to exclude leaves, a large-diameter shallow bowl with an anti-vortex plate surrounded by baffles, which have the effect of restricting the volume of air carried into the system and a small diameter outlet to carry the discharge to waste. When rain falls, the bowl is filled, the air is excluded and a suction is induced, causing the outlet to flow at full bore.

Maintenance may be required at more frequent intervals, depending on the location of trees and potential detritus (eg from green roofs) to ensure the drainage outlets are cleared of debris and perform as intended.

EAVES DROP SYSTEMS

These are gutterless systems that simply discharge rainwater to the surrounding ground. They are commonly used on thatched roofs where a good overhang is provided, but the design should address issues of protection of the building from splashing, protection of entrances from falling water and protection of foundations from concentrated discharges.

SNOWFALL

Gutters that are designed for rainfall will be perfectly satisfactory for the drainage of melting snow. Gutters and outlets can become blocked by frozen snow and this danger can be mitigated by the use of trace heating or snowboards.

Snowguards can be fitted to the eaves of a pitched roof where sliding snow is a risk to people or any structure below. They can be necessary for roof pitches up to 60° from the horizontal and need not be higher than 300 mm. Fixings should be strong enough to withstand forces calculated in accordance with BS 6399-3[7].

SPLASHING OF RAINFALL

Rain, whether wind driven or vertical, is conventionally expected not to splash up more than 150 mm from horizontal hard surfaces. This is the height normally used for siting damp proof courses (dpcs) above paving as well as for flashings above roofing level. However, detritus carried by splashes has been found on vertical surfaces up to 300 mm above horizontal surfaces or paving, indicating that the 150 mm guide height is insufficient to prevent this phenomenon in many circumstances.

HAILSTONES AND SNOW

The impact of small hailstones up to 10 mm in diameter does little or no damage to buildings. On rare occasions hailstones over 75 mm across can fall in severe local storms. These can weigh over 100 g and can smash roof lights and patent glazing, and dent metal roof coverings. Even smaller sized hailstones have been known to cause extensive damage to horticultural glasshouses. On average, one of these severe hailstorms may occur at some place in southern Britain about once in five years, but they are very localised and the likelihood of one occurring at any given place is very small indeed.

Dry, wind-blown snow can penetrate tiny gaps in roof coverings, and accumulate, sometimes in drifts, within the roof void. The correct installation of sarkings is important in preventing penetration (Figure 2.15).

RAINWATER RECOVERY SYSTEMS

Guidance on rainwater recovery systems is covered in Approved Document (AD) H2[4], BS 8515[26] and two WRAS guidance documents[27, 28]. Where rainwater run-off could occur from shingled or timber-boarded roofs, the harvesting of rainwater may not be appropriate because substances such as tannins leached from the timber can discolour the harvested water.

RAINFALL AND DRIVING RAIN

Periods of higher rainfall rates are not considered to be a direct problem for most roof coverings. Removing the water at a sufficient rate is considered in the earlier section on siphonic roof drainage systems.

Figure 2.15: A plain tile roof with no sarking and no torching. This roof can be penetrated by fine, wind-blown snow, though a roof clad with machine-made tiles is less vulnerable than one with irregular handmade tiles

Wind-driven rain can enter a tiled or slated roof through capillary leakage. To assess the risk of water ingress the wind-driven rain index is normally used. Although the index varies throughout the UK, the highest driving rain index could be adopted to safeguard against this risk due to climate changes.

At the time of writing, a CEN standard for a test method for wind-driven rain on roof coverings with discontinuous laid small elements is being developed[29]. This test method will be used to determine the resistance of pitched roof coverings to wind-driven rain. Four test conditions are given for a number of climatic areas. The UK would be classed as Northern Europe coastal. Four sub-tests are also given which vary the wind speed and rainfall rate.

The following are relevant to the UK and are both worst-case scenarios for a return period of 1 in 50 years:
- *Sub-test B* is for wind-driven rain and a maximum amount of water leakage.
- *Sub-test D* is for a deluge and no leakage is allowed during the test.

The lower the pitch of the roof, the higher the risk of water ingress due to driving rain. BS 5534[30] limits the pitch for double-lap clay tiles and concrete plain tiles to 35°, single-lap clay and concrete tiles to 30° and double-lap fibre cement and natural slates to 20°. It is anticipated that these limitations will provide adequate resistance to climate change. Any slate/tile/pitch combination below these limits should be tested to ensure their performance is satisfactory for the anticipated conditions.

A minimum pitch of 25° is used to minimise the risk of exceeding the driving rain penetration limit of the system. The manufacturer's recommendation for the head lap should be a minimum of 100 mm.

MASONRY PARAPET WALLS

Driving rain can be a problem for masonry parapet walls. These are typically on flat or low-sloped roofs. Any rain penetration is normally dealt with by the provision of dpcs. However, these are not always installed correctly. Any dpc should be located so that it drains to the outer wall. There is more information in *Understanding dampness*[31].

2.4 ENERGY CONSERVATION AND VENTILATION

There have been substantial amendments to the energy conservation requirements for both domestic and non-domestic buildings, whether newly built or refurbished, in the last few years. These requirements play a major role in the design of all types of roofs, for example, the introduction of the new Approved Document (AD) L in England and Wales (April 2006)[4], Section 6 of the Technical Handbook in Scotland[5] and Technical Booklet F in Northern Ireland[6].

Protected buildings (listed buildings, buildings in conservation areas and ancient monuments) are exempt from the energy-efficiency requirements of the latest building regulations, but this still leaves the vast majority of the UK's existing stock of buildings which arguably needs to be brought up to acceptable/required standards.

AD L will play an increasingly significant role in the design and construction of roofs of both existing and new housing and non-housing in England and Wales, particularly in insulation levels, the effects of airtightness and the installation of rooflights and roof windows, ie AD L 'controlled fittings'. However, since building regulations are not prescriptive, designers are free to adopt different solutions in meeting the whole-building carbon emission targets for new buildings or elemental performance standards for existing buildings where carbon targets are not appropriate. Complying with the maximum U-value for a roof and other elements, will not necessarily be adequate to satisfy the requirements for the whole building.

Table 2.7: Limiting standards in AD L (2006) for roofs in new buildings (domestic and non-domestic)

	U-value
Area weighted average	0.25
Individual element	0.35
For existing buildings having pitched roofs:	
Insulation at ceiling level	0.16
Insulation between rafters	0.20
For flat roofs:	
Between rafters	0.20
When the element is in an extension	0.20
When the element is in an existing building	0.25

Limiting standards for roofs in new buildings, both domestic and non domestic, are given in Table 2.7.

Note that the Code for Sustainable Homes (CSH)[32] and BREEAM[33] assessments require enhanced performance for energy conservation and therefore higher standards of thermal performance for roofs and other envelope construction.

The thermal performance of the roof is a very significant item in the consideration of energy conservation of any building, particularly where that building is single-storey. Although the roof, arguably, is the easiest element to upgrade thermally, there may be side-effects. BRE has published guidance[34] which, amongst other things, draws attention to the potential risks in roof design and maintenance.

If energy conservation is to be taken into account in assessing the heat balance of the building, care will be needed to ensure that energy received over the lifetime of the building is considered in relation to the energy embodied in manufacture, erection and running costs, including maintenance of the glazed areas of the roof and its ultimate reclamation.

PREVENTION OF HEAT LOSS

All the materials of a roof, and even the spaces between layers of the materials (cavities), contribute to the thermal performance of a roof. The contribution of the different elements is assessed by testing, but for many purposes values are rounded to standardised published values which are used in calculations (eg for sizing heating plant and for building regulation requirements). For pre-formed materials the usual requirement is for a specified density, supported by test data on thermal transmission characteristics. For materials formed in situ, such as blown insulation, British Standards, or third-party certificates, as appropriate, specify both the material properties and the workmanship levels to be attained. For more information, refer to

The Red Book[35] and organisations offering third-party certification.

The thermal insulation performance of materials reduces with increased moisture content so materials that do not absorb water are needed where prolonged wetting (eg through condensation) is inevitable. Some thermal insulation materials are more vulnerable than others. In most buildings, insulants are kept at acceptable moisture contents by protecting them from rain and detailing to avoid the build-up of condensation. The controls range from materials to reduce water vapour flow, such as vapour control layers used towards the inside surfaces of buildings, to materials which are waterproof but vapour permeable, such as slater's felt and other breather membranes used towards the outside surfaces. Condensation build-up can also be prevented by ventilating cavities adjacent to the insulation, especially where the roof contains layers with high vapour resistance on the outside of the insulation. Interstitial condensation calculations will be needed to confirm that the design will prevent water vapour condensing out inappropriately[36, 37].

The insulation value of the roof can be degraded by thermal bridges where high thermal transmission material penetrates layers of low thermal transmission material. Heat losses due to thermal bridges are often ignored in calculations, especially where thin sections are involved; but these and other materials such as concrete roof beams become more important as thermal insulation standards increase. Some thermal bridges are also important because they produce inside surface temperatures below the dewpoint of the air, leading to selective condensation on parts, for instance, of a ceiling. Careful design can overcome or reduce thermal bridging to acceptable levels.

Air movement into and within the roof, and especially through layers of low-density insulation material or gaps between insulation panels, can reduce thermal efficiency considerably (see also the next section on Airtightness). Sealing at joints and around areas where services penetrate the insulation is important. Unintended air movement within the roof void may also carry water vapour to areas where condensation can cause problems.

It is beyond the scope of this book to discuss the detailed requirements for thermal insulation in terms of thermal transmittance (U values). Indeed, the economic values will vary according to building type, fuel used, its relative cost, and a host of other factors. But what is important is where that insulation goes within a roof void, ie:
• whether it is below or above the deck,
• that it is laid consistently, filling every space and avoiding thermal bridges where moisture may condense, and
• that it is located on the correct (cold) side of any item of construction which functions as a vapour control layer.

The terms cold deck and warm deck are described in more detail in Chapter 2.11 since they are more normally used in relation to flat roofs.

In most of the diagrams which follow in this book and which show thermal insulation, a U-value of at least 0.25 W/m² °C has been assumed. However, thermal insulating material which is susceptible to moisture intake (eg from condensation or rainwater ingress) may suffer from reduced values of thermal insulation.

Where the thermal layer is intended to provide support for a weatherproof surface, it will need to have certain minimum structural and other properties, including:
• resistance to loads, both positive and negative (eg wind uplift),
• a degree of movement compatible with the weatherproof layer,
• chemical compatibility with the weatherproof layer,
• appropriate fire resistance, where necessary.

AIRTIGHTNESS

Airtightness has crucial implications for energy conservation and is now an important issue for the construction industry. Designers are responsible for designing an airtight building. Contractors are responsible for building an airtight building, arranging for an airtightness test to be carried out after completion, and also for any necessary rectification and re-testing.

It is important that operatives are instructed on appropriate details and are well supervised, since in most situations remedial work to rectify hidden deficiencies exposed by post-completion testing will be much more expensive than getting it right in the first place.

Buildings need to be ventilated to provide a healthy environment for the occupants by supplying fresh air, and getting rid of excess heat, moisture (humidity), smells and other pollutants. Many existing buildings are excessively leaky and this leads to uncontrolled heat loss, draughts, interference with the performance of designed mechanical and natural ventilation systems, uncontrolled polluted air entering the building and reduction in the performance of the building's insulation.

Air permeability is defined in Box 2.1. CIBSE's Technical Memoranda *Testing buildings for air leakage*[38] recommends air leakage indices and air permeabilities for some building types. Tests are carried out by pressurising or depressurising the building using

---

Box 2.1: Definition of air permeability

Air permeability is the measure for airtightness of the building fabric. It measures the resistance of the building envelope to inward or outward air leakage.

It is defined as the average volume of air (in cubic metres per hour) that passes through the unit area of the structure of the building envelope (in square metres) when subject to an internal or external pressure difference of 50 Pa.

The envelope area of the building is defined as the total area of the external floor (eg a flying floor over a passageway between two dwellings), walls and roof, ie the conditioned or heated space.

a fan fixed to an opening such as the front entrance. Tracking down the cause of the leaks is not easy but using smoke pencils or puffers can help. The building needs to be practically complete before it can be tested so fixing the problem post construction can be costly.

AD L requires a maximum permeability of 10 m^3/(h.m^2) @50 Pa; better than this will usually be required if the CO_2 Dwelling Emission Rate (DER) or CO_2 Building Emission Rate (BER) is to be greater than the CO_2 Target Emission Rate (TER).

Post-construction testing of air leakage rates is required. Generally, poor performance in this respect is associated with adventitious ventilation through:

- gaps between walls and windows/ doors,
- poorly fitting windows/doors,
- lack of draught-proofing,
- gaps round service pipes,
- gaps where joists are built into the inner leaf of cavity walls, and
- leaks around loft hatches.

A continuous membrane over a flat roof will in principle be airtight, but it still has vulnerable edge joints. However, the same is not necessarily true for pitched roofs covered with a flexible sarking and overlapping slates or tiles that can result in a leaky roof.

If the roof has been designed as a warm roof, the airtight but vapour-permeable membrane will normally be on the slope under the slates or tiles.

Although cold roofs are expected to become less common in future, they still may be needed in certain circumstances. The 'traditional' cold roof has a vapour control layer and then insulation laid over it at ceiling level. The underlay to the roof tile or slate system has typically in the past been a type '1F' bituminous felt which is not vapour permeable although some air movement at laps can be expected. The roof space requires ventilation and this is achieved through openings at the eaves and ridges. This means that special attention is required at ceiling/wall junctions and around loft hatches to achieve the

airtightness required (using seals and well-lapped and sealed membranes).

A more recent development is the cold sealed pitched roof. The vapour control layer and insulation are at ceiling level but a vapour permeable underlay to the slating and tiling system is used. Ventilation is through the vapour permeable membrane and no other provision for ventilation is required. The roof design should be checked according to third-party certification of the vapour permeable membrane when used for this type of roof.

Sheet material overlaps will need to be properly sealed, as will all traps and service perforations into the roof void.

Roofs and services in roof spaces

The detailing necessary for a roof will be largely determined by the location of the air barrier within it. The air barrier can either use the roof structure or, especially if the roof is ventilated, it can be located at ceiling level, in which case the roof structure would fall outside the air barrier.

In cases where the roof structure is used to house the air barrier, it will probably mean using a vapour control layer as the air barrier. As with all construction relating to the air barrier, all penetrations for services, vents, flues and chimneys will require sealing. In addition, attention will be needed to ensure continuity of the air barrier at eaves where the barrier transfers from the roof to the external walls.

It is common to use the ceiling as the line of the air barrier with pitched roofs. While it is generally easier to make the ceiling airtight, the roof structure service penetrations can pose problems. Large roof voids are obvious places to route or locate services. Water storage cisterns have traditionally been located in the loft space. These, together with electrical cables, soil and vent pipes and, more recently, ventilation ducts, all present a number of locations for air leakage through the holes and gaps around these services that pass through the ceiling into the loft space (Figure 2.16).

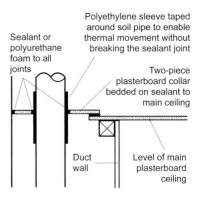

Figure 2.16: Forming an airtight seal around a soil pipe

Loft hatches provide another location for air leakage. Warm moist air entering the roof space through gaps around services and around the loft hatch can result in condensation occurring on the cold surfaces within the roof so it is important to seal these gaps. Seals should be fitted around the loft hatch frame, the hatch should be draught-stripped and catches should be provided so that the hatch compresses the draught-strip (Figure 2.17).

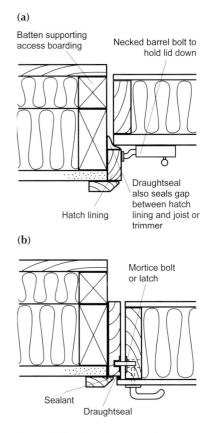

Figure 2.17: Minimising air leakage at the loft hatch: (a) lift-up hatch, (b) hinge-down hatch

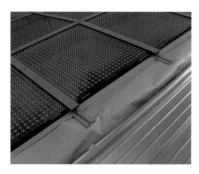

Figure 2.18: Vapour control layer sealed to rooflight framing before installation of cover panel

All service penetrations should be sealed including light fittings. Gaps around cables serving ceiling roses should be sealed from above, after the rose has been wired and before loft insulation has been laid. This prevents sealant displacement during wiring up when cables are often pushed back up into the roof space.

The number of service penetrations through ceilings and roofs should be minimised.

Rooflights and solar pipes

It is important to seal around large penetrations such as rooflights and solar pipes. Rooflights, while weathertight, may not be completely airtight. The manufacturer's details should be checked. Where a vapour control layer is used as the air barrier it can be dressed around and sealed to the rooflight frame (Figure 2.18).

Vapour control layers

Vapour control layers in the form of membranes and reinforced sheets are used in a wide range of construction types and are recognised as an effective air barrier material. The manufacturer's literature should indicate whether a particular product will be suitable as an airtightness layer. Polyethylene sheet is widely used as a vapour control layer or airtightness layer in timber-frame construction. Care is needed to ensure that the polyethylene sheet is not accidentally damaged by following trades. Reinforced sheets provide additional robustness.

CONDENSATION

The need to minimise the risk of condensation in roofs is covered in the supporting documents of the national building regulations:
- England and Wales: AD C[4],
- Scotland: Section 3 of the Technical Handbooks, Domestic and Non-domestic[5],
- Northern Ireland: Technical Booklet C[6].

The Scotland and Northern Ireland guidance to the regulations suggest that cold, level-deck roofs should be avoided.

Condensation occurs on a surface when its temperature falls below the dew point of the air for a sustained period. In these conditions, the relative humidity of the air in direct contact with the cold surface rises to 100% and the moisture in the air condenses on the surface.

On a clear night, the outer roof covering radiates heat to the sky and its temperature may fall even below that of the surrounding air. If the relative humidity of the air inside the building is greater than about 60%, there may then be a risk of condensation on the underside of the roof unless precautions are taken to avoid it. The risk is especially high in profiled sheet roofs; this is dealt with in Chapter 5.1.

In existing roofs, especially flat roofs where the diagnosis of the cause of dampness is often complicated, it may be worth calculating the risk of condensation occurring (Figure 2.19). The calculation can be done using the procedure outlined in BS 5250[39]. This calculation should always be made for new flat roofs of all kinds and for warm-deck pitched roofs. Further guidance is given in BRE Information Papers IP 2/05[36] and IP 5/06[37].

Interstitial condensation occurs out of sight within the thickness of a roof or other part of the building structure when water vapour diffuses through the internal fabric (including any insulation) and condenses on a cold surface beyond. Interstitial condensation can rot timber and lead to deterioration in other organic materials; it also encourages corrosion of metal.

There are three main ways in which the risk of condensation in roofs can be reduced:
- installing thermal insulation to reduce heat losses,
- installing a vapour control layer (with a vapour resistance greater than 200 MNs/g) on the warm side of the thermal insulation to restrict moisture which diffuses through the insulation from condensing on any colder outer surface,
- ensuring cross-ventilation of any cavity or void to remove excessive moisture.

Vapour control layers are notoriously difficult to install without creating gaps at joints between sheets or at interfaces with other parts of the structure. For this reason, they are best fixed against continuous stiff sheet materials (although not fully bonded to them in case movements at joints fracture the vapour control

Figure 2.19: Evidence of severe condensation and fungal growth on the underside of a plywood-decked flat roof

layer). Joints between sheets should be glued with a suitable adhesive; welting is less effective.

Thermal bridges, which may be so localised as to make little contribution to the total heat loss from the building, can lead to surface temperatures low enough to promote the growth of troublesome moulds. Roofs and their associated ceilings should be designed with a continuous layer of thermal insulation, with special consideration being given to junctions between elements such as occur at eaves and abutments where more dense materials may overlap or penetrate thermal insulation.

Where a group of buildings is showing problems due to deficiencies in the thermal insulation, it may be worthwhile carrying out a survey using infrared thermography to indicate parts of the structure needing attention.

Some evidence of condensation in the lofts of dwellings having pitched roofs was given in the English House Condition Survey 1991[9]. Of the 14 million or so houses in the pitched roof category, 208,000 (ie roughly 1 in 50) displayed signs of condensation having occurred in the roof space. Interestingly, the age of the property did not make a great deal of difference to the incidence of condensation. The ratio of dwellings with condensation in the roof space ranged from approximately 1 in 50 of those older than 1944 to about 1 in 85 for those built after 1980. The latter was caused perhaps by blockage of the eaves ventilation in cold deck roofs insulated to higher standards. Of dwellings dated before 1850, 1 in 3 had no sarking (presumably many within this age group had sarking installed when re-roofing took place) but it is a matter of concern that just under 1 in 100 post-1980 houses had

no sarking. At least these roofs will not suffer from condensation. This detail is no longer gathered in the EHCS, so the information cannot be updated. However, since the annual demolition rate of dwellings is about 0.1% of the total stock, many of these dwellings still exist. Comparable data from Scotland, Wales and Northern Ireland are not available.

Unpublished information from Housing Association Property Mutual shows that a significant number of new houses built from 1991 to 1994 had cold roofs insulated at ceiling level where the ventilation did not comply with the then current building regulations. An even higher proportion of houses insulated on the slope had unsatisfactory provision for ventilation. Comparatively new houses therefore are not immune to the risk of condensation.

PRESSURE DIFFERENTIALS CAUSING WATER 'LEAKS'

Although condensation and rain penetration through holes should always be investigated first as causes of any water lying under sheet roof coverings, there may be another phenomenon at work. Enclosed voids in roof constructions which provide only limited and perhaps inadvertent contact with the outside air (as in a roof constructed with a welted seam sheet material) may be subjected to pressures and suctions which take up water lying against the open edge of the welt and transfer it to the inside of the roof. In addition, water vapour in the atmosphere may similarly be transferred to the roof interior and condense on relatively cold surfaces near the outside of the structure. The phenomenon is loosely referred to as 'pumping'. Pumping can be attributed to two main mechanisms:
• external wind pressures and

suctions forcing air to and from the inside of the roof void (this tends to occur, given appropriate conditions, where the voids are relatively large),
• temperature changes affecting the air inside the void (tending to occur, given appropriate conditions, where the voids are relatively small).

In both cases, the resultant movement of air through the gaps carries with it some liquid water or water vapour from outside to inside. The water or water vapour then cannot easily escape if the void is not ventilated.

Significant volumes of water can be transported in some combinations of circumstances; indeed, water could well be present in sufficient quantities to confuse diagnosis of the cause with rainwater penetration through holes or other damage to the covering.

Where the problem has been correctly diagnosed, effective remedies may be found, in the case of the first mechanism, by providing better air seals at the welts or joints, effectively making them less responsive to external air pressure; with the second mechanism, by better ventilation to the immediate underside of the membrane via special weatherproof vents situated away from the joints. This should remove water vapour without admitting rain. Specialist advice may need to be sought.

WATER IN CONSTRUCTION MATERIALS

In sealed constructions, the moisture stored in materials during construction, either by hygroscopic absorption from the atmosphere or from rainfall, will not be able to escape. Under solar heating and radiative cooling, this moisture will move to other areas, possibly generating problems there.

2.5 CONTROL OF SOLAR HEAT AND AIR TEMPERATURE

It can be anticipated that under the influence of global warming, higher temperatures will occur than are currently being experienced. They will occur both more frequently and also for longer periods. At present, it is not possible to predict with any precision what will happen in the UK. However, the performance of roofing materials in hot climates is well known and this experience can be drawn on if required. Further information on temperatures is available from the UKCIP website[1] and in Chapter 1.

In summer, a roof will gain heat through solar radiation. Thermal insulation in the roof will reduce the transfer of this heat to the interior of the building, but there will be an attendant rise in the surface temperature of the outer covering. In many cases, it will help with the durability of this external covering if it is a sheet material to provide a highly reflective top surface to reduce heat absorption.

SOLAR ENERGY

The intensity of solar energy at the earth's surface is about 1 kW/m². Solar radiation is conventionally divided into:
- short wave, 0.29–4 μm,
- long wave, 4–100 μm.

(1 μm = 1/1000 mm)

The short waveband contains about 99% of the solar energy reaching the surface of the earth. About half of this is in the visible range with wavelengths of 0.4–0.7 μm (Figure 2.20).

Solar radiation affects surface materials in the outer envelope in two distinct ways:

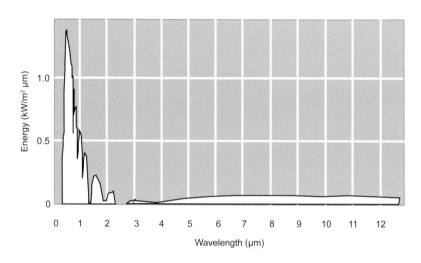

Figure 2.20: Radiation from the sun and from a low temperature source. The curve shows the distribution of energy in the solar spectrum at sea level. The dips in intensities are due to selective absorption by substances (eg particulates) in the atmosphere

- ultraviolet (UV) radiation disrupting specific chemical bonds between materials,
- solar energy on surfaces causing their temperature to rise.

UV radiation, with its high wave energy, can cause profound chemical change in certain materials by disrupting specific chemical bonds. This is particularly significant for organic materials such as plastics and coatings for sheet metals. UV degradation may cause changes in the appearance or light transmission characteristics of organic materials, and ultimately lead to surface erosion and loss of strength. Components exposed to sunlight behind glass may undergo fading of their original colours. For any given material, the wavelengths responsible are generally quite specific. Measures are usually taken in the design of such materials to lessen the effect of UV radiation, for

example by introducing chemical stabilisers.

Solar energy falling on surfaces causes their temperatures to rise significantly above air temperatures. This can occur on indoor surfaces if the rays have been allowed to pass through rooflights into rooms. Glass is transparent to short-wave solar energy and opaque to long-wave radiation, thus exacerbating the situation internally.

Some of this absorbed heat is re-radiated from the surface, but, because surface temperatures are quite low, re-radiation takes place in the long waveband. Figure 2.20 shows the relative dispositions and intensities of received radiation from the sun and re-radiation from a surface that has become heated to 80 °C. There is a buildup of surface temperature as solar energy is absorbed in excess of long-wave emission from that surface.

During design, it is important to be able to predict the extreme temperatures that materials or components may reach so that due allowance can be made for chemical or physical changes that may take place, eg:

- thermal expansion and contraction,
- softening or embrittlement,
- loss of plasticiser from plastics, and
- photo-oxidation of plastics.

To assess the magnitude of this effect, knowledge is required of the solar radiation absorptance of a material and its long-wave emittance. Values for these quantities are widely tabulated in standard textbooks.

It will be important to assess the risks of differential movement (eg between a membrane and thermal insulation) due to temperature changes which may have the capacity to tear the membrane.

The difference between the two quantities gives some initial impression of the solar heating to be expected. For example, since visible energy accounts for about half of the total solar energy, it is not surprising that light-coloured materials have high reflectance and therefore absorb less solar energy. But light surfaces also have good emittances for long waves. This gives the best combination for reducing solar gain and hence maintaining a cool surface.

Prediction of temperatures

Bright metal sheet materials (that is to say, not coated) also tend to have high reflectance but in their case long-wave emittance is low. This leads to the perhaps surprising result that they can experience significant solar heating.

Calculations can be used to predict the maximum temperatures that roof surfaces will reach. The equations assume a balance between incoming short-wave radiation, emitted long-wave radiation, convection and conduction into the structure behind. A sol air temperature is often used, for example in Section A2 of the *CIBSE Guide A*[40], it is the hypothetical outdoor temperature that would give the same net heat flows in the absence of radiation. In effect, the sol air temperature is the same as the temperature that would be reached by the surface if there was no heat conduction into the space or structure behind. So sol air temperatures approximate very closely to measured temperatures on highly insulated panels. Allowance in the calculation for conduction behind the surface can be made fairly simply, but the sol air temperatures may be regarded as the safest basis for estimating for design purposes.

Figure 2.21 illustrates the sol air temperatures for different surface finishes on a hot summer's day in the UK (eg an air temperature of 30 °C). It shows the large difference that could be expected between, say, two highly insulated roofing panels: one with a white and one with a black finish. Because maximum values rather than long-term averages are used, Figure 2.21 is likely to be equally valid for roofs or walls. It is also likely to apply equally to most locations in temperate or equatorial zones.

There is a certain amount of experimental data on measured surface temperatures, though most of it refers to walls rather than roofs. For example, in measurements on wall sandwich panels taken by BRE over a two-year period, maximum surface temperatures of about 78 °C were recorded for vertical surfaces. They would have been correspondingly greater for sloping surfaces directly exposed to the sun.

The obverse of solar overheating is cooling by night sky radiation. This occurs when there is a clear night sky so that significant long-wave radiation can continue to take place from surfaces, but with no incoming solar contribution. In these circumstances, surface temperatures as much as 8 °C below the outdoor air temperatures can commonly occur. Condensation will result, and this phenomenon is examined in relation to the various types of roofs in later chapters.

The standard height for air temperature measurements is 1.2 m. Air temperature, often described as the 'screen' or 'shade' temperature, is measured by thermometers shielded from direct radiation and precipitation by a louvred box or Stevenson Screen.

Measurements of air temperature are made very widely in the UK. At some locations only 24-hour maxima and minima are recorded; at others, temperatures are read every hour, while at others, continuous records are kept. There is obviously an enormous number of ways of analysing and representing these data, and meteorological records are available over many years to anyone who wishes to undertake a particular statistical treatment appropriate to their needs.

Air temperature at any one place is continually varying. Variations over short time intervals, often for seconds or even minutes, may be ignored for most purposes, but variations from one hour to another can be significant in their effect on buildings.

Regular shifts in temperature occur on an annual cycle and, superimposed on this, a diurnal cycle. One way of representing this is shown in Figure 2.22.

Measurements at any given hour of the day for, say, each January are averaged and plotted. The same is done for each hour of each month, and isotherms drawn to show how, on average, the temperature varies with time of day and time of year. Diagrams of this kind can be drawn for any location in the UK and will show generally similar patterns, though changing in detail according to latitude, distance from the sea and altitude.

The variations in air temperature from one part of the country to another are best shown as maps. Examples are shown in Figures 2.23 and 2.24 which illustrate the average minimum and maximum air temperatures for January and July, respectively.

Protection from solar heat gain by means of a dressing or paint may need to be considered with certain types of material used in both pitched and flat roofs.

Principles of modern building[3] discussed effective treatments,

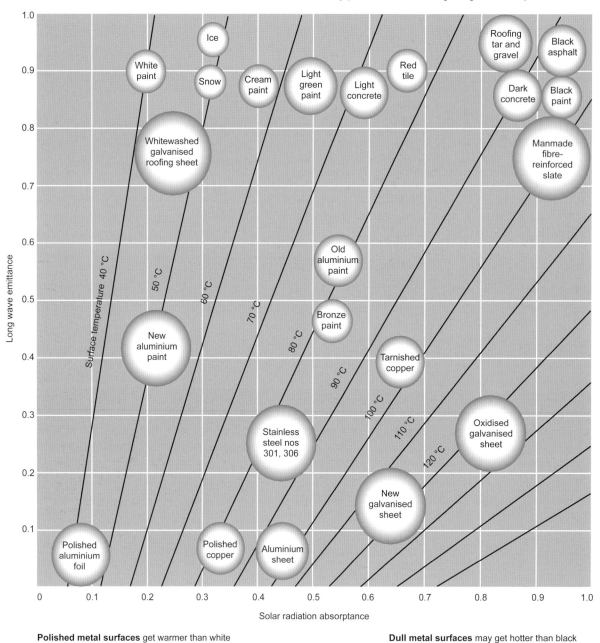

White materials are cool. They do not absorb much heat from the sun

Black materials get much hotter than white materials, and a black surface may experience a very large temperature swing between night and day (more than 50 °C for a lightweight black roof)

Polished metal surfaces get warmer than white surfaces because, though they absorb a similar amount of heat, they re-radiate less of it

Dull metal surfaces may get hotter than black surfaces. They absorb less heat from the sun but re-radiate little of it

Figure 2.21: Calculated extreme surface temperatures for air temperature of 30 °C

concluding at that time (the early 1960s) that a matt white surface of lime–tallow wash was much more effective than any other surface treatment. Moreover, it was the only coating that had been found, up until that time, not to cause deterioration in asphalt or built-up bitumen felt. The recipe depends on the availability of quicklime to melt the 10% by weight tallow content. However, treated surfaces need yearly replacement. White spar or other finishes are now

available that are equally effective and more durable (BRE Information Paper IP 26/81[41]). It is common for reinforced bitumen membranes to incorporate a mineral surface finish to the capsheet without the need for additional chippings to be applied.

It can be seen that the areas of highest maximum temperatures are in south east England, and areas of lowest minimum temperatures are in the central areas of Scotland, northern England, Wales and the Midlands. Temperatures in coastal

areas are generally more equable than those inland so that the interior part of the UK is coldest in winter and warmest in summer. Maps of this type usually give data for sea level. The temperature at a specific site can be calculated by subtracting 0.6 H/100 °C (where H is the height above sea level in metres) from the sea level temperature. This represents the fall in temperature with height within the UK.

Long-term average temperatures are important for some purposes,

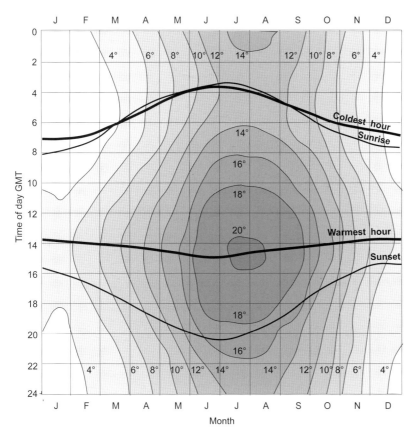

Figure 2.22: Diurnal and annual variation of air temperature at London Airport (Heathrow), 1957–1966, prepared from average hourly values

eg in calculating seasonal heat losses through roofs. For other purposes, shorter term averages or average peak values may be important, eg for sizing of heating, ventilating and air-conditioning plant. Finally, threshold or extreme values may sometimes be important, eg where material or mechanical failure may occur. Sufficient meteorological data are now available to allow analysis for any of these purposes and for almost anywhere in the UK.

Seasonal accumulated temperature difference (ATD) or commonly called degree-day totals, varies by a factor of about 1.6 when values are reduced to sea level over the UK between south west and extreme north (Figure 2.25). The map illustrates just how much more important it is to achieve adequate thermal insulation in the colder northern latitudes of the UK. Totals are considerably increased at higher altitudes. Standard ATD totals give a general indication of the geographic variation in energy needed for space heating but make only a small fixed allowance for solar gains. Thus, they do not reflect the full potential of the sun to warm buildings and

the spaces around them, or of wind-induced losses. The use of ATD totals to building-specific base temperatures can take account of some of these factors.

The temperature of the exposed surface of a building is rarely the same as the air temperature. One reason for this is that surface temperature is continuously modified by heat exchange between the outdoor air and the occupied internal space of the building. In winter, for example, heat loss from inside the building leads to a slightly elevated temperature for the outer surface of the roof, depending on insulation values. In addition, structures with significant thermal mass will always tend to lag behind any short-term fluctuations in temperature and smooth out diurnal variations.

Perhaps a more important consideration than either of these is radiated heat to or from the external surface. It is this which, in conjunction with already high or low air temperatures, is often responsible for the extreme values experienced by many surfaces.

As well as withstanding the extreme temperatures described above, roofs are also subjected to rapid swings in temperatures caused, for example, by sudden showers on an already hot surface. Method of Assessment and Test No 27[15] of the European Union of Agrément indicates that repeated rapid reversals of the order of 60 °C should be used in carrying out an assessment of any new material for which a certificate is being prepared.

Frost on saturated construction

Deterioration of building components frequently happens when several meteorological parameters act together. One example of this is frost damage to porous materials. This occurs when freezing conditions coincide with high moisture content, near saturation. The air temperature at which frost damages roof materials usually needs to be substantially below freezing since the roof often receives heat energy from inside the building of which it forms a part. It will also have a certain thermal capacity. Furthermore, although water in the largest pores of the materials may freeze at a little below 0 °C, the freezing point in the smallest pores may be depressed by several degrees.

A number of factors affect the susceptibility of a given roof to frost damage, including:
- the precise porous characteristics of the material,
- the speed and duration of freezing,
- the number of faces frozen simultaneously,
- the moisture content and its distribution, and
- the internal strength of the porous tile or other unit and its parts.

Each of these factors is itself influenced by a complex interaction of subsidiary factors.

Frost testing

It is scarcely surprising then that it has proved difficult to devise a reliable and representative test for assessment of frost resistance. Any test method must select particular

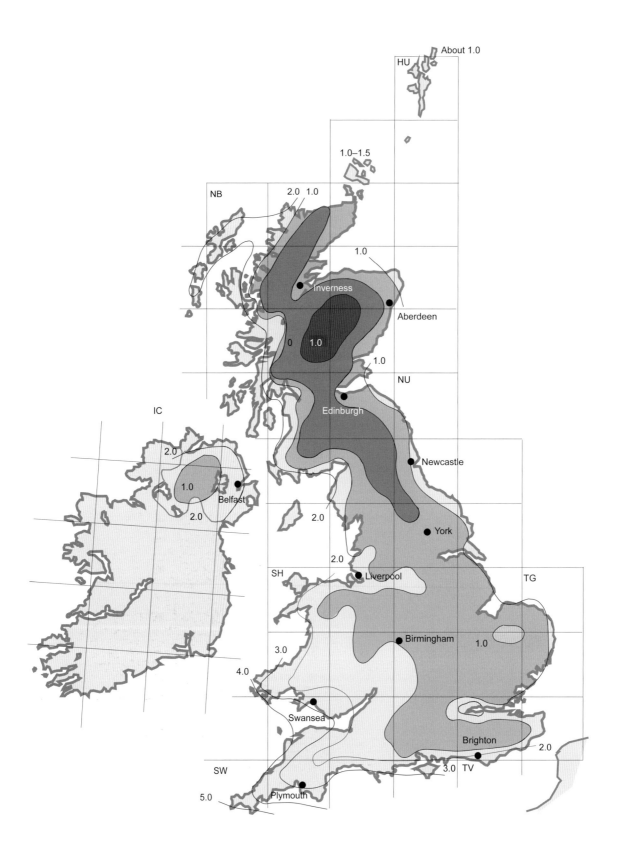

Figure 2.23: Average January minimum daily temperatures in the British Isles

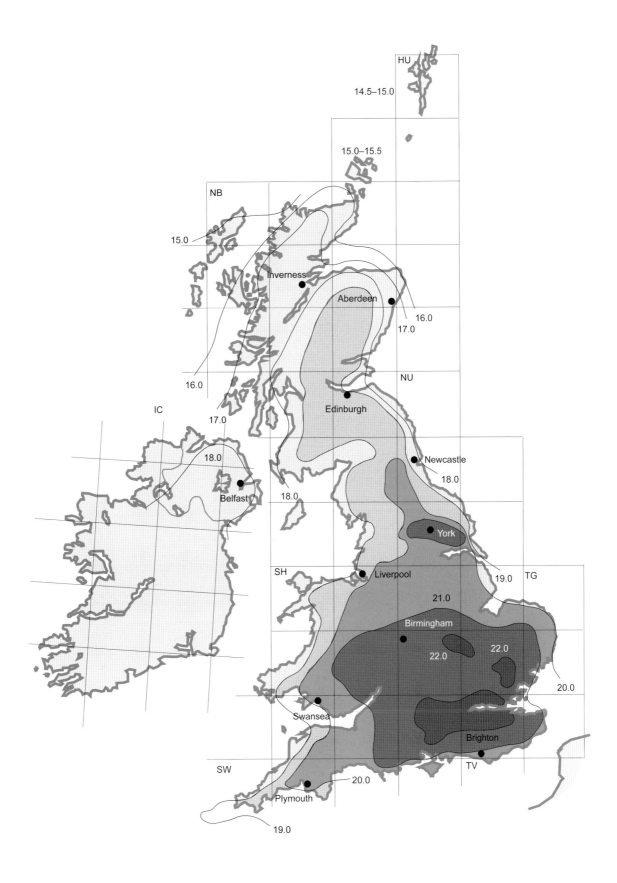

Figure 2.24: Average July maximum daily temperatures in the British Isles

Figure 2.25: Accumulated temperature difference (degree-day) totals, September–May average, 1957–76: data reduced to sea level (Reproduced from the *CIBSE Guide A*[40] by permission of the Chartered Institution of Building Services Engineers, 222 Balham High Road, London SW12 9BS)

conditions which may then not apply to a specific application of the material. For example, freeze/thaw tests on individual tiles are easy to carry out and give very quick results. However, they represent unique conditions which rarely occur in practice in exactly those same circumstances.

Despite these difficulties, a number of types of test are used, including:

- natural exposure tests,
- laboratory freeze/thaw cycling,
- crystallisation tests,
- measurements of absorption and strength characteristics, and
- characterisation of pore structure.

For each, the procedures and criteria for durability vary greatly according to the material under test and the country or even the organisation in which the test is carried out.

It is worth noting that there is no current British Standard test for durability to frost damage that can be applied to all materials subject to freezing. Even so, there are a number of tests that are well established and quite widely used for similar purposes; for example, salt crystallisation tests are used in the UK for the assessment of limestones. Reference samples are included in test batches. These tests allow prediction of suitability for different locations on a building; that is, they determine different susceptibilities to frost attack. Crystallisation tests may also be used for other types of stone, although interpretation is less certain (BS EN 539-2[42]).

2.6 FIRE SAFETY AND PRECAUTIONS AND LIGHTNING PROTECTION

The designer of the roof of a building will need to know a great deal about the conditions that will be imposed on that structure in service. Factors such as loading, environment and durability all have to be understood and assimilated into the design process, and all aspects considered in relation to behaviour in fire.

The main safety requirement for roofs is to ensure the external fire performance of roofs is such that they do not burn through quickly and that fire does not spread from one compartment or property to another and that they do not allow the flames to spread across the roof quickly (Box 2.2).

At the time of writing this third edition, the requirements for limiting penetration of the external surfaces of roofs by fire, and limitation of surface spread of flame, are subject to both UK standards and recently introduced European Norms. BS 476-3 (1958) has been in use for many years despite being withdrawn officially by BSI in the 1970s. In order to maintain a consistent approach it was revised and re-issued as BS 476-3[43]. This enabled the UK to put forward this standard as DD ENV 1187[44], Test 4.

The new European test method for penetration by fire described in DD ENV 1187[44] deals only with the first of the two criteria, namely penetration, while the question of surface spread of flame is not at present covered. Four methods are described for testing penetration by fire, based on those developed in France, Germany, Scandinavia and the UK, and it is intended that the acceptance criteria for products will in future be specified in BS EN 13501-5[45]. Therefore, at the

moment, products, and in particular membranes, for use in the UK, will be tested in accordance with DD ENV 1187[44], Test 4 (the method based on BS 476-3[43]).

The external fire performance is defined as penetration and surface spread of flame and is assessed by the test methods in BS 476-3[43]. The level of performance is designated by two letters: the top designation is AA, AB or AC. AD B[4] recognises that not all roof coverings require testing and that some are given a rating without testing.

> **Box 2.2: Performance of roof coverings**
>
> The designation for performance of roof coverings in BS 476-3[42] consists of two letters (AA, AB, AC, etc.). The first letter relates to performance in respect of penetration by fire:
>
> - A no penetration in 1 hour,
> - B no penetration in half an hour,
> - C penetration in less than half an hour,
> - D penetration in a preliminary test,
>
> and the second letter relates to performance in respect of flame spread:
>
> - A no spread of flame,
> - B spread of flame not exceeding 533 mm,
> - C spread of flame exceeding 533 mm,
> - D continuation of burning for 5 minutes after the withdrawal of the test flame or spread more than 381 mm across the region of burning in the preliminary test.

Building Regulations in England and Wales recognise that all roof coverings do not necessarily need to be tested for penetration by fire, and some, such as slate and tile and flat roofs with a minimum cover of 12.5 mm stone chippings, have in the past been allocated designations without testing, using a minimum classification of AC (see Box 2.2) as the criterion.

How to conform with the Building Regulations (England & Wales) is advised in Approved Document (AD) B which was revised in 2006 and came into effect in April 2007[4]. It includes reference to both the traditional UK approach of classification as given in BS 476-3[43] and also to the European approach in BS EN 13505-5[45]. It is anticipated that these will co-exist for the immediate future. It is likely that true harmonisation of fire safety tests throughout Europe will occur only when the regulations in the member states assume a closer relationship.

In the future, the absence of a measure of surface spread of flame in a harmonised European Standard may cause industry to promulgate its own standard to cover this particular aspect of fire performance.

FIRE PRECAUTIONS

The latest versions of the national building regulations and their accompanying guidance documents contain much material that is relevant to roofs and their fire protection, particularly in relation to loft conversions, means of escape, compartmentation, concealed spaces and protection of openings.

The requirements are given in the supporting documents of the national building regulations:

- England and Wales: AD B1 & B2[4],
- Scotland: Section 2 of the Technical Handbooks, Domestic and Non-domestic[5],
- Northern Ireland: Technical Booklet E[6].

Passive fire protection measures are those features of the fabric (such as cavity barriers) that are incorporated into building design to ensure an acceptable level of safety. These features are dealt with in outline in this book. Measures that are brought into action in the event of a fire (such as fire and smoke detectors, sprinklers and smoke extraction systems) are referred to as active fire protection and are not dealt with here (see BRE's companion book on building services[46]).

The objectives of fire precautions include the following:
- reducing the number of outbreaks of fire,
- providing adequate facilities for the escape of occupants,
- minimising the spread of fire, both within the building and to nearby buildings.

Relevant practical aims arising from the last of these objectives are:
- limiting the size of the fire by controlling fire growth and spread, and by dividing large buildings, where practicable, into smaller spaces,
- avoiding conflagration by attention to the building boundaries and space separation from neighbouring structures.

Cavity barriers within roof voids are intended to prevent the growth and development of a fire within an enclosed space, though that is not their only purpose under building regulations. They are also used to prevent fire spreading round the edges of fire-separating elements.

Recent developments to assist in the prevention of the spread of fire in roof voids include mineral wool curtains which can give up to two hours' fire resistance, and are comparatively light in weight.

Where refurbishment is being undertaken on buildings controlled under the Fire Precautions Act

1971[47], the fire certificate issued by the fire authority will list conditions that will need to be maintained in the building, provided the use has not changed. So far as roofs are concerned, this will affect roof escape routes and roof ventilation required to assist means of escape. Therefore, all those concerned with refurbishment of controlled buildings should make themselves aware of the contents of this certificate and, indeed, with all other legislation affecting occupied premises.

Fire can attack a roof from both the outside (see Box 2.2) and the inside. All roof designs must take into account the risk from the outside, but some designs may also need to take into account the risk from the inside, and these factors are explored further in the following sections.

Fire from outside

The performance of the roof as a whole will depend on the degree of combustibility of the constituent materials, their softening or melting temperatures, the fixing and jointing methods used, and the thermal properties of the insulation and the surrounding materials.

Roof coverings have been the cause of large conflagrations in the past, particularly in urban environments, and, following public concern, a method of test was published in 1958 (there was a later revision, but this has not been adopted) to assess the performance of roofs to an external fire exposure hazard. That test has been used in connection with the building regulations for England and Wales to relate the acceptability of different types of roof coverings to their proximity to the boundary, and schedules of notional designations of roof coverings have been included in the different building regulations. Roof coverings of different kinds have expected values for properties in fire when tested, and several of them are listed under the types of materials given in later chapters.

The Loss Prevention Council's Code of Practice for the protection of industrial and commercial buildings[48] may require a different or higher standard of construction

than would be needed to comply with building regulations.

The main principle used is that the roof covering, including roof lights, is required to achieve a standard which is set in relation to the distance of the roof from the relevant boundary (relevant boundary, as broadly interpreted from AD B[4], being the actual boundary of the land belonging to the building represented by, say, hedges, fences or walls; or, where the land abuts a road, railway, canal or river, the relevant boundary may be the centre line of that road, railway, canal or river).

Under AD B1 & B2[4], boundaries may be actual, as defined above, or notional. A notional boundary may apply in the case of any building within the residential, assembly and recreation purpose groups described in the Approved Document which is on the same site as any other building.

The various levels of performance expected to be achieved by the more common roof coverings are given in the appropriate chapters that follow.

Where a separating (ie compartment) wall is carried up only so far as the underside of the roof covering, the roof covering will need to be AA, AB, or AC for a distance of 1.5 m each side of the wall (Figure 2.26a). If the wall separates dwellings or other residential, office, assembly or recreational buildings, and is not more than 15 m high, boarding or tiling battens may be carried over the wall provided that they are fully bedded in mortar or other no less suitable material where they pass over the wall. Mineral wool quilt will comply. Alternatively, the wall must be extended 375 mm above the upper surface of the roof, measured vertically (Figure 2.26b).

If the provisions of the Approved Documents are being followed, roof coverings of certain performances cannot be placed nearer to site boundaries than indicated in Figure 2.27, though there are permitted exceptions.

Roof lighting

The rules that govern the selection and use of roof lights are complex,

(a)

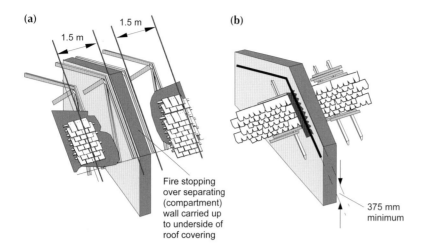

Fire stopping over separating (compartment) wall carried up to underside of roof covering

(b)

375 mm minimum

Figure 2.26: Distances given in the Building Regulations 2000 AD B[4] for coverings adjacent to separating (ie compartment) walls. AA, AB, AC designation, while being required for the configuration shown in (a), is not required where the separating wall is carried through the covering (b)

and reference must be made to the provisions of national building regulations. Some general points in relation to the Building Regulations 2000 (England & Wales) AD B1 & B2, and which have been used in BRE site inspections, are made below.

The undersides of roof lights are considered to be parts of the ceilings of rooms and spaces, and Class 0 is called for over the circulation and common areas of flats and maisonettes unless certain area and spacing limitations are observed.

Class 0 is defined in the Approved Documents as a combination in part of BS 476 Parts 6 and 7[43]. These limitations apply more particularly to those roof lights made from thermoplastic materials.

The behaviour of roof lights in relation to external fire also needs to be considered. Unwired glass at least 4 mm thick can be regarded as having an AA classification. However, thermoplastic roof lights may be made from a variety of materials (eg PVC or polycarbonate) and may be single or multi-skinned.

It is these properties which, in the main, govern their selection and positioning in roofs.

AD B states that thermoplastic roof lights may be classified as:

- **TP(a) rigid:** products comprising rigid and solid PVAC (polyvinyl acetate) sheet, 3 mm solid polycarbonate, multi-skinned sheet of PVC-U or polycarbonate having a Class 1 surface spread of flame, or any other rigid thermoplastic product that conforms when tested to BS 2782, Method 508A[49].

- **TP(a) flexible:** products not more than 1 mm thick that comply with type C requirements of BS 5867-2[50].

- **TP(b):** rigid and solid polycarbonate sheet less than 3 mm thick, or multi-skinned products of polycarbonate which do not qualify for TP(a) by test, or other products that conform when tested to requirements specified in Appendix A of AD B.

Thermoplastics roof lights must not be placed over protected stairways. However, they may be used in situations where their positioning meets requirements for distance from relevant boundaries; these dimensions are:

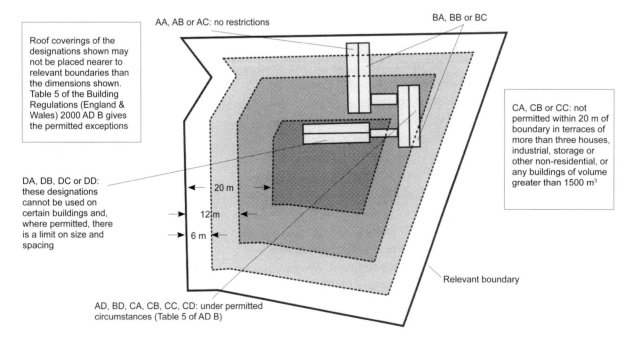

Figure 2.27: Permitted distances from site boundaries for roof coverings of various performances following the provisions of the Building Regulations 2000 AD B[4]

- **for TP(a) rigid:** 6 m over any space except a protected stairway (or no limit in the case of spaces over a balcony, veranda, carport, covered way or loading bay which has at least one longer side wholly or permanently open, detached swimming pool, conservatory, garage or swimming pool with a maximum floor area of 40 m²).
- **for Class 3 surface spread of flame or TP(b) and AD, BD, CA, CB, CC, CD or TP(b) external surface:** 6 m, or 20 m in the case of DA, DB, DC or DD, over spaces as shown in brackets above for TP(a) rigid.
- **over circulation spaces (except a protected stairway) and rooms:** 20 m, and not within 3 m of another plastics roof light, maximum area 5 m².

In addition, roof lights constructed of thermoplastics materials having a Class 3 or TP(b) lower surface, each not amounting to more than 5 m² (or a group of roof lights having an area no more than 5 m²), may not be placed nearer to each other than 3 m. Even then the roof covering has to be non-combustible or of limited combustibility, as defined in BS 476-11[43], for the same distance (Figure 2.28).

Fire from inside

So far as fire from the inside of a building is concerned, where the

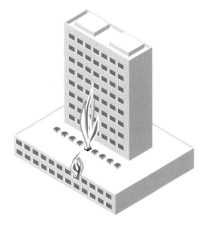

Figure 2.29: In some circumstances (eg where it acts as a floor), a roof may need to have the ability to contain a fire

roof forms part of an escape route, or is required to give protection from fire spreading from a podium to a tower, then it must be constructed to achieve particular standards in just the same way as floors (Figure 2.29). A number of fires have occurred which involved foamed plastics boarding underdrawn on ceilings.

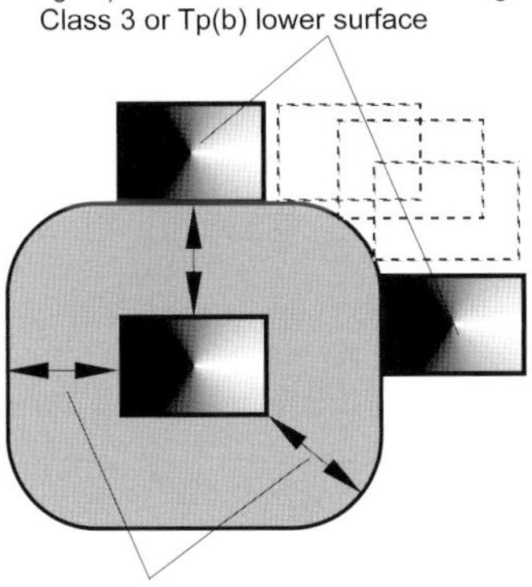

Where a roof is required to contain a fire, the roof in question must achieve the necessary period of fire resistance when the test specified in BS 476: Parts 20–23[43] is carried out from the underside. Where there is a fire-protecting suspended ceiling, the construction as a whole is subjected to test. The roof must satisfy, for the specified time, the performance requirements in relation to loadbearing capacity (freedom from collapse), integrity (no holes through which flames can pass) and insulation (temperature rise of the top surface contained to a mean temperature of 140 °C above initial temperature, and not more than 180 °C at any point on the surface).

Fire damage

Inspection of fire-damaged structures should be carried out by competent and experienced persons. It should be noted that tests carried out to BS 476 specifications on building materials and structures are not intended to give guidance on the serviceability of such structures after exposure to a fire. Absence of spalling or absence of

change of colour after a fire should not be taken as evidence that any reinforced concrete members are in satisfactory structural condition since different aggregates vary in their behaviour in fire.

FIRE SAFETY

The main requirement for roofs regarding ADs B1 & B2 is related to external fire performance. The roof has to provide sufficient protection for a suitable period of time to allow the occupants to escape.

It is not envisaged that climate change will be a major influence regarding the fire performance requirements for roofs.

LIGHTNING PROTECTION

The roof carries the main provision for terminals for any lightning protection system (Figure 2.30). As a general rule, if lightning protection is to be effective, no part of the roof should be more than 10 m from a conductor; 5 m for high risk buildings. It will also be necessary to connect metal parts of a roofing system to any lightning protection system.

The risk of a building or part of a building being struck by lightning should be calculated (see BS EN 62305[51]). BS EN 62305[51] includes a map of the UK indicating the number of lightning strikes to the ground per square kilometre

Figure 2.30: Lightning conductors on a domestic roof. The positioning and support of the conductors need improvement

Plastics roof light (or group of roof lights) of maximum area 5 m² having a Class 3 or Tp(b) lower surface

3 m minimum spacing. Also, roof covering to be of limited combustibility of 3 m

Figure 2.28: Criteria for distances between roof lights (given in the Building Regulations 2000 AD B[4])

per year, and this then should be multiplied by the effective collection area of a structure to give the probable number of strikes to the structure per year.

Weighting factors, including the use of the structure, type of construction, the contents of the building, the degree of isolation, and the type of country in which the building is situated, are used to obtain the overall risk. A decision then needs to be taken on whether the risk is considered to be acceptable without protection, or whether some measure of protection is thought to be necessary.

A lightning protection system generally consists of the following networks:
- air termination,
- down conductors,
- earth electrodes,
- earth termination.

Air termination

This provides the point of interception of the lightning strike. Terminals may take the form of vertical rods (Figure 2.30) or horizontal conductors generally placed at the eaves of flat roofs where the risk of a strike is greatest (an example is shown in Figure 5.38). Parts of the construction and especially its services, such as roof-mounted air-conditioning units, should be connected to the air terminal system if they are not within the zone of protection of a higher terminal. The metal parts of the structural frame of a building should also be connected to the air termination network.

Down conductors

The purpose of the down conductors network is to provide a low impedance path safely to conduct the lightning strike from the air termination network to the earth termination network. The route should be as short and as direct as possible. Wherever possible, too, the conductors should be routed directly down an external wall. If they bend to follow recesses in the building's surface, there is an increased risk of side flashing with consequent damage to the structure. On pitched roofs, where conductors are sited on chimneys, as in Figure 2.30, conductors may need to be positioned over sloping surfaces, and care should be taken to avoid bends as much as possible.

Joints in the conductor network should be kept to a minimum, and joints between dissimilar metals should be avoided.

There is further information on routing of down conductors on walls in the companion book in BRE's Building Elements series, *Walls, windows and doors*[52].

Inspection and testing

The lightning protection system should be inspected by a competent person every 12 months, and as soon after a strike as possible. Inspections should include:
- conductors for signs of displacement, damage or corrosion,
- clamps and other supports for continued mechanical effectiveness,
- earth termination for electrical effectiveness.

Modifications to either the lightning protection system or the structure on which it is carried should lead to a re-assessment of the system by a competent person.

There is further description of lightning protection, as it affects larger roofs, in Chapter 4.1.

2.7 DAYLIGHTING AND CONTROL OF GLARE

The provision of daylighting through a roof will be considered in detail in this book only for larger spans where it is not possible to provide sufficient light through windows. Area for area, quantities of light received at the working plane from above can be considerably greater than from the side; it is also especially important to consider the effects of glare when considering the design of roof lights.

Light tubes, also called light pipes, are a modern alternative for the traditional skylight. They are designed to transfer diffused light through a large diameter pipe (200 mm to approximately 500 mm diameter stock sizes) through the ceiling. As with any penetration of the roof cover, due attention must be given to the flashing to ensure a watertight joint for flat roofs, a weathered detail for pitched roofs and avoidance of condensation and breaches in air tightness.

To exclude glare, it may be desirable to screen any low-angle direct view of the sky through roof glazing, if necessary by louvres or baffles, and to increase the brightness of the surrounding surfaces. This should certainly be considered in spaces that are likely to contain display screen equipment.

DAYLIGHT FACTOR

Daylight factor is the usual way of indicating the amount of daylight received under any roof lighting. It is simply expressed as a proportion of the available light outside the building that can reach the working plane (Figure 2.31). The calculation of daylight factors is explained more fully in Chapter 5.1, since it is for the medium-span pitched roof category that calculations will usually be prepared.

As a general rule, however, it can be assumed that an average daylight factor of around 5% should be sufficient for most tasks to proceed without the need for additional artificial lighting. For spaces that are lit only by roof lights giving less than 2% daylight factor, the permanent use of artificial light may have to be considered.

REFLECTION

It is not only the amount of daylight reaching the inside of the building that is important. The reflectance value of the surface of the roofing of a low building may affect the amount of daylight reaching adjoining higher buildings. Glare or dazzle can occur when sunlight is reflected from a glazed roof light or from specular, light-coloured roof surfaces. For flat roofs, this problem usually occurs only when the sun is high in the sky; but some types of modern design incorporate sloping glazed or metallic roof surfaces that can, under certain circumstances, reflect unwanted high altitude sunlight into the eyes of motorists, pedestrians and people in nearby buildings (BRE Information Paper IP 3/87[53] and Figure 2.32).

An important development has been the production of new forms of coated low-emissivity glazing that can be used to conserve heat. These types of glazing have found widespread application (it is unlikely that the requirements of AD L will be met without using low-emissivity glazing). Low-emissivity glazing does not reflect significantly more light (within the visible spectrum) than normal glazing; the main differences occur in the longer wavelengths of the total spectrum.

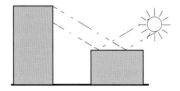

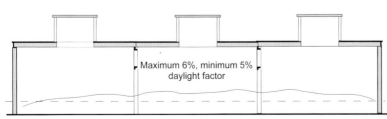

Monitors with vertical glazing: glass area 30% of floor area

Figure 2.31: Typical daylight factors (maximum 6%, minimum 5%) under a roof light in an industrial building. Other profiles are given in Chapter 6.2

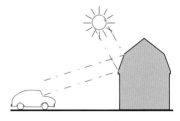

Figure 2.32: Bright sunlight can reflect from polished metal surfaces on low flat roofs and dazzle people in adjacent buildings. The same can happen to motorists from reflective surfaces on sloping roofs

It is important that the possible effects of unwanted external solar reflection are considered at the design stage; failing that, remedial measures may be needed. From simple input data, the times of the year and of each day at which reflected sunlight might occur, and their duration, can be calculated. It is necessary to obtain the relationship between the angles of incidence and reflection to derive the sun positions at which solar dazzle may be a problem in each particular case.

For a sloping roof design where solar dazzle has been identified as a potential problem, the geometry of the roof may have to be altered. Initial experience suggests that, in Europe and the USA at least, the greatest problems occur with surfaces facing within 90° of due south, sloping back at angles between 5° and 30° to the vertical. Where the roof surface slopes at more than 40° to the vertical, solar reflections are likely to be less of a problem. It is very unlikely that surfaces which slope forward, so that the top of the building forms an effective overhang, will cause problems in this respect.

ROOF LIGHTS

Roof lights take many forms, and can be made of many materials. Glass roof lights are normally either of the lantern variety or are made of patent glazing: these are dealt with in Chapters 5.3 and 6.3, respectively.

There are four basic types of plastics glazing used for roof lights:
• corrugated sheets,
• flat sheets,
• moulded roof lights,
• ETFE pillows.

Corrugated sheets

These are designed to match the profiles of metal or cementitious sheeting. In the UK, they are most commonly made of PVC or glass-fibre-reinforced polyester (GRP). It is normal practice to fabricate double-skinned units having improved thermal insulation characteristics. The profiles do not always match profiles of sheet materials sufficiently closely to prevent rain penetration, particularly where fixings are over- or under-tightened; special care should therefore be taken in replacement work. Such sheets are also known to flex in the wind, producing a pumping action at seals. These topics are further explored in appropriate later chapters.

Flat sheets

These are usually of high light transmission, but may be tinted. The principal types of thermoplastic used are PVC (sometimes reinforced with wire mesh) or polycarbonate.

Acrylic sheet is not now widely used in the UK because of its negative fire properties, but is used elsewhere in Europe.

Polycarbonate, in particular, is available in the form of double- or triple-skinned box section sheets. These offer improved thermal insulation and stiffness, and hence improved spanning characteristics when compared with plain sheet.

Plain sheet is widely used to span barrel vaults because it is readily cold bent. The minimum permissible bending radius depends on polymer type and sheet thickness, and should be checked with manufacturers.

GRP flat sheet is also available, and capable of being cold bent. High light transmission can be achieved by closely matching the refractive indices of the resin and the glass-fibre used. These sheets are usually installed with gaskets and glazing beads. It is important to ensure that the system is chemically compatible with its surroundings.

Both thermoplastic and GRP sheets can be surface coated to improve their durability to weathering.

Moulded roof lights

These units are most often fitted to upstands inserted in flat roofs and are generally made of thermoplastics such as PVC or polycarbonate. Shapes can also be moulded in clear GRP; rounded or pyramidal shapes are thermoformed from suitable grades of flat sheet and can be made up as double-glazed units if required. Most types of surface-coated thermoplastics sheet are not suitable for thermoforming; this point needs to be checked with manufacturers before specification. Care is needed to avoid overheating during thermoforming as this can reduce durability.

Large areas of roofing have in recent years been covered with ethylene tetrafluoroethylene (ETFE) pillows, which offer a much reduced weight when compared with more conventional glazing systems.

Safety

Where roof lights are fitted, HSE regulations require that safe access is provided from inside and outside so that the units can be opened and cleaned safely. These regulations do not apply to individual dwellings. Roof lights also need to be guarded where the roof forms a means of escape in case of fire, or where regular access is required. Glazing with proprietary self-cleaning glass may reduce the frequency with which cleaning is needed but may not eliminate it entirely.

2.8 SOUND INSULATION

Roofs may be required to protect their occupants from noise just as much as walls. This book does not deal with protection of the outside from noise generated within the building, but rather with protection within from noise generated from the outside.

In addition to meeting the requirements for resistance to noise from adjacent dwellings, other aspects to be considered include:

- Weather-generated noise (eg drumming of rain or hail on thin metal sheets, whistling at ventilation slots, movements between metal to metal joints or from thermoplastics gutters, slack sarking membranes vibrating in windy conditions). BRE Information Paper IP 2/06[54] gives guidance on assessing the likely effect of rain noise from lightweight roofs and roof elements on the indoor ambient noise levels in rooms.
- Noise travelling from room to room (failures tend to be concentrated on missing elements of separating walls in roofs, unfilled joints in masonry, or holes around services and joists).

SOUND AND VIBRATION CHARACTERISTICS

Sound is energy travelling through the air, while vibration is associated with energy travelling through the ground or building structure. Generally, sources external to the building produce more sound than vibration, even though this may not be obvious to the occupants. For example, a heavy vehicle passing a house may cause ornaments and loose fittings to rattle even though the sound level is not very high. The most probable cause of this is not ground-borne vibration, but low-frequency sound waves set up by the vehicle displacing a large volume of air which causes the building structure to vibrate.

Vibration can damage a building structure but, in practice, it happens rarely. Usually, the level of vibration likely to cause even cosmetic damage, such as plaster cracking, is significantly greater than the level that would be clearly perceptible (and probably unacceptable) to the occupants.

SOUND DESCRIPTORS

Sounds differ in their level (loudness), frequency content (pitch), and they may vary with time. Consequently, different units have been developed to describe different types of sounds.

SOUND LEVEL

Sound level is described on a logarithmic scale in terms of decibels (dB). If the power of a sound source is doubled (eg two compressors instead of one) the level will increase by 3 dB. Subjectively, this increase is noticeable but not large. An increase of 10 dB doubles the perceived loudness of a sound.

FREQUENCY CONTENT

The ear can respond to sounds over a wide frequency range (roughly 20 Hz to 20 kHz), but most environmental sounds lie between 20 Hz and 5 kHz. The ear is more sensitive to sounds of some frequencies than others, and is particularly sensitive between about 500 Hz and 5 kHz. The 'A' weighting now in wide use results from an electronic circuit being built into a sound level meter to make its sensitivity approximate to that of the ear. Measurements made using this weighting are expressed as dB(A). Sounds with tonal components (such as fan noise) are particularly annoying, and sometimes this is recognised by adding 5 dB(A) onto a measured level. However, where the sound level varies with time, a simple dB(A) level does not describe it adequately and special advice will be necessary.

NOISE SPECTRA

Noise spectra from common sources are shown on the graph, Figure 2.33. Road traffic has most of its energy around 80 Hz, whereas aircraft have considerable energy up to 2 kHz. Trains are also a broad band source, although with less energy in the 100–400 Hz range.

WEATHER-GENERATED NOISE

Noise from the roof itself, caused by wind, sun, rain or hail, may need to be considered.

The factors include:
- hail or rain-induced drumming on thin sheets,
- wind-induced whistling through ventilation slots at eaves or within ventilators placed on the roof slope,
- sun-induced 'oil canning' of edge-stiffened thin sheet metals (particularly self-supporting sheets), and
- 'stick-slip' intermittent movements on metal-to-metal joints in fixings and from thermoplastics gutters intermittently exposed to sunlight.

Drumming of thin sheets can best be reduced by coating the underside with a heavy layer of 'soft' material: much the same principle as used to reduce noise from car body panels.

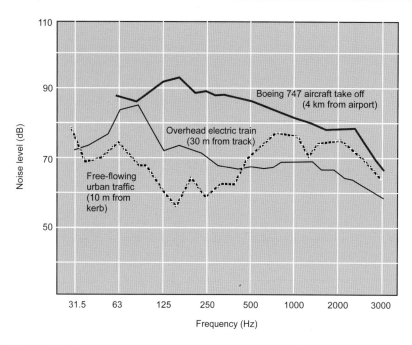

Figure 2.33: Noise spectra for a jet aircraft, an electric train and road traffic

An alternative which has been used in Europe, is a non-adhered layer of sponge-like material. Sheets made of salvaged rubber particles have recently become available, but there is no information on their longevity under metal roofing. 'Oil canning' is far more difficult to deal with and is best considered at the initial design stage; damping, however, may reduce its worst effects. Sliding joints in metal can be isolated with PTFE washers. Nothing can be done about noisy thermoplastics gutters, save making sure that no detritus has built up in the sliding joints.

More recently, BRE investigators on site have occasionally discovered evidence of slack sarking felts over cold roof spaces vibrating in windy conditions and slapping the under surface of the tiling battens.

INTERNALLY GENERATED NOISE

When complaints of sound transmission through a separating wall have been investigated, that part of the separating wall within the roof space has been found often to be a major contributing route. In some cases, BRE investigators have found no separating wall at all in the roof space. The material of a separating wall, or its thickness, may change above ceiling level, or there may be unfilled joints in masonry, or even holes. Lath and plaster ceilings, being heavier than a single thickness of plasterboard, are likely to help prevent this problem from arising. In other cases, rough rendering the separating wall within the roof space may help to reduce noise transmission. Guidance is contained in the relevant parts of national building regulations.

2.9 DURABILITY, EASE OF MAINTENANCE AND WHOLE-LIFE COSTS

There is a British Standard on durability, BS 7543[55], which applies to roofs and roofing. It gives general guidance on required and predicted service life, and how to present these requirements when preparing a design brief. BS ISO 15686 describes the methodology[56].

When selecting products, systems, installers or maintenance companies, BRE strongly recommends those that have been independently approved by a third-party certification body. The performance of an excellent product can be severely undermined by poor installation or maintenance. BRE produces lists of approved products and services which can be viewed free of charge on *The Red Book* web site[35].

One of the most important influences on durability is the quality of the original work that is carried out. Specific points to bear in mind on workmanship are listed later in the book under headings appropriate to each kind of roof. However, no roof, even if constructed of the highest quality materials, will give good service if crucial aspects of workmanship are not properly carried out (Figure 2.34).

Other significant factors for the durability of roofs are the types and amounts of pollution and organisms that attack roofing materials. These agents of deterioration are examined first.

POLLUTANTS IN GENERAL

Whether any particular form of deterioration will affect the life of roofing depends for the most part on:

• what materials are used,
• in what combination, and
• where the building is situated.

The combined effects of pollutants and weather is also important. A comprehensive review of the effects of pollutants on buildings can be found in a report of the Building Effects Review Group (BERG)[57]. However, in some circumstances it will be necessary for the specifier to evaluate the long-term performance of roofing materials in specific environments.

INDUSTRIAL POLLUTANTS

The major industrial pollutants considered to have a potential effect on roofs are sulfur dioxide, nitrogen oxide, ozone, smoke, chlorides and hydrocarbons. Industrially derived chlorides and ozone are described separately in later sections since they supplement significant natural background levels.

Sulfur dioxide, produced by the burning of fossil fuels containing sulfur, is the pollutant which, with its associated secondary pollutant, sulfuric acid, has been most strongly associated in the public mind with the concept of 'acid rain' since at least the middle of the 19th century (Figure 2.35). Acid rain is now defined as rainfall with a pH below 5.0 (a neutral solution has a pH of 7.0). This value makes allowance for acidity due to natural sources such as carbon dioxide and the natural sulfur cycle.

Emissions of sulfur dioxide increased steadily with advancing industrialisation until the early 1970s. Since then there has been a significant decline. For urban environments, the decline of sulfur dioxide concentrations has been even more dramatic due to the proportional shift of emissions from low- and medium-level urban sources to high-level emitters, often located outside urban areas.

Pollutants can be transferred to roof surfaces by wet deposition (ie in rain or snow) or by dry deposition (ie direct deposition in damp conditions or absorption by the material surface). The relative effects of wet and dry deposition on decay of materials depends on geographical location and on the material involved. Recent evidence indicates that dry deposition accelerates the corrosion of steel.

Available information on the wet and dry deposition of sulfur compounds suggests that wet deposition tends to dominate where rainfall is high and the sources of pollution are at some distance, for example, the west coast of Scotland. Dry deposition tends to dominate where rainfall is relatively low and sources nearby, such as in eastern England. However, dry deposition can lead to damage only when the material becomes wet and run-off occurs, mobilising and removing the reaction products.

Figure 2.34: Opening up this blister reveals a complete lack of adhesion between layers. There can be several possible causes (see Chapter 4.1)

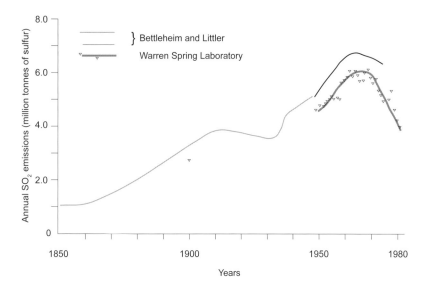

Figure 2.35: Sulfur dioxide emissions in the UK (from the BERG report[57])

It is known that, in sufficient concentration, sulfur dioxide plays an important part in the deterioration of a number of building materials, the most sensitive being calcareous stones and certain metals. There is evidence that the considerable reduction in emissions between 1960 and 1990 has resulted in a reduction of weathering rates for steel, but little evidence that other metals have benefited from the lower emissions.

For porous materials such as clay or concrete roof tiles it is difficult to isolate current from historic damage. It is difficult also to distinguish damage caused by the present day atmosphere from that initiated by past atmospheres which are still able to cause damage by recrystallisation of reaction products held within the pores (ie the memory effect).

Furthermore, available evidence does not indicate if 'safe' levels exist, or which atmospheric components are now the major causes of damage: moisture, frost, salts, sulfur dioxide, oxides of nitrogen, chlorides or carbon dioxide (Figure 2.36).

Oxides of nitrogen are produced during combustion processes when high temperatures result in the combination of nitrogen and oxygen from the air. The primary pollutants are nitric oxide and nitrogen dioxide, and the secondary pollutant is nitric acid. Emissions of oxides of nitrogen have approximately doubled since 1945, chiefly due to increased motor transport. Typical annual mean values for central London are 100–145 μg/m^3 with peak hourly concentrations three to four times higher than this.

Smoke emissions are often considered in parallel with sulfur dioxide. They are perhaps the most visible evidence of a polluted environment. Since 1960 there has been a major decrease in overall smoke emission (Figure 2.37). This has been reflected in urban smoke concentrations where the average urban mean in 1988 was about 17 μg/m^3 compared with 140 μg/m^3 in the early 1960s. The levels are still reducing.

A major source now of smoke emission in urban areas is particulate emission from diesel engines in vehicles. There are also localised emissions from building heating systems.

NATURAL POLLUTANTS (INCLUDING CARBON DIOXIDE AND MARINE POLLUTANTS)

A number of chemicals which occur naturally in the atmosphere can affect the performance of roof materials. Some, such as water vapour, rain and oxygen, are not usually regarded as pollutants. But others, such as carbon dioxide, ozone and chlorides, can be regarded as pollutants insofar as, arguably, human activities have raised their values sufficiently to increase their environmental impact. In the case of chloride, sulfate and nitrate levels in precipitation, some distinction has to be made between marine and non-marine sources.

CARBON DIOXIDE

Carbon dioxide is present in the atmosphere at a concentration of about 350 parts per million (ppm). It dissolves in pure distilled water to give a slightly acid solution, carbonic acid, with a pH of 5.6. Dissolved in rainwater, carbon dioxide is able to react slowly with the carbonate component of calcareous roofing stone slates, leading to long-term deterioration.

Carbon dioxide can also cause deterioration of concrete: in particular, reinforced concrete. Steel reinforcement is protected against corrosion by the alkaline environment of fresh concrete. However, carbon dioxide gradually penetrates from the concrete surface inwards, neutralising the alkalinity

Figure 2.36: A clay roof tile attacked by salt crystallisation. The nibs on the reverse have all but disappeared

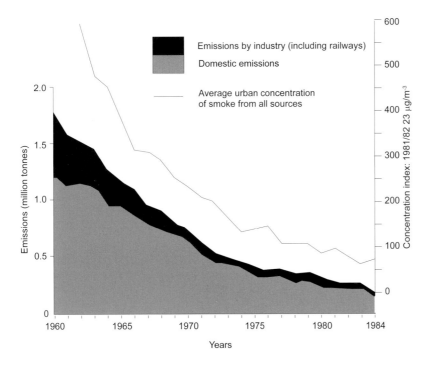

Figure 2.37: Trends in sources of smoke emission and air quality: emissions from coal combustion (from the BERG report[57])

as it progresses (carbonation). When carbonation reaches the reinforcement, the steel is much more vulnerable to corrosion and, in bad cases, spalling of the concrete cover can result.

The chief source of carbon dioxide from human activities is the burning of fossil fuels. It has been suggested that, at their peak, urban concentrations reached 3000 ppm.

The effect of carbon dioxide as a greenhouse gas is not an issue relevant to the durability of roofing. However, the climatic changes implicitly resulting from an increase of carbon dioxide would be likely to modify the value of several other agents affecting performance, most obviously temperature.

CHLORIDES

Systematic measurements of particulate chloride or gaseous hydrogen chloride over extended periods in the UK are limited. An 18-month survey in Leeds indicated almost equal contributions to total particulate chloride from marine and non-marine sources.

In the industrial context, chlorides are emitted from a number of sources, mainly as hydrogen chloride from the burning of coal. It should be noted that national manmade emissions of chlorides are of an order of magnitude less than those of sulfur dioxide or oxides of nitrogen considered earlier, and they have shown a decline parallel to sulfur dioxide concentrations.

Apart from their effect on metals, chlorides are likely to be significant for other materials if they are present in sufficient concentration to form unsightly or perhaps disruptive crystallisation products. Hydrogen chloride can add to the total acidity of rain or ambient air, but in this role it is thought to be far less important than emissions of sulfur dioxide.

In a maritime climate, as that in the UK, rain almost anywhere may at times contain sufficient chloride derived from sea salt to cause increased rates of corrosion of metals such as iron, copper, zinc and aluminium. Natural concentrations fall with distance from the coast. The highest concentrations are found in sea spray from breaking waves (up to 3 mg/m^3). Average values near to the sea but away from spray are about 0.1 mg/m^3 and, well inland, 10–30 μg/m^3. Figure 2.38 shows average rates of deposition as a function of distance from the tide mark for a large land mass. However, there are many other factors, other than airborne chlorides, that can significantly affect corrosion rates.

Elevations facing the sea are normally washed during rain storms and deposition of chlorides will be lower than on landward elevations.

OZONE

Ozone is produced naturally by photochemical reactions at various levels in the atmosphere. The natural background level in the atmospheric boundary layer (the lowest 1–2 km) can be increased by photochemical episodes. This additional ozone is a secondary pollutant.

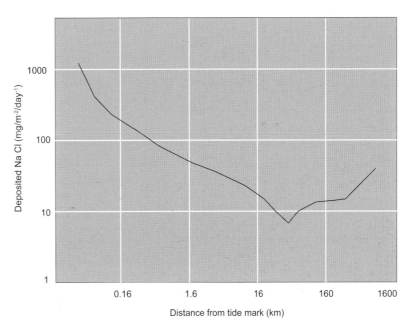

Figure 2.38: Deposition of sodium chloride aerosol as a function of distance from tide mark

Measurements of ozone concentrations in the UK show rural concentrations in the range 40–60 $\mu g/m^3$ with peak hourly values at 100–175 $\mu g/m^3$. Urban levels are lower by about one-third.

Ozone is implicated in the deterioration of polymer materials including plastics and paints. Its precise role is difficult to quantify since it cannot be distinguished from other decay processes: in this case, the action of heat and ultraviolet light.

Ozone does not affect inorganic materials directly, but it plays an important part in converting sulfur and nitrogen oxides to their respective acids, sulfuric and nitric, which can then affect inorganic materials.

INSECTS

Two types of problem occur in roofs as a result of insect action:
- the degradation of materials, and
- infestation and harbourage.

Degradation of timber and some timber-based board materials can result from the action of wood-boring beetles; common furniture beetle (*Anobium punctatum*) is the most common species found in the UK. Though damage is potentially of structural significance, the risk of significant damage in roof structures is low because the normal designs and conditions generally discourage initiation and development of attack.

In a survey carried out in 1993, BRE found that there had been a marked decrease in infestations of the common furniture beetle in UK properties up to 30 years old, though just under one-third of buildings older than this can still be affected, with slightly greater numbers recorded for properties built before 1900.

House longhorn beetle (*Hylotrupes bajulus*) infestations can cause severe structural damage to softwood roof timbers, but are mainly concentrated in a few areas to the south west of London, with the highest levels occurring in properties built between 1920 and 1930 (Figure 3.60 shows an attack on domestic pitched roof timbers). Areas of England where treatment

is mandatory are defined in the building regulations for England and Wales.

Guidance is available in two BRE books for distinguishing insects that damage wood in buildings in the UK from those that are harmless (*Recognising wood rot and insect damage in buildings*[58]), and for the treatment of insect attack (*Remedial treatment of wood rot and insect attack in buildings*[59]).

Infestation by pest insects can occur in suitably sized voids in roof constructions. Where these voids have openings into the internal space of the building this can make pest control operations both complex and expensive. Cockroaches and pharaoh's ants are usually found only in larger voids whereas book lice (psocids), bed bugs and silver fish use small crevices on internal surfaces. Many garden insects will intrude into buildings on a casual basis where suitable openings exist. In particular, wasps may build substantial nests in roof voids. The majority of the more troublesome larger varieties of insects can be excluded by provision of a 3–4 mm mesh over ventilation slots. Care should be taken to ensure that any smaller mesh size introduced as a remedial measure does not reduce the ventilation opening size below that required by building regulations.

RODENTS, BATS AND BIRDS

Although rodents such as squirrels may damage soft materials (eg insulants and timber) by gnawing, animals in general can be regarded as presenting only an infestation hazard and rarely do any significant damage to the structural parts of buildings. However, infestations may cause nuisance and fouling of wall and floor surfaces, and even have health and hygiene implications.

Rats and mice will use voids within roofs as nesting areas as well as runways to gain access to room areas. Fouling of roof voids and ceiling surfaces with droppings and nesting materials has obvious health implications. Once established in voids infestations may be difficult to eradicate. Access by rodents

normally occurs through external openings at eaves.

Infestations of roofs by edible dormice (Glis glis) may be a problem in parts of the Chilterns. Their removal requires a licence issued under the Wildlife and Countryside Act 1981.

In certain areas of the UK, grey squirrels have been known to infest roofs, gaining access by shinning up rainwater pipes and enlarging holes pecked by birds in wooden fascias. Once inside, extensive damage can ensue, not least to electric wiring insulation. They have also been known to cause considerable noise at night. Replacing timber fascias has sometimes failed to cure the problem, the squirrels gnawing through the new timber. Removal of the first two courses of tiles and folding a sheet of expanded metal over the exposed part of the fascia before replacing the tiles has proved to be one successful remedy.

Bats are known to colonise cavities, usually gaining access through openings at wall heads. Although bats are not usually regarded as having health implications, their presence is often regarded as a nuisance. Remedial measures are difficult as bats are protected from any interference under the Wildlife and Countryside Act. There is an obligation under the Act to consult the Nature Conservancy Council about any building work which might affect bats or their roosts. Guidance is available in *Bats in houses*[60] and BRE Good Repair Guide 36[61] (Figure 2.39).

Birds of many kinds will nest in roofs and it may be a constant

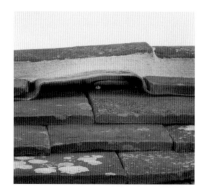

Figure 2.39: Close-up of a bat crawl entrance point under ridge tiles

battle to prevent this happening in certain kinds of construction using soft materials or those which are subject to decay. Various parasitic mites carried by the birds may enter buildings through openings and can cause dermatological problems for human occupants.

Seagulls in particular have been known to peck gaskets, but some success has been achieved in reducing this form of damage with proprietary bird-scaring devices.

MOULD, FUNGUS, ETC.

There are two kinds of building fungi: those that cause wood rot, and those that do not (Figure 2.40).

Wet rot can be quite a problem in roofs, though it is more likely to be in evidence in the fascias and bargeboards than in the structure. To give one indication of its frequency, the English House Condition Survey 1991[9] showed that, on average, around 1 in 4 houses had faults in its fascias, with the majority of cases probably associated with decay. Age did not seem to affect the rate of occurrence until after 1980 when the incidence in houses showed a fall to around 1 in 10; this is more likely to be due to the time delay required for decay to become significant rather than the growth in the application of preservative treatment or the use of alternative materials such as plastics.

Wet rot decay of timber and timber-based board materials can take place only where these are maintained in persistently damp conditions. Initiation of attack generally results from microscopic airborne spores, but can also occur where pre-infected timber has been used; in this situation, very rapid decay can occur in new construction. Under appropriate conditions damage may be rapid and severe, and therefore of structural significance; damage is almost always restricted to the timbers adjacent to the wall head. The dry rot fungus, *Serpula lacrymans*, is less common in roofs.

Surface moulds occur on external and internal building surfaces such as ceilings, usually preceded by persistent condensation. Externally, moulds are unsightly and may also cause premature failure of paint films. On internal surfaces, as well as being unsightly, mould is thought to cause respiratory problems in susceptible individuals. Problems of

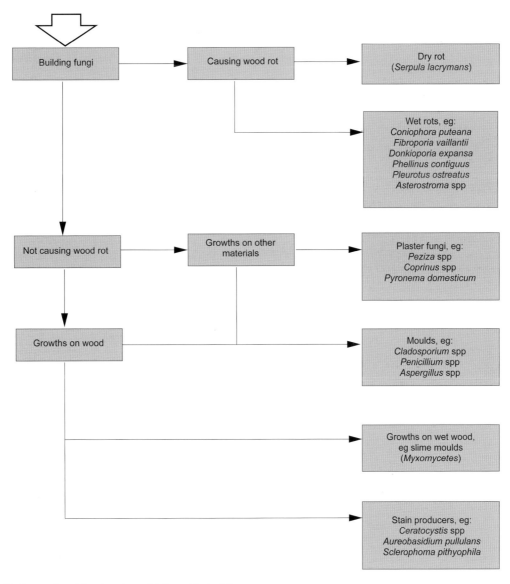

Figure 2.40: Categories of fungi and moulds found in buildings

appearance and paint failure can occur.

Algal growths occur in the main only on external surfaces where high moisture levels persist for lengthy periods.

Guidance is available on the recognition of fungi and moulds in buildings[58] and on remedial treatment[59].

Algae, lichens and mosses are commonly seen on roofs, and the resulting appearance is often regarded as beneficial. However, these growths can increase the risk of frost damage to coverings such as porous tiles and fibre cement sheets, and they may interfere with drainage occasionally; for that reason they may need to be removed. Growths up to 100 mm deep have been reported to BRE, though these are rare. Since the Clean Air Act of 1956, growths have occurred in urban areas as well as in rural. BRE Digest 370[62] explains how to treat affected surfaces. Even the less porous surfaces such as slate are not immune, and growths have been seen to begin in as little as two or three years on some kinds of materials.

If growths of these organisms need to be removed, there are a number of proprietary solutions available that have been found to be satisfactory. *Only those that have been given a number by HSE should be used.* Alternatively, brushing or, in the worst cases, hosing with a power jet may be used, provided care is taken not to damage the

surface or to drive water between the laps of sheeted or tiled surfaces. Future growth may be inhibited by fitting small diameter copper pipe or wire above the area to be protected, though the visual acceptability should be checked as some disfiguring stains will result. It will be seen that copper chimney flashings inhibit organic growth only in the areas immediately under the flashing.

An HSE leaflet[63] gives guidance on cleaning old asbestos cement roofs. One of the most important considerations is that asbestos cement must not be brushed when it is in a dry state. It needs to be thoroughly wetted with an approved biocide to minimise fibres being released from the material, and dust masks of the appropriate rating should be worn. Where moderate levels of organic growths are wanted, they may be encouraged by washing the surface with a dilution of cow dung in water.

PLANTS

Plants can grow on both pitched and flat roofs if the roofs are not properly maintained. Weeds will grow in gutters if the detritus is not cleared regularly (Figure 2.41). The roots of plants may also be a problem with inverted roofs, and this is dealt with in Chapter 3.3.

It is possible that climbing plants can have detrimental effects on buildings. There is some evidence that Virginia creeper, *Vitis quinquefolia*, which clings by

Figure 2.42: Ivy has destroyed this plain tiled roof on an outbuilding (most of the growth has been cut away). The moss and lichen on the adjoining asbestos cement slate roof do not necessarily accelerate deterioration, although the slates are old and need to be replaced (Photograph by permission of A Penwarden)

sucker pads on tendrils, does not in general damage building materials. However, the common ivy, *Hedera helix*, which clings by short adventitious roots on the stems, can penetrate masonry and the joints between slates and tiles, forcing them apart (Figure 2.42); growths up to 3 m long have been observed by BRE within roof voids, the length depending on the amount of light reaching the plant. Cutting the bases of stems does not always kill the plant. In removing growths of ivy from affected roofs, the plants have to be removed with care to avoid damage to the roof. Any attempt to pull off the growths wholesale will also tend to pull off the covering of the roof.

Apart from the obvious effect of blocking gutters and perhaps harbouring pests, climbing plants which cling by suckers probably do little harm to roofs (Figure 2.43). The Royal Botanic Gardens, Kew, believe that it would be most unlikely for Virginia creeper to root above ground level. Vigorous climbing roses will occasionally enter roof voids through gaps in open eaves,

Figure 2.41: Weeds growing in the detritus of a gutter on a farm building. The gutter will overflow and the wall will become saturated

Figure 2.43: A creeper growing on a plain clay tile roof. In this case, the plant is of a species which is unlikely to cause damage

but the shoots will die back if no light reaches them.

Most roofs need periodic maintenance if they are to continue to perform well. Recommendations are made in later chapters dealing with particular kinds of roofs, but one factor which applies to all kinds of roofs, without exception, is the annual clearing out of gutters and rainwater hoppers (BS 12056-3)[24].

STRUCTURAL SURVEYS OF ROOFS

It will be a matter of judgement when the building surveyor or architect needs to call in the assistance of a structural engineer to carry out a full structural survey. All long-span roofs should be subjected to regular inspections in any event, and building owners should be advised of this.

Any preliminary examination and assessment of a roof to determine whether or not to call in a structural engineer will need to consider the items listed in Box 2.3.

Where it is thought necessary to call in an engineer to appraise a structure, the appraisal will normally be carried out in accordance with the framework described by the Institution of Structural Engineers' *Appraisal of existing structures*[64].

The report from the engineer to the client, or the client's other professional advisers, should include recommendations on at least the following points:
- items (including the numbers of structural members) requiring immediate repair or replacement for reasons of safety, stability or serviceability of the structure either as a whole or substantial parts of it (the assessments will need to be supported by calculations in appropriate cases),
- items likely to require attention within the short term (say two years),
- items requiring attention within the medium term (say five years),
- items provided for access to the roof for maintenance purposes,
- a suitable maintenance programme if one is not already in place,
- a date for the next structural survey and appraisal.

SAFETY

Compliance with all relevant items of safety legislation is outside the scope of this book but reference should be made to Digest 493[65] for a list of relevant information services.

The Health and Safety Commission's *The Construction (Design and Management) Regulations* 2007[66] describe responsibilities of clients, designers and contractors for the health and safety of construction workers. The regulations also require the appointment of a CDM coordinator and principal contractor for most design and construction work.

2.10 FUNCTIONS PARTICULAR TO PITCHED ROOFS

Pitched roofs can be categorised as one of four kinds for the purposes of this book.

- *Dual-pitch roof* where the structure is triangulated to span between supporting exterior walls, and where the structure is entirely above ceiling level (Figure 2.44). This general category includes both gabled and hipped configurations.
- *Portal frame* in which the wall and roof are supported by a single structural member (Figure 2.45).
- *Single-pitch roof* where the slope is formed either by a simple beam or purlin laid to slope and no tie is provided at the lowest level of support, or where a single-pitch truss is carried on both walls (Figure 2.46).
- *Troughed dual-pitch roof* where the valley runs in the centre of the building, normally at right angles to the span of the supporting members. Again, no tie is usually provided though a valley beam is (Figure 2.47).

Pitched roofs offer an opportunity for using an efficient means of spanning large spaces without necessarily imposing on the interior space available; with flat roofs, downstand beams are required, depending on the length of span. For large spans, the general form of the structure must contribute to its capacity for supporting loads; to achieve this, the triangulated truss has been the most common form adopted.

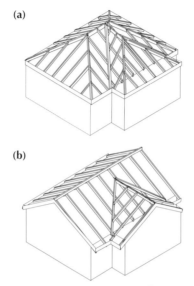

(a)

(b)

Figure 2.44: Dual-pitched triangulated roof structures: (a) hipped, (b) gabled

Figure 2.45: A portal frame in timber

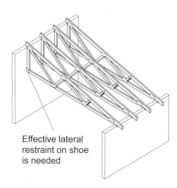

Effective lateral restraint on shoe is needed

Figure 2.46: A single-pitch roof

Figure 2.47: A troughed dual-pitch roof

PITCHED ROOF COVERINGS: PRINCIPLES

The outer covering for a pitched roof is normally one of two kinds: permeable or impermeable to air (Figure 2.48). The former category includes components normally relatively small in size, such as tiles or slates, whereas the latter includes relatively large sheets of various materials. Limiting the effect of wind on roofs in the former category is carried out by secondary lines of defence such as underlays of various kinds, sarkings, etc., whereas with the latter it is the sealed outer covering that performs this function.

The principle underlying the performance of overlapping units on a pitched roof is that all direct paths from the underside to the outer surface are lapped by the adjacent units. There is relatively little tolerance with the positions of some units (eg interlocking tiles) if performance is to be up to expectations, but other materials have more tolerance. Also, any broken or displaced unit that reveals an unprotected joint in the surface below will increase the risk of rain penetration. Vigilance is always needed on the part of those

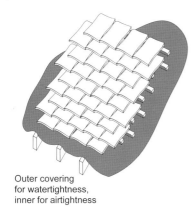

Outer covering
for watertightness,
inner for airtightness

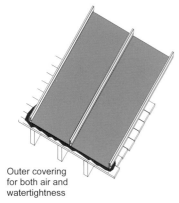

Outer covering
for both air and
watertightness

Figure 2.48: First and second lines of defence in roof coverings

responsible for maintenance so that missing or damaged units are replaced before rain penetrates to the building interior.

Small units may be either single or double lap. Single-lap units are normally larger in size than double lap and have frequently been of all-round lap but with interlocking edge profiles. However, there are exceptions: pantiles for example. Provided all units, whether large or small, interlocking or not, are laid to manufacturers' recommendations, there should be little difference in their weathertightness properties.

A number of materials from which tiles are made are porous, but this does not affect their rain exclusion function as the pores are relatively small, and water is not forced through by the wind or gravity.

The natural flow of water on a pitched roof is governed, of course, mainly by gravity. The ridging of double-lap tiles and the vertical rolls of pitched metal roofs help to direct water flow downwards. In plain tile and slate, verges can be canted upwards to assist in diverting water

away from the verge (which also has the added effect of countering a potential optical illusion of sagging where the roof surface runs through without such canting). However, where relatively smooth surfaces are used, wind can and will blow water sideways across roofs. BRE experiments, measuring rainwater run-off on both real buildings and fully instrumented natural exposure rigs, have shown that relatively large quantities of water can be moved sideways under wind pressure (BRE Current Paper 90/74[67]). It is clear that water can move sideways on smooth pitched surfaces, given the right conditions. In metal covered roofs with standing seams or batten rolls, the water will be driven sideways until it encounters the roll, where it will be diverted downwards. The water load on the joint is considerable in these circumstances. Therefore, in areas of the UK with the highest driving rain indices it may be necessary to evaluate the risk of sideways flow, the potential use of the building and its proximity to walkways, and to provide verge gutters where appropriate.

If the span of a pitched roof is large, the quantities of rainwater reaching the eaves can be considerable and it may be necessary to take special precautions to prevent discharge of rainwater overtopping the gutter. Asymmetrical

gutters with large upstands on the outer rim are feasible, though they may need to be specially fabricated. The Standards make provision for discharge weirs at the ends of gutters to direct overload run-off where the flows exceed the design rate (Figure 2.49).

SARKINGS AND UNDERLAYS

Information on the kinds of sarkings and underlays used in the pitched roofs of dwellings in England is given in the English House Condition Survey 1991[9]. Taking all ages together, in approximate terms the following were noted:

- 1 in 6 of all houses had no sarking whatsoever,
- 1 in 2 had material coming under the general description of felt,
- 1 in 50 had timber boarding,
- 1 in 50 had polyethylene sheet,
- 1 in 500 had sprayed coatings,
- 1 in 150 had an internal lining of some other unspecified form, and
- the lining on the remainder of those houses with pitched roofs went unrecorded.

It is understood that this kind of information is not currently obtained, so the figures cannot be updated. Nevertheless, since less than 0.1% of the existing stock of houses is demolished each year it is likely that many older houses which have not been re-roofed in the intervening years will still

Figure 2.49: Valley gutter extended; profiled weir discharges overload

embody these features. There is more information in the House of Lords Select Committee on Science and Technology, second report, July 2005[68].

Rain and snow may be forced past the edges of lapped tiles and through the joints by wind action. In theory, at least, tiled and slated roofs with boarded sarking, as is common practice in Scotland, will be at less risk of rain penetration than roofs with flexible sheet sarkings since the boarded sarking performs much better as a wind barrier. Flexible sarkings can billow in the wind. Following the principle of the ventilated but closed cavity, in which air pressure can be equalised between exterior and cavity, there would normally be very little air pressure difference between the upper and lower faces of the tiles in a boarded roof to provide the driving force for rain to penetrate.

Sarking membranes should have a water vapour permeability in the range 0.1–2.0 MNs/g.

WIND LOADING EFFECTS

Dual-pitch roofs forming either a ridge or a valley, with pitches greater than 30° to the horizontal, in general have positive pressures only from the wind. Below that pitch, suction increases as pitch drops. Unequal pitch (asymmetrical), dual-pitch roofs present problems in calculating wind loads, and specialist advice should be sought.

It can be seen from Figure 2.50 that suctions at eaves, verges and ridges are generally greater than over the remainder of the roof, placing great importance on the fixing of

coverings at these positions. Local high suctions are averaged out to give uniform values for the purposes of design, and, in a typical house, such zones extend for around 1.2–1.5 m from the relevant edge of the roof.

With hipped roofs, vortices form along each of the ridges of the hips, although the vortices are greater over a gable and the total loads on a hipped roof are therefore less than with a gabled dual pitch.

It has become clear that aircraft landing and taking off generate vortices, in some cases causing tiles to be stripped from houses near airports. This phenomenon has been investigated by BRE, concluding that installers and those responsible for the maintenance of roofs close to airports need to check carefully the provision of tile and slate nailing and clipping. This subject is again referred to in Chapter 3.1. BRE Digest 467[69] also covers the problem.

Figure 2.51: Snow sliding from a relatively shallow pitched slate roof

SNOW

On pitched roofs, accumulated snow melting under heat from the sun or as a result of heat loss from poor roof insulation, will of course slide off – even from relatively shallow pitches (Figure 2.51); the danger to people below and the potential damage to vehicles is obvious. Falling snow, depending on wetness, is assumed to be unlikely to accumulate to such a critical mass on pitches of more than 60°. Icicles formed on rainwater gutters have been known to fall and puncture or damage roof coverings below – and, of course, falling icicles can cause serious injury to people.

Snow boards or guards should be provided where a risk is identified. Guards need to be around 300 mm in height, preferably slatted or of wire mesh to allow the wind and powdery snow through (Figure 2.52). Since the weight of wet snow can be considerable, the guards need to be fixed firmly, and kept clear of the gutters. Best practice, where the roof covering allows, is to use specially fabricated angle brackets screwed firmly to rafters over the top of sarking felt and protruding under the laps of the covering (Figure 2.53). An idea of the relative mass of snow to be allowed for is given by some of the design criteria in BS EN 12056-3[24] and for sliding snow in BS 6399-3[7].

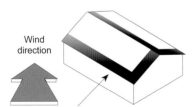

Wind direction

Zones of highest pressure coefficients nearest the windward eaves and verges and the leeward ridge (darkest = highest)

Figure 2.50: Wind loads on a simple symmetrical, dual-pitch roof (for more details see BRE Digest 346, Part 6[13])

Figure 2.52: A mesh snow guard. The fixing of the upstands to the fascia may prove inadequate for the heavier loads of snow and ice

Figure 2.53: Snow boards resting on L-shaped brackets fixed to rafters under the slates. These boards are insufficiently deep to hold substantial snow loads on the roof slope

FIRE

Roofs are not treated as elements of structure according to the guidance given in ADs B1 and B2 of the Building Regulations for England and Wales[4] and therefore do not need to be fire resisting, unless they serve the function of a floor. In the case of pitched roofs the need for fire resistance does not occur.

Cavities in external walls should be closed where the wall meets the roof construction unless the cavity is filled totally with thermal insulation.

Spaces or cavities within the fabric of a building provide routes for the spread of smoke and flame; of particular importance in this respect are the cavities within roof voids. The maximum distances between cavity barriers in roof voids of non-domestic buildings is 20 m in any direction, irrespective of the class of surface exposed. No provisions are made for maximum spacing of cavity barriers in the roofs of domestic buildings.

However, in dwelling houses, flats and maisonettes, and other residential buildings, cavity barriers are required within roof voids above protected escape routes of fire-resisting construction which is:
- not carried to full storey height, or
- not carried to the underside of the roof covering in the roof void (Figure 2.54).

ADs B1 and B2 also contain provisions relating to cavities in profiled metal decking.

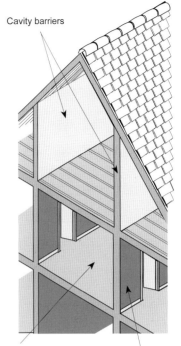

Protected escape route   Fire-resisting construction

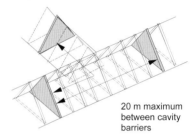

20 m maximum between cavity barriers

Figure 2.54: Cavity barriers are required within a roof void, particularly over a protected escape route

2.11 FUNCTIONS PARTICULAR TO FLAT ROOFS

For the purposes of this book a flat roof is taken to be a roof of slope less than 10°. For convenience, however, certain materials that can be used on both flat and pitched roofs have been considered in the earlier chapters dealing with pitched roofs. Any additional points in relation to their use on pitches under 10° are made in the chapters on flat roofs.

As mentioned earlier, the main point of concern with flat roofs is that they are not normally flat. BRE recommends a minimum fall is always maintained. It is the lack of maintenance of this minimum fall that gives rise to many problems, particularly in long-span roofs where deflections (eg under snow loading) can be significant. A design fall of 1 in 40 to achieve a minimum of 1 in 80 when built has been the usual practice in the past. Minimum actual falls for particular materials in this standard include:
- aluminium, copper and zinc: 1 in 60,
- lead: 1 in 80,
- built-up felt: 1 in 80,
- semi-rigid fibre reinforced bitumen sheet:
  - roll cap, 1 in 60,
  - ribbed, 1 in 75.

DECKS
Roofs of very small span may consist of little more than a structural deck or slab with suitable weatherproofing above. But as the span increases, so the need for greater depth of primary members to support the increased loads imposed by the weatherproofing layers without significant deflection, leads to the use of beams of various kinds. Some flat roofs can be rather complicated (Figure 2.55).

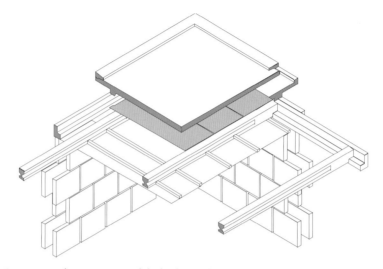

Figure 2.55: The concrete roof deck of an Orlit system house: a relatively complicated structure

Balconies are a form of flat roof. Where they are inset into the building they need to perform the full range of functions of a roof in addition to carrying traffic, and they are not therefore dealt with separately in this book. Where they project from the building they should be checked to ensure they do not provide cooling fins to the interior, leading to condensation risk. At least 1 m width of insulation should be provided on the inner soffits and upper surfaces of contiguous floors.

The dimensional stability of substrates and decks is crucial to the subsequent behaviour of the covering, particularly where the covering is bonded, especially fully bonded, to the deck. Movements of the deck due to thermal and moisture content changes have been shown in the past to be a major factor in the breakdown of coverings. For example, if a membrane is fully bonded to a deck across a joint only 1 mm wide, and that joint subsequently opens

a further 1 mm, the covering is subjected to a 100% extension. Very few materials, if any, can withstand that degree of movement.

WEATHERTIGHTNESS
Flat roofs, so far as weathertightness is concerned, are of three kinds:
- seamless,
- virtually seamless with adhered joints to sheets,
- sheets jointed with welts or folds.

The weatherproofing layer is held down against wind-induced suction by one of three main systems:
- ballasted (Figure 2.56a),
- mechanically fastened (Figure 2.56b),
- adhered (Figure 2.56c).

Adhered plastics or rubber membranes may be backed with a layer of resilient material (eg a polymer fleece) to help overcome small height differences encountered between adjacent thermal insulation sheets and also to provide an improved key for the adhesive. They

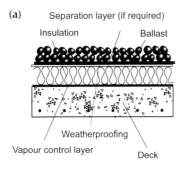

(a)

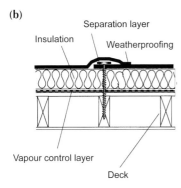

(b)

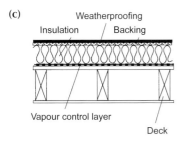

(c)

Figure 2.56: Three main roof-covering systems for holding down the weatherproofing layer: (a) gravel ballasted, (b) mechanically fastened, (c) adhered

Figure 2.57: Ponding has occurred because the outlet has been set too high

may also provide some flexibility in overcoming slight movement in use. Such sheets with suitable backing can also be used as overlays to refurbish old bitumen sheet roofs (BS EN 13956[70]).

Coverings applied in liquid form in situ, which are the most common form of seamless covering, depend on the quality of workmanship for their integrity. For membranes that are welded or otherwise supposedly fully stuck together, it is better if joints are made where free water will not lie in quantity in case the joints have imperfections (Figure 2.57); in practice, this ideal cannot always be met. Welted or folded seams operate on the principle

that all joints lie above the highest standing water level, and this principle must be achieved for the roof to remain watertight.

The amount of lap and upstand on joints should be related to possible wind pressures since water can be driven uphill by wind pressure. *Principles of modern building*[3] noted that a 26 m/s wind will support a column of water 50 mm high. If joints in roofs have to withstand such wind speeds without leaking, leaving aside the effects of continuous capillaries, the seams will in general need to stand at least that height above the slope of the roof.

Flat roof coverings, where these are light in weight, are also very prone to suction caused by the wind. Furthermore, the scouring action of the wind on gravel ballast over inverted roofs can lead to considerable movements. Unbonded ballast should be at least 20 mm in size, though it is better to relate the size to the particular wind conditions expected (BRE Digest 311[71]).

A form of construction sometimes seen is the reservoir roof: ie a roof that is flat, having no pitch, and which is designed to trap and hold rainwater with the excess flowing over a weir. In the past, these have invariably given trouble, although, with the newer materials now available, durability may become less of a problem. The load of water on a reservoir roof can be considerable, and lead to deflections, becoming progressively more serious.

MODIFICATIONS FOR CLIMATE CHANGE

The main waterproofing membrane on a flat roof has to have an upstand (recommended minimum of 150 mm) at edge details to ensure splashback from heavy rainfall does not enter the building and any temporary flooding can be coped with. This upstand could be carried up the entire height of the parapet wall, which will significantly reduce the rain penetration and hence the risk of water ingress. The increased cost depends on the height of the wall.

WIND LOADING

Wind flow over flat roofs is, in some respects, similar to the flow over walls; that is to say, the flow separates at the upwind edge and may then converge downwind. Vortices form from the upwind corner. The maximum suctions occur at verges (Figure 2.58), especially where these are associated with shallow slopes.

THERMAL INSULATION

The position of the principal layer of thermal insulation in a flat roof defines the types of flat roof which came into general use in the 1980s. These roofs comprise:
- cold deck (Figure 2.59a),
- warm deck (Figure 2.59b),
- inverted warm deck (Figure 2.59c).

The thickness of insulation required depends on the thermal conductivity of the insulant and the requirements

Wind direction

Zones of high pressure coefficients nearest the windward eaves (darkest = highest)

Figure 2.58: Wind pressures on a flat roof (for more details see BRE Digest 346, Part 6[13]

of regulations. The regulatory requirements (AD L) have been expressed as a U-value where the lower the value the more insulation is required. Although AD L allows different ways to comply, the thickness of insulation for a foamed plastics material is likely to be about 200 mm.

Cold deck roofs

Cold deck roofs have the thermal insulation below the deck so that the deck is not warmed by the building (Figure 2.59a). Cold decks are normally characterised by an air space above the insulation. A vapour control layer beneath the thermal insulation reduces the risk of condensation occurring on the underside of the weatherproof layer, but the difficulty of ensuring the integrity of a vapour control layer leads to the need to ventilate the air space to reduce the condensation risk.

The cold deck flat roof is considered a poor option in the temperate, humid climate of the UK where sufficient ventilation may not be achieved in sheltered locations or in windless conditions, even where the roof is correctly designed. *Cold deck flat roofs should be avoided in new work, and existing cold deck roofs should be converted to warm deck wherever possible.* The procedure for conversion given in Chapter 4.1 should be used.

The following clause from the requirements of the Building Standards in Scotland means that it is unlikely that 'cold roof' constructions satisfy regulatory requirements:

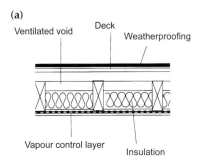

(a)

Ventilated void   Deck   Weatherproofing

Vapour control layer   Insulation

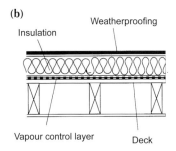

(b)

Insulation   Weatherproofing

Vapour control layer   Deck

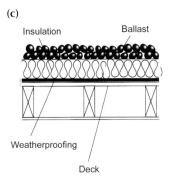

(c)

Insulation   Ballast

Weatherproofing

Deck

Figure 2.59: Flat roof systems: (a) cold deck, (b) warm deck (sandwich), (c) inverted warm deck

'Condensation
3.15 Every building must be designed and constructed in such a way that there will not be a threat to the building or the health of the occupants as a result of moisture caused by surface or interstitial condensation.'

Warm deck (sandwich) roofs

Warm deck (sandwich) roofs have the insulation over the top of the deck (in effect the main flat surface of the roof which supports both weatherproofing and access loads) so that the deck is warmed by the interior of the building (Figure 2.59b). A vapour control layer (in effect acting as a barrier) is normally required beneath the insulation to reduce the risk of condensation forming on the underside of the

weatherproof layer. This forms the so-called sandwich from which this type of roof gets its name. Ventilation of the interior of the roof structure is not required. In the process of converting cold to warm deck, existing thermal insulation may need to be removed entirely before re-insulating.

Inverted warm deck

Inverted warm deck (or 'upside down') roofs have the thermal insulation laid over the top of the weatherproof layer (Figure 2.59c). The function of vapour control to reduce the risk of condensation occurring on the underside of the thermal insulation is carried out by the weatherproofing layer. If the thermal insulation is light in weight it will usually need ballasting to provide resistance to wind suction, though mechanical fixing is possible for some materials. The fact that the thermal insulation is above the deck means that it protects the deck and the membrane from the extremes of temperature, so movements can therefore be less than where the insulation is below the deck. There is much to recommend in this form of flat roof, with some indication that thermal performance can be better even than the warm deck sandwich.

There are, however, some risks with this form of roofing:
- grit may be washed down underneath the thermal insulation,
- degradation of the insulant may occur if not chosen correctly,
- condensation may occur,
- gaps in insulation may decrease thermal performance.

When grit is washed down underneath the thermal insulation, subsequent thermal movements may abrade the waterproof layer causing leaks. A geotextile filter membrane should overcome this risk.

To prevent degradation of insulants, foamed plastics insulants should not be exposed to ultraviolet degradation.

If condensation occurs, the most likely cause is cold rainwater percolating below the thermal insulation (as it must for drainage) chilling the deck or the top part of the rainwater downpipe and taking it below the dewpoint. The problem is solved by ensuring that the thermal resistance of the construction between the weatherproofing and the internal space is at least 0.15 m²K/W. This phenomenon is examined in more detail in Chapter 4.3.

ACCESS

There are two kinds of flat roof from the point of view of designing for access (BS 6399-3[7]), ie as a means of providing safe access to parts of the same roof or to other parts and elements of the building:
- the roof which provides access only for repair and cleaning,
- the roof which provides access for amenity, means of escape or for other purposes.

The essential differences to take into account include loads imposed on the roof structure, including impacts, and the provision of guarding. (This book does not cover, in all respects, roofs that provide for vehicle access.)

Where pedestrian access is needed, the compressive strength of thermal insulation boards used over decks should have a minimum value of 175 kPa and 300 kPa for crowd loads. Care should be taken that the membrane is not deformed under any of the loading criteria and that paving supports are of sufficient area to avoid crushing the insulation boards.

If a roof is used for recreational facilities where the public have access, ramps and doorways will need to be provided to ensure disabled access. The requirements are given in the supporting documents of the national building regulations:
- England and Wales: AD M[4],
- Scotland: Section 4 of the Technical Handbooks, Domestic and Non-domestic[5],

- Northern Ireland: Technical Booklet H and R[6].

Parapets or balustrades also need to comply with national design guidance:
- England and Wales: AD A[4],
- Scotland: Section 1 of the Technical Handbooks, Domestic and Non-domestic[5],
- Northern Ireland: Technical Booklet H[6].

FIRE

Where used as a floor (eg as a car parking level), a flat roof is treated as an element of structure and therefore needs to have the same standard of fire resistance as the elements of structure for that building. The various regulations or accompanying technical documents also indicate the need for any supporting element of structure to have fire resistance of not less than the element supported which, in practice, means that any supporting wall or other supporting members will also need to achieve the same standard of fire resistance as the roof.

Where a flat roof is used as part of a means of escape, the relevant part of the roof will need to have specific fire resistance from the underside. This will consist of 30 minutes for loadbearing capacity, integrity and insulation if the roof is not also serving as a floor which requires a higher period. There are also other restraints on the design of roof escapes; for example, the proximity of roof lights and vents that do not have fire resistance, and through which flames and smoke might affect an escape route over the roof.

Cavities within the roof structure are required to be closed at all junctions with other elements of construction.

2.12 EXTENSIVE LIGHTWEIGHT GREEN ROOFS

Colonisation by mosses and lichens on roofs is sometimes unavoidable. However, the 'green roof' is designed to accommodate and encourage plant growth both on pitched and flat roofs (Figure 2.60). Roof gardens have been built for many years but if trees and shrubs are to be used, then the structure will need to support an increased load. In Germany, where the construction of green roofs is more common than in the UK, green roofs are divided into two categories: 'intensive' and 'extensive'.

The 'intensive' green roof is essentially a garden and is accessible to people. Planters are required for trees, a deep earth layer is needed for shrubs and artificial irrigation is necessary. Landscaping and paths all add to the weight so the structural implications of this need early consideration. Snow loading should not be forgotten. In this book this topic is covered in Chapter 5.4.

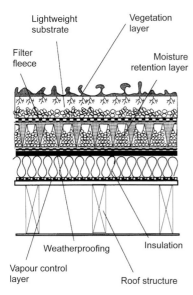

Figure 2.60: Structure of a green roof

The 'extensive' green roof is simpler in concept with the planting selected to have low management requirements and no artificial irrigation. The thin soil layer has only to support low-growing plants or turf. However, the structure will usually be more substantial than for a tiled or membrane-covered roof.

The reasons for selecting a green roof design include:
- visual appeal,
- controlling rainwater,
- improving the thermal performance, and
- providing a habitat for insects and other wildlife.

On greenfield sites they can, to some extent, replace lost habitat. There is more guidance in BRE Digest 486[72] and *Green roofs and façades*[73].

ROOF SLOPES
Although it is not necessary for a roof to be flat to support vegetation, a flat roof will inherently provide better water retention than a sloping roof, and thereby reduce the need for irrigation. A flat roof will also avoid potential problems with vegetation slipping down the slope. It may be necessary to provide an anchoring mechanism such as a metal grid to reduce this risk. Slopes of up to 40° have been used successfully.

LAYERS
A lightweight green roof will consist of several layers, each performing a defined function.

Structural deck
Many kinds of structural deck can form a suitable base for a lightweight green roof. However, it is important that there is no possibility of deflections which would cause

excessive ponding of rainwater to take place over the membrane, since this will have a detrimental effect on some kinds of vegetation. The risk of this occurring may be reduced by using a moulded plastics board which willl ensure free drainage, and a slight slope will also be advantageous in this respect.

It is also necessary to ensure that the deck is sufficiently robust to resist wind suction.

Protection of the waterproof layer
If the waterproof layer is vulnerable to puncturing from uneven deck surfaces, it is usual to lay a protective fleece.

Waterproofing
The most crucial layer, as with all kinds of roofs, is the waterproof layer. This usually consists of a sheet material such as bitumen, polyurethanes, polyvinylchloride (PVC), butyl rubber (polyisobutylene) and ethylene propylene diene monomer (EPDM). The integrity of the sheet can often be tested after installation. Membranes applied in liquid form to the deck, including bitumen and asphalt have also been used, but integrity is vital to satisfactory performance.

The waterproofing may also need to have a level of root resistance depending on the plant species selected. Specialist membranes are available which have an additional layer (eg copper) incorporated specifically for root resistance. Otherwise, an additional layer may be required.

Root resistance test methods are contained in BS EN 13707[74] for bituminous products and BS EN 13956[70] for plastics and rubbers. The test takes 2 years to perform.

Geotextile membrane

Since the roof will gather detritus from birds attracted to the habitat provided by the vegetation, and also from the residue of windblown leaves and dust from the covering itself, this needs to be prevented from entering the drainage routes. The most usual provision is by covering the waterproofing with a non-woven geotextile filter fleece or blanket made from a suitable plastics such as polyester, polypropylene or polyethylene. This fleece may also be water retaining, which will help the vegetation to survive prolonged dry spells.

Growing medium

The growing medium for a lightweight green roof can consist of a range of proprietary materials, many of which are largely inorganic in origin, such as perlite or calcined clay, with a small proportion of organic material.

PLANTING

The most frequently used plants on 'extensive' green roofs are sedums or stonecrops of the *Crassulaceae* family which are classed as succulents or water-storing plants able to withstand prolonged dry spells. Sedums used in green roof planting are small, low-growing plants with white or yellow flowers whose leaves effectively shut down in high temperatures, and open during the cool night air to absorb the moisture and carbon dioxide essential for growth.

Sedums can be established by seeding or by cuttings, and can also be grown on blankets of growing medium which makes their establishment relatively straightforward. Sedums give colour to a roof during most of the year, although that colour is not always green (Figure 2.61). Russet colours are common, and the plants can change colour to red or dark brown during periods of drought.

Stonecrop flowers do attract butterflies and other insects, but if the roof is intended to maximise bio-diversity, specialist advice should be sought to take into account local conditions.

Mosses and lichens, depending on local conditions, will grow on almost any roof covering, and a sedum mat is no exception. They tend to die back in drought conditions, but revive again when rain falls. They provide a habitat for insect life.

Figure 2.61: Sedums make a colourful and attractive roof covering having a combination of green, red and brown colours

2.13 MODERN METHODS OF CONSTRUCTION

While there is no precise definition for this topic, a simple one might be: *"dwellings whose structural units are wholly or in part manufactured off site, or on site by contemporary methods other than 'traditional' methods such as brick and block cavity masonry"*.

Only recent systems (eg from the mid 1990s) are regarded as 'modern methods' although some of the current systems have been used, mainly in other countries, for much longer. There is a growing demand for housing so the UK Government is encouraging the house building industry to innovate and develop new solutions and ideas.

In 2003 the Housing Corporation published a construction classification system that it used for its own purposes (Box 2.4) and which has subsequently been adapted by others.

Box 2.4: Housing Corporation construction classification system for dwellings

• **Off-site manufactured:**

Volumetric construction
Three-dimensional units produced in a factory, fully fitted out before being transported to site and stacked onto prepared foundations to form the dwellings.

Panellised construction
Flat panel units built in a factory and transported to site for assembly into a three-dimensional structure or to fit within an existing structure (Figure 2.62).

Hybrid construction
Volumetric units integrated with panellised systems.

Sub-assemblies and components
Larger components that can be incorporated into either conventionally built or MMC dwellings.

• **Non off-site manufactured modern methods of construction**

Figure 2.62: A house being constructed using structurally insulated panels (SIPs) (Courtesy of Kingspan Tek)

Roof cassettes are prefabricated panels designed specifically for pitched roofs. The panels are very stiff, spanning from eaves to ridge, often requiring no intermediate structural support. They are useful for room-in-the-roof constructions where the absence of structural timbers is beneficial (Figure 2.63a).

Pre-assembled roof structures are constructed at ground level before constructing the shell of the building. On completion of the substructure, the whole unit is craned into place (Figure 2.63b).

Other components that are used include prefabricated chimney stacks and dormers (Figure 2.64).

For more information see references [75]–[77].

Figure 2.64: Prefabricated dormers

Figure 2.63 (a, b): Prefabricated roof cassettes being installed (Courtesy of Milbank Roofs Ltd)

2.14 ROOF-MOUNTED PHOTOVOLTAIC SYSTEMS

Many different types of roof-mounted photovoltaic (PV) systems are in use in the UK. Some have been designed specifically for UK roof construction and practice, while others are imported systems originally designed for use in other countries. This can lead to confusion and in some cases inadequate installation has resulted in failures due to wind action or rain penetration. Guidance exists for electrical installation of PV systems but there is no equivalent guidance for mechanical installation.

The PV system should be fully defined at the planning stage, including all system components. This includes coordination of the assembly sequence for individual system components. The overall design, assembly sequence and detail design solutions should be mutually consistent. Special consideration should be given to the adequacy of the fixings or anchors to ensure that they can withstand wind forces. It should be noted that wind forces vary throughout the UK and are affected by roof height, roof pitch, orientation, etc. Therefore, a design that is suitable for one roof shape might not be suitable for a differently shaped roof, or a fixing system used in the London area might not be adequate for use in northern England or Scotland. BRE Digest 489 gives further information on calculating wind loads[78].

The specifications of the roof covering, roof weatherproofing system or external substrate should be taken into account when planning the installation of a PV system. In particular, thermal insulation, structural stability and weathertightness of the existing roof should not be compromised by the installation of a PV system. Installation should not reduce the air space beneath the roof covering to such an extent that it affects roof ventilation. The ideal free air gap beneath the roof covering is 50 mm, although 25 mm is acceptable in some situations: a similar air gap should be maintained beneath PV modules (see BS 5534[30] and *Thermal insulation: avoiding risks*[34] for more information).

The PV modules optimally should be located on the roof with respect to the sun in terms of both compass orientation and angle of inclination (ie ideally, south facing and at 30–40° for the UK). Avoid shading where possible. The PV system supplier will be able to provide further advice. Design and installation requirements for many PV systems are covered by third-party certificates or by the manufacturer's recommendations. Long-term durability should be considered since the PV components could need replacing long before the remainder of the roof and there is no guarantee that replacement parts will be available.

TYPES OF PV SYSTEM

There are two main types of PV system:
- systems integrated into the roof surface,
- systems mounted over the top of the existing roof covering, generally known as stand-off systems.

Integrated systems are usually installed in new buildings and stand-off systems are generally retrofitted to existing roofs, although this is not always the case. There are no technical barriers to the use of either system in new or existing roofs.

Integrated systems

There are six main categories of integrated systems:
- shingles,
- interlocking tiles,
- slates (non-interlocking),
- large format tiles (using standard frameless laminates),
- weathertight laminates,
- module systems joined/flashed with surrounding tiles or slates, and bonded to a metal pan.

Stand-off systems

There are fewer varieties of stand-off systems, probably due to the constraints of retrofitting. There are two main categories:
- laminates,
- modules.

These can be combined with one of several possible fixing methods.

INSTALLATION

Connection of items that penetrate through the waterproofing layer (eg cabling and piping) should be made using appropriate standard manufactured parts or special mountings. At the same time, these should be connected to the roof covering, roof weatherproofing system or external wall, or roof facing in accordance with Standards or the manufacturer's recommendations. The connection heights and widths, as detailed in the manufacturer's fixing instructions, should be maintained.

The British Standards for roofing and roof fixings will also be generally applicable to the installation and

Figure 2.65: Small format PV tiles integrated into a pitched tiled roof array (Courtesy of Marley Building Materials Ltd)

fixing of PV systems on pitched and flat roofs.

Pitched roofs

Integrated systems
Integrated systems can comprise fully integrated modules or modules that are intended to replace roof tiles or slates. They are installed as integrated components into the roof covering and should maintain a weatherproof seal.

Stand-off systems
Stand-off systems usually comprise PV modules attached to a frame that is mounted above the roof surface. The frame is constructed from prefabricated elements designed to transfer the applied forces (eg self-weight, wind and snow) to the supporting structure.

Various types of PV modules can be used for stand-off mounting using track and loadbearing units made of various non-corroding materials such as stainless or galvanised steel. The mechanical fixings should be manufactured from non-corroding metals and consideration should be given to avoid galvanic corrosion.

The framing system should be appropriate to the roof covering, the inclination of the roof and to conditions of the site.

The distance between the upper surface of the roof covering and the underside of the modules should generally not be less than

60 mm. It should be possible to dismantle the system which may be necessary in the event of repairs to the roof covering or if dirt or debris accumulates under the modules.

Flat roofs

Integrated systems
The installation of integrated systems into flat roofs is rare. However, if they are used then they should be installed into the roof weatherproofing system in the same manner as integrated components like roof lighting domes. They should be fitted into the weatherproofing system to maintain weathertightness.

They consist of modules and mounting frames. When selecting the mounting frames, attention should be paid to the compatibility of the materials. In some cases, PV modules can be integrated directly into the weatherproofing materials to form the PV system.

Stand-off systems
PV modules on flat roofs are generally mounted on a sloping frame. They can be mounted on bases or supports and incorporated into the roof weatherproofing system or attached to load-distributing bearers such as pans loaded with ballast (eg gravel or tile shards). The roof weatherproofing system beneath the bearers should be protected and the accumulation of water should be avoided. Attention should be paid to the static load capacity of the load-bearing design and the roof structure (eg the compressive strength of the thermal insulation). It might be necessary to fix the frame mechanically to the roof structure, though this should be avoided wherever possible because penetrations are hard to seal and make inspection difficult. In all instances, the functionality of the roofing beneath stand-off mounted systems should be maintained.

It should be possible to dismantle the system which may be necessary in the event of repairs to the roof covering or if dirt or debris

Figure 2.66: Flat roof-mounted PV stands (Courtesy of PV Systems Ltd)

Figure 2.67: Installation of PV system (Courtesy of solarcentury.com)

accumulates under the modules. For more information on care, maintenance and inspection, see BRE Digest 495[79].

DESIGN

PV systems can be susceptible to wind damage and rain penetration. They should not reduce the weather resistance of the existing or surrounding roof to which they are fixed or integrated. For stand-off systems, the primary areas of concern are the hook fixings or anchor points which should be strong enough to resist wind loads. Where these hooks and fixings penetrate through tiled or slated roofs, they should not lift the tiles to allow rain penetration. Stand-off systems should also be stiff enough to resist wind load which can cause lifting of the tiles. There are no test methods available specifically for determining the uplift resistance of modules and their fixings. However, a test method designed for determining the uplift resistance

of tiles and slates has been shown to be applicable to many types of integrated and stand-off systems (BS EN 14437[80]). This test method can be used to determine the resistance to wind uplift of modules and their fixings. Wind loads can be determined using BRE Digest 489[78].

For integrated systems, the area most at risk from rain penetration is the interface with the existing roof covering. Adequate weatherproofing can often be achieved by the use of suitable flashings. These can either be integrated or externally applied.

There are no test methods specifically for determining the resistance of roof-mounted PV systems to wind-driven rain. However, an EN test method is currently being developed for roof tiles and slates[29]. This draft test method has been used successfully to test a wide range of systems. It should be used where test evidence of weathertightness performance is required.

2.15 REFERENCES

[1] **UK Climate Impacts Programme (UKCIP).** Climate change scenarios for the United Kingdom. The UKCIP briefing Report. Norwich, The Tyndall Centre for Climate Research, 2002. See www.tyndall.ac.uk

[2] **Saunders GK.** Designing roofs for climate change. Digest 499. Bracknell, IHS BRE Press, 2006

[3] **Building Research Station.** Principles of modern building, Volume 2: Floors and roofs. London, The Stationery Office, 1961

[4] **Communities and Local Government (CLG).** The Building Regulations 2000.
Approved Documents:
 A: Structure, 2004
 B: Fire safety, 2006
 Volume 1 Dwellinghouses
 Volume 2: Buildings other than dwellinghouses
 C: Site preparation and resistance to contaminates and moisture, 2004
 F: Ventilation, 2006
 H: Drainage and waste disposal, 2002
 L: Conservation of fuel and power, 2006
 L1A: New dwellings, L1B: Existing dwellings
 L2A: New buildings other than dwellings, L2B: Existing buildings other than dwellings
 M: Access to and use of buildings, 2004
Available from www.planningportal.gov.uk and www.thenbs.com/buildingregs

[5] **Scottish Building Standards Agency (SBSA).** Technical standards for compliance with the Building (Scotland) Regulations 2009
Technical Handbooks, Domestic and Non-domestic:
 Section 1: Structure
 Section 2: Fire
 Section 3: Environment
 Section 4: Safety
 Section 6: Energy
Edinburgh, SBSA. Available from www.sbsa.gov.uk

[6] **Northern Ireland Office.** Building Regulations (Northern Ireland) 2000. Technical Booklets:
 C: Preparation of site and resistance to moisture, 1994
 D: Structure, 1994
 E: Fire safety, 2005
 F: Conservation of fuel and power.
 F1: Dwellings, F2: Buildings other than dwellings, 1998
 H: Stairs, ramps, guarding and protection from impact, 2000
 K: Ventilation, 1998
 R: Access and facilities for disabled people, 2000
London, The Stationery Office. Available from www.tsoshop.co.uk

[7] **British Standards Institution (BSI).** BS 6399: Loading for buildings. London, BSI
 Part 1: 1996 Code of practice for dead and imposed loads (AMD 13669)
 Part 2: 1997 Code of practice for wind loads (AMD 13392) (AMD corrigendum 14009).
 Part 3: 1988 Code of practice for imposed roof loads (AMD 6033) (AMD 9187) (AMD 9452)

[8] **Walker D.** Certification of building structural design in Scotland. Proceedings of the Institution of Civil Engineers: Structures and Buildings: 2007: **160** (3): 125–128

[9] **Department of the Environment.** English House Condition Survey: 1991. London, The Stationery Office, 1993 (Analyses of so far unpublished data were carried out especially for Roofs and roofing. Further information may be obtained from BRE Technical Consultancy)

[10] **Communities and Local Government.** English House Condition Survey 2005. London, CLG, 2007

[11] **Northern Ireland Housing Executive.** Northern Ireland House Condition Survey 1991. First report of survey. Belfast, Northern Ireland Housing Executive, 1993

[12] **BRE.** Resistance of pitched lightweight roofs to wind uplift (leaflet). Watford, BRE, 1990

[13] **BRE.** The assessment of wind loads. Digest 346, 8 Parts. Bracknell, IHS BRE Press, 1989–1992
 Part 5: Assessment of wind speed over topography
 Part 6: Loading coefficients for typical buildings

[14] **Health and Safety Commission.** Workplace (Health, Safety and Welfare) Regulations 1992. Approved Code of Practice and Guidance L24. London, The Stationery Office, 1992

[15] **European Union of Agrément (UEAtc).** General directive for the assessment of roof waterproofing systems. Method of Assessment and Test (MOAT) No 27. Paris, UEAtc, 1983

[16] **Currie D.** Roof loads due to local drifting of snow. Digest 439. Bracknell, IHS BRE Press, 1999

[17] **Currie DM.** Handbook of imposed roof loads: a commentary on British Standard BS 6399 'Loading for buildings': Part 3. BRE Report BR 247. Bracknell, IHS BRE Press, 1994

[18] **BRE.** Estimation of thermal and moisture movements and stresses: Part 1. Digest 227. Bracknell, IHS BRE Press, 1979

[19] **BRE.** Estimation of thermal and moisture movements and stresses: Part 2. Digest 228. Bracknell, IHS BRE Press, 1979

[20] **BRE.** Estimation of thermal and moisture movements and stresses: Part 3. Digest 229. Bracknell, IHS BRE Press, 1979

[21] **British Standards Instiutution (BSI).** BS 5268-3: 2006 Structural use of timber. Code of practice for trussed rafter roofs. London, BSI, 2006

[22] **British Standards Instiutution (BSI).** BS 8110-2: 1985 Structural use of concrete. Code of practice for special circumstances. London, BSI, 1985

[23] **BRE.** Disposing of rainwater. Good Building Guide 38. Bracknell, IHS BRE Press, 2000

[24] **British Standards Instiutution (BSI).** BS EN 12056-3: 2000 Gravity drainage systems inside buildings. Part 3: Roof drainage, layout and calculation. London, BSI, 2000

[25] **British Standards Instiutution (BSI).** BS 8490: 2007 Guide to siphonic roof drainage systems. London, BSI, 2007

[26] **British Standards Instiutution (BSI).** BS 8515: 2009 Rainwater harvesting systems. Code of practice. London, BSI

[27] **Water Regulations Advisory Scheme (WRAS).** Reclaimed water systems. Information about installing, modifying or maintaining reclaimed water systems. WRAS Information and Guidance Note 09-02-04. August 1999. Available from www.wras.co.uk

[28] **Water Regulations Advisory Scheme (WRAS).** Marking and identification of pipework for reclaimed (greywater) systems. WRAS Information and Guidance Note 09-02-05. August 1999. Available from www.wras.co.uk

[29] **British Standards Institution (BSI).** prEN 15601: Hygrothermal performance of buildings – Resistance to wind-driven rain of roof coverings with discontinuously laid small elements – Test method. London, BSI

[30] **British Standards Institution (BSI).** BS 5534: 2003 Code of practice for slating and tiling (including shingles). London, BSI, 2003

[31] **Trotman P, Sanders C & Harrison H.** Understanding dampness. BR 466. Bracknell, IHS BRE Press, 2004

[32] **Communities and Local Government (CLG).** The Code for Sustainable Homes. Available from www. communities.gov.uk/thecode or from www.planningportal.gov.uk

[33] **BREEAM Centre.** BREEAM: BRE Environmental Assessment Method. www.breeam.org

[34] **Stirling C.** Thermal insulation: avoiding risks. BR 262. Bracknell, IHS BRE Press. 2002 edition

[35] **BRE Certification.** www. RedBookLive.com

[36] **Sanders C.** Modelling and controlling interstitial condensation in buildings. Information Paper IP 2/05. Bracknell, IHS BRE Press, 2005

[37] **Sanders C.** Modelling condensation and air flow in pitched roofs. Information Paper IP 5/06. Bracknell, IHS BRE Press, 2006

[38] **CIBSE.** Testing buildings for air leakage. Technical Memoranda TM23. London, CIBSE, 2000

[39] **British Standards Institution (BSI).** BS 5250: 2002 Code of practice for control of condensation in buildings. London, BSI, 2002

[40] **CIBSE.** Environmental design: CIBSE Guide A. 7th edition. London, CIBSE Publications, 2006

[41] **Whiteley P & Gardiner D.** Solar reflective paints. Information Paper IP 26/81. Bracknell, IHS BRE Press, 1981

[42] **British Standards Institution (BSI).** BS EN 539-2: 2006 Clay roofing tiles for discontinuous laying. Determination of physical characteristics. Test for frost resistance. London, BSI, 2006

[43] **British Standards Institution (BSI).** BS 476: Fire tests on building materials and structures. London, BSI

   Part 3: 2004 Classification and method of test for external fire exposure to roofs

   Part 6: 1989 Method of test for fire propagation for products

   Part 7: 1997 Method of test to determine the classification of the surface spread of flame of products

   Part 11: 1982 Method for assessing the heat emission from building materials

   Part 20: 1987 Method for determination of the fire resistance of elements of construction (general principles)

   Part 21: 1987 Methods for determination of the fire resistance of loadbearing elements of construction

   Part 22: 1987 Methods for determination of the fire resistance of non-loadbearing elements of construction

   Part 23: 1987 Methods for determination of the contribution of components to the fire resistance of a structure

[44] **British Standards Institution (BSI).** DD ENV 1187: 2002 Test methods for external fire exposure to roofs. London, BSI, 2002

[45] **British Standards Institution (BSI).** BS EN 13501-5: 2005 Fire classification of construction products and building elements. Classification using data from external fire exposure to roofs tests. London, BSI, 2002

[46] **Harrison HW & Trotman PM.** Building services: Performance, diagnosis, maintenance, repair and the avoidance of defects. BRE Building Elements. BR 404. Bracknell, IHS BRE Press, 2000

[47] **HMSO.** The Fire Precautions Act 1971 (Modifications) (Revocation) Regulations 1989. Statutory Instrument 1989 No. 79. London, HMSO, 1989

[48] **Loss Prevention Council (LPC).** The LPC Design Guide for the Fire Protection of Buildings 2000. A Code of Practice for the Protection of Business. Available from www.thefpa.co.uk/Resources/ Design+Guide/

[49] **British Standards Institution (BSI).** BS 2782-0: 2004 Methods of testing. Plastics. Introduction. London, BSI, 2004

[50] **British Standards Institution (BSI).** BS 5867-2: 2008 Fabrics for curtains, drapes and window blinds. Flammabililty requirements. Specification. London, BSI, 2008

[51] **British Standards Institution (BSI).** BS 62305: 2006 Protection against lightning. 4 Parts. London, BSI, 2006

   Part 2: Risk management

[52] **Harrison HW & de Vekey RC.** Walls, windows and doors: Performance, diagnosis, maintenance, repair and the avoidance of defects. BRE Building Elements. BR 352. Bracknell, IHS BRE Press, 1998

[53] **Littlefair PJ.** Solar dazzle reflected from sloping glazed facades. Information Paper IP 3/87. Bracknell, IHS BRE Press, 1987

[54] **Hopkins C.** Rain noise from glazed and lightweight roofing. Information Paper IP 2/06. Bracknell, IHS BRE Press, 2006

[55] **British Standards Institution (BSI).** BS 7543: 2003 Guide to durability of buildings and building elements, products and components. London, BSI, 2003

[56] **British Standards Institution (BSI).** BS ISO 15686: Buildings and constructed assets. Service-life planning. 8 Parts. London, BSI, 2000–2008

[57] **Building Effects Review Group (BERG).** The effects of acid deposition on buildings and building materials in the United Kingdom. BERG Report. London, The Stationery Office, 1989

[58] **Bravery AF, Berry RW, Carey JK & Cooper DE.** Recognising wood rot and insect damage in buildings. 2nd edition. BR 232. Bracknell, IHS BRE Press, 1992

[59] **Berry RW.** Remedial treatment of wood rot and insect attack in buildings. BR 256. Bracknell, IHS BRE Press, 1994

[60] **Natural England.** Bats in houses: guidance for householders in England. Westbury-on-Trym, Bristol, Natural England Wildlife Managment and Licensing Service, 2007. Available from www.naturalengland.org.uk/

[61] **Collins J.** Bats and refurbishment. Good Repair Guide 36. Bracknell, IHS BRE Press, 2009

[62] **BRE.** Control of lichens, moulds and similar growths. Digest 370. Bracknell, IHS BRE Press, 1992

[63] **Health and Safety Executive (HSE).** Cleaning weathered asbestos cement (AC) roofing and cladding. a12: asbestos essentials. London, HSE, 2007

[64] **Institution of Structural Engineers (ISE).** Appraisal of existing structures. 2nd edition. London, ISE, 1996. (NB: 3rd edition is due for publication in 2009)

[65] **Saunders G.** Safety consideration in designing roofs. Digest 493. Bracknell, IHS BRE Press, 2005

[66] **Health and Safety Commission.** The Construction (Design and Management) Regulations 2007. London, The Stationery Office, 2007

[67] **Herbert MRM & Harrison HW.** New ways with weatherproof joints. Building Research Establishment Current Paper 90/74. Watford, BRE, 1974

[68] House of Lords Select Committee on Science and Technology, second report, HMSO, July 2005

[69] **BRE.** Slate and tile roofs. Avoiding damage from aircraft wake vortices. Digest 467. Bracknell, IHS BRE Press, 2002

[70] **British Standards Institution (BSI).** BS EN 13956: 2005 Flexible sheet for waterproofing. Plastic and rubber sheets for roof waterproofing. Definitions and characteristics. London, BSI, 2005

[71] **BRE.** Wind scour of gravel ballast on roofs. Digest 311. Bracknell, IHS BRE Press, 1986

[72] **Saunders GK.** Reducing the effects of climate change by roof design. Digest 486. Bracknell, IHS BRE Press, 2004

[73] **Grant G.** Green roofs and façades. EP 74. Bracknell, IHS BRE Press, 2006

[74] **British Standards Institution (BSI).** BS EN 13707: 2004 Flexible sheets for waterproofing. Reinforced bitumen sheets for roof waterproofing. Definition and characteristics. London, BSI, 2004

[75] **NHBC Foundation.** A guide to modern methods of construction. NF 1. Amersham, NHBC, 2006

[76] **Ross K.** Modern methods of house construction: A surveyors guide. FB11. A BRE Trust publication. Bracknell, IHS BRE Press, 2005

[77] **Stirling C.** Off-site construction: an introduction. Good Building Guide 56. Bracknell, IHS BRE Press, 2003

[78] **Blackmore P.** Wind loads on roof-based photovoltaic systems. Digest 489. Bracknell, IHS BRE Press, 2004

[79] **Blackmore P.** Mechanical installation of roof-mounted photovoltaic systems. Digest 495. Bracknell, IHS BRE Press, 2005

[80] **British Standards Institution (BSI).** BS EN 14437: 2004 Determination of the uplift resistance of clay or concrete tiles for roofing. Roof system test method. London, BSI

3 SHORT-SPAN DOMESTIC PITCHED ROOFS

Short-span roofs are normally defined as being of less than 8–9 m span. Pitched roofs are conventionally defined as those roofs with slopes greater than 10°, whereas roofs of slope 10° or less are defined as flat.

For the purposes of this book, pitched roofs have been categorised into those covered in relatively small overlapping units, dealt with in Chapters 3.1, 3.2 and 3.6, those covered in sheet materials, dealt with in Chapters 3.3–3.5, and thatch, which forms a category of its own, dealt with in Chapter 3.7.

Small overlapping units consist of the following types:
- clay tile,
- concrete tile,
- natural slate,
- manmade slate,
- shingles.

Figure 3.2 shows the current market share held by the first four categories. Following increased appreciation of the need to protect the environment, there has been limited use of recycled materials such as rubber. The increased popularity of PV systems is envisaged which will affect the type of roofing chosen to support them.

Some general information about defects in pitched roofs in housing can be found in *Assessing traditional housing for rehabilitation*[1].

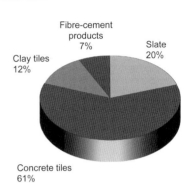

Figure 3.2: Roof tile market by value, 2006. Data from *Roofing Market Report*[2]

Figure 3.1 This steeply pitched plain tiled roof was built in 1880 but re-covered after bomb damage in the 1939–45 war

3.1 CONCRETE AND CLAY TILES

Information on the basic forms of roof coverings and their condition are given in the UK national house condition surveys, but are not always included in the published summaries. Special analyses have been carried out for this book using the 2005/6 data for England[3] and Northern Ireland[4].

Roof coverings

Clay and concrete tiles
Table 3.1 shows data for dwellings in the UK in 1991 having roofs covered wholly or partly by clay or concrete tiles, and the proportion that showed faults.

By 2005/6 in England, some 3.9 million (18.4%) of all dwellings had clay tiles covering the whole or part of their roofs, and some 12.5 million (59%) had concrete tiles. This represents a significant change over the intervening 15 years, with a considerable reduction in the numbers of clay tiled roofs, and a concomitant increase in the numbers of those covered in concrete tiles.

Information on the proportions showing faults were not itemised, but experience indicates that they are unlikely to differ greatly from the numbers recorded 15 years earlier.

An analysis of comparable data for Wales was not possible in the early 1990s, but a review of the limited data available pointed to figures very similar to those for England.

Current data for Wales were not available for this book.

The Scottish data in 1991[7] did not differentiate between clay and concrete tile roofs. For the two kinds of tiles, there did not appear to be much difference between the

Table 3.1: Reported problems with clay and concrete tile roof coverings in the UK. Data from national House Condition Surveys circa 1990[5–8]

	Number of roofs (whole or part) covered	Proportion with a problem (approx.)
England:		
Clay tiles		
Houses	4.2 million (25% of total stock)	1:3
Flats	571,000	1:3
Concrete tiles		
Houses	8.1 million (50% of total stock)	1:12
Flats	1.5 million	1:10
Wales:	No data	—
Scotland:		
Clay and concrete tiles	1.2 million	1:5
Northern Ireland:		
Clay tiles	33,000 (6% of total stock)	No data
Concrete tiles	271, 000 (50% of total stock)	No data

condition of the stock in Scotland and the rest of the UK.

Current data for Scotland were not available for this book.

Similar proportions of the total housing stock in Northern Ireland as in England in 1991 used concrete tiles to partly or wholly cover roofs.

By 2005/6 in Northern Ireland, just over 5000 dwellings (0.7%) of all dwellings had clay tiles covering the whole or part of their roofs, and just under 500,000

(68.2%) had concrete tiles. As is the case in England, this represents a considerable change over the intervening 15 years, with a massive reduction in the numbers of clay tiled roofs, and a concomitant increase in the numbers of those covered in concrete tiles.

Information on the proportions showing faults were not itemised, but experience indicates that they are unlikely to differ greatly from the numbers recorded 15 years earlier.

Other roof coverings
No data were available for the numbers of roofs covered in pressed metal tiles, but in 2005/6 the totals of those predominantly covered in glass, metal and laminates was 0.7% in England and 0.8% in Northern Ireland — a very small proportion of the total.

Roof type
Tables 3.2–3.4 give data for structural faults (not necessarily due to loading) for different roof types over the whole or part of their area for dwellings in the UK. Additional data showed that 291,000 houses in England had faults in dormer windows.

Comparable information to that in Tables 3.2 and 3.4 for 2005/6 in England and Northern Ireland is not available for this book, but experience indicates that figures quoted for faults are unlikely to differ by more than one or two percentage points. However, some indication of faults in all roof coverings which required urgent attention was recorded, giving 6.7% in England and 3.6% in Northern Ireland.

Structural faults are not necessarily age dependent. For

Table 3.2: Reported problems with roof structure. Data from English House Condition Survey 1991[5]

Type of roof	Number of roofs	Proportion with a problem (approx.)
Dual-pitch		
Houses	14 million	1:8
Blocks of flats	2.7 million	1:10
Mansard		
Houses	71,000	1:10
Blocks of flats	107,000	1:10
Chalet-style (Room-in-the-roof)		
Houses	123,000	1:20
Flats	—	—
Single-pitch		
Houses	214,000	1:20
Blocks of flats	112,000	Negligible

Table 3.3: Reported problems with roof structure. Data from Scottish House Condition Survey 1991[7]

Type of roof	Number of roofs	Proportion with a problem (approx.)
Mansard or half-mansard	32,000	1:5
Single-pitch	33,000	1:3

Table 3.4: Reported problems with roof structure. Data from Northern Ireland House Condition Survey 1991[8]

Type of roof	Total no.	Proportion with a problem (approx.)
Dual-pitch	410,000	1:20
Mansard, Single-pitch, Chalet-style	4,500	Negligible

example, the National House Building Council (NHBC) reported that structural failures in relatively newly built pitched roofs cost them about £400,000 in the 1993 claims year[9].

The great majority of domestic pitched roofs in the UK have been constructed as cold roofs. In the future, however, surveyors should increasingly come across warm pitched roofs following the recent increases in thermal insulation requirements in all the UK building regulations. Accordingly, most of the details which follow in this chapter deal with cold roofs, while the additional precautions to be taken with warm roofs are given in Chapter 3.8 on thermal insulation in lofts.

CHARACTERISTIC DETAILS

Basic structure

The more complex the geometry of the structure, the more likely it is that faults will ensue (Figure 3.3). For very short spans (eg over porches or single room extensions) where the loads from the roofs are small, roofs may consist of rafters without ties or with collars halfway up the slope. Single roofs (ie those without purlins to support the rafters, and without collars or ties at wall plate level) are called couple roofs, and those with ties at wall plate level are called

close-couple (or couple-close) roofs (Figure 3.4). Sometimes, a close-couple roof will have rod or thin-section batten hangers to support the ceiling ties. Sizes of rafters were given in the past in, for example, Approved Document (AD) A to the Building Regulations for England and Wales but are now to be found in the revised AD issued by TRADA[10].

In a couple roof, the whole of the horizontal component of the thrust of the rafters is taken by the walls, whereas the introduction of a collar or a tie will reduce the horizontal forces that are tending to overturn the wall. Although couple roofs figure in the standard construction textbooks, none has been seen, other than over porches, on the BRE site inspections carried out in the 10-year period, 1980–90[11]: not even on garages. If they had been found, they would have been recorded as faults.

In the so-called double roof, though, purlins are used to support the rafters which can thereby be made of smaller section. The larger spans of older houses and blocks of flats are, of course, to be found with timber trusses supporting the purlins, though comparatively few of these have been encountered in the BRE quality inspections.

Single pitches have occasionally been encountered in BRE inspections, employing appropriately

Figure 3.3: Old single-lap tiled roofs with many features giving rise to flashing complications. There is also evidence of movement in the roof structures; in particular, of tiles over a separating wall in line with the chimney on the left-hand side of the photograph

(a)

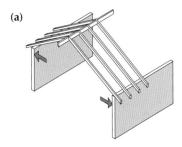

(b)

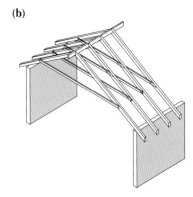

(c)

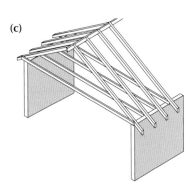

Figure 3.4: (a) Couple, (b) collar, (c) close-couple pitched roofs

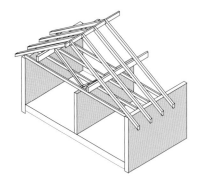

Figure 3.5: Double-cut roof: a common form of strutted purlin roof

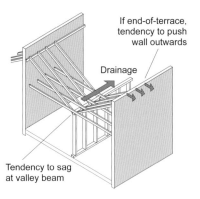

Figure 3.6: A 'butterfly' roof with provision for an internal valley gutter

shaped trussed rafters in similar configurations to the symmetrically pitched. Lateral stability to the walls has been given in most of the cases seen by hooked shoes taking the heels of the trusses. This type of connection has proved sometimes to be a source of problems; it is described in the section on strength and stability later in this chapter.

By far the most common form of roof for small dwellings seen in the BRE site studies is the double-cut roof with strutted purlins supporting the rafters and carrying tiles of one form or another (Figure 3.5). The struts, at or near mid-points of the purlins, are usually carried on internal partitions and formed either from a continuation of the masonry or by a timber prop. In the case of a hipped roof, the purlins are usually mitred at the hips.

BS 8103-1[12] deals with the structural design of low-rise buildings; it covers stability requirements for roofs of maximum clear span 12 m with no part of the roof higher than 15 m above lowest ground level, and design wind speeds not exceeding 44 m/s.

BRE investigators on site inspections have occasionally come across pitched roofs over older terraced houses which slope from the separating wall to an internal valley beam ('butterfly' roofs), sometimes strutted from internal partitions (Figure 3.6). The whole roof is usually hidden behind a parapet through which the rainwater drains out via a hopper head or box receiver. This kind of roof is prone to three types of faults:

- sagging of the valley beam caused by inadequate support from struts on a stud partition founded on a suspended timber floor,
- a leaning gable in an end-of-terraced house caused by the thrust of the rafters on the outer wall not being counterbalanced,
- poor discharge of rainwater into the hopper which saturates the wall.

During the 1970s and 1980s, the fink and fan roofs (or gang-nail truss) built with stress-graded timber sections held together with punched plates of mild steel became very common in new construction (Figure 3.7). Occasionally, roofs of other types of design may be encountered (Figure 3.8). These forms of roof permit only limited use of the roof voids for storage. The attic truss roof, which permits clear

headroom within the roof space, is an alternative.

Where they are to be placed in the roof space, the support of cold water tanks needs specific consideration (Figure 3.9). BRE does not recommend siting cold water storage tanks in cold roof spaces susceptible to freezing. Although there may be fewer instances in the future where a storage tank needs to be positioned in the roof space (owing to the increased use of boilers that do not require domestic water storage), precautions against freezing should be taken in those cases where there are no reasonable alternatives.

Bracing and binders are critically important in trussed rafter roofs: properly positioned and fixed, they convert a collection of individual trusses into a single three-dimensional structural unit (Figure 3.10). A list of faults relating to trussed rafters found during site inspections appears in Box 3.3). Roofs covered in rigid sarking boards, as is common in Scotland, do not appear to suffer from structural deficiencies in the same way as the unboarded roofs described above. Further information may be found on these points in BRE Defect Action Sheets 83[13] and 84[14]. Where wide span roofs are employed in housing, there are indications that in about 1 in 10 cases chevron bracing has been unsatisfactory. There are further notes on chevron bracing in Chapter 5.1.

In the 1950s and 1960s, many narrow fronted terraces were roofed very economically with trussed

(a)

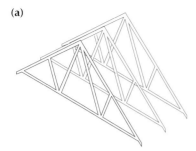

(b)

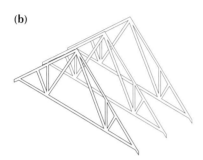

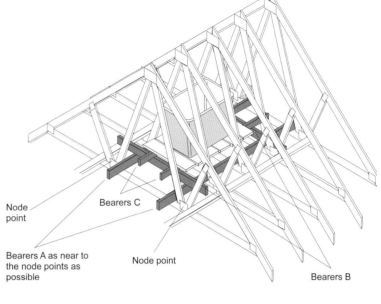

Node
point

Bearers C

Bearers A as near to
the node points as
possible

Node point

Bearers B

Figure 3.9: Cold water storage tanks must have adequate support. *Note:* bracing
and binders are not shown

(c)

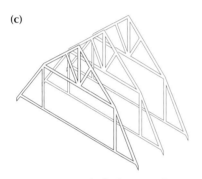

Figure 3.7: (a) Fink, (b) fan trussed
rafters, (c) attic mansard trusses.
Note: no bracing or binding is shown

purlins of depths varying to suit
the slope, spanning between the
cross-walls. Such purlins bore on
the separating walls, and could be a
source both of sound transmission
and potential fire hazard where a
separating wall was penetrated.

Main roof areas

Clay tiles should conform to BS EN
1304[15] and concrete tiles to BS EN
490[16].

Pitch

Recommended pitches for tiles vary
enormously and it is not practicable
here to list the various types and
their performance in slopes of
various pitches. Manufacturers'
recommendations should be
followed. As a general rule, however,
plain tiles should not be used on
pitches of less than 40°. Laps should
be not less than 65 mm, or 75 mm
in exposed areas. Single-lap tiles
generally should not to be laid at
pitches of less than 35°, though they
have been laid on pitches down to
30° provided they have anti-capillary
clearances built into the design.
Modern designs of single-lap tiles
may be laid at lower pitches (some
are claimed to perform adequately
at very low pitches, even 12.5°),
though it might be a wise precaution
to seek evidence of tests at these low
pitches before specifying.

Gauge

Some single-lap tiles may be suitable
for laying either to a fixed or to
a variable gauge: manufacturers'
recommendations should be
checked. Occasionally, a warped-
plane roof covered in non-
interlocking plain tiles or pantiles
will be seen, usually over a non-
rectilinear building in the congested
centre of a village. The more the
deviation from a flat plane, the
higher the premium placed on the
integrity of the underfelt to ensure
a dry roof. Naturally, this shape of
roof ought not to be attempted in
interlocking tiles because they will

Figure 3.8: A roof of handmade scissor trusses under construction. Although unusual
in that the structure for such a roof is normally prefabricated off site, there is no
particular reason to suspect its adequacy simply because it has been handmade

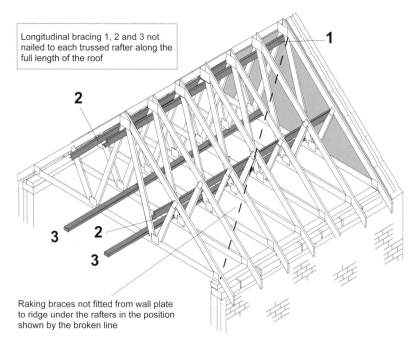

Longitudinal bracing 1, 2 and 3 not nailed to each trussed rafter along the full length of the roof

Raking braces not fitted from wall plate to ridge under the rafters in the position shown by the broken line

Figure 3.10: Bracing and binding to domestic trussed rafters have frequently been found by BRE to be sources of faults. *Note:* truss clips and lateral restraint straps are not shown

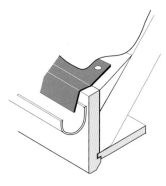

Figure 3.11: A strip of dpc material should be dressed into the gutter

not interlock.

It is often difficult to construct porch and dormer roofs in single-lap tiles, and BRE investigators on site have seen many examples of inadequate work on such small roofs with, in some cases, almost every tile needing to be cut. Plain tiles are much more adjustable to gauge.

Fixings

Nibbed plain tiles should be twice-nailed at every fifth course on pitches below 60° (though some manufacturers recommend that each tile be nailed with at least one nail in pitches over 45°), and every tile twice-nailed on pitches of 60° or steeper. Fixings are covered in BS 5534[17]. Fixing requirements for single-lap tiles vary, but, in general, end-tiles in all rows, and the top and bottom rows, should be fixed, and clips and nails used at verges in accordance with manufacturers' instructions. Every peg tile, naturally, needs fastening, whatever the pitch.

It is customary in some parts of the UK not to nail heavy pantiles.

Eaves

Overhanging

Overhanging eaves may be of 'closed' or 'open' configuration. These terms refer to whether the rafters are boxed in with fascia and soffit, or left exposed with the wall surface carried up to the underside of the roofing.

To achieve adequate weather protection to the wall below, BRE recommends an overhang at eaves of at least 300 mm (in contrast to some construction textbooks which recommend a much smaller minimum overhang at eaves), irrespective of the kind of roof covering and angle of pitch. It has been estimated that a pitched roof with a pitch greater than 20° and an eaves overhang of 350 mm gives approximately the same protection to the wall below as if the wall were rendered or clad with a covering of some kind. There is a presumed advantage if the overhang can be provided with a drip: say a tile with a sharp arris to the under-surface. It is also advantageous to continue the underlay into the gutter (Figure 3.11).

In exposed areas, however, a large overhang can be something of a mixed blessing. BRE experiments on a natural exposure rig, using overhangs of various dimensions, have shown that under some conditions of driving rain, water load reaching the wall under the eaves can actually be increased (BRE Current Paper 81/74[18]). Taken

over the whole range of conditions, however, the experiments showed that generous overhangs do reduce water loads on walls. It should also be remembered that overhanging eaves provide a lower surface on which the wind pressures can act on the roof, adding to the suction that occurs on the upper surface above the eaves.

Large overhangs at both eaves and verges should be securely anchored with straps back to the roof. Some examples have been seen on site where sagging in eaves and verges has occurred.

With some kinds of roof structure over thick external walls and roof pitches steeper than 45°, there may be difficulty in obtaining good overhangs at the eaves and, at the same time, providing adequate clearance for the window heads. This may have given rise to the use of sprockets, cantilevered from the rafters, to give a suitable overhang with flatter pitch at the eaves. The fixings of these sprockets seem to be the first part of these roofs to show distress. Nailing or bolting sprockets to the sides of rafters, where the fixings operate in shear, would probably in most circumstances give better reliability than those nailed to the tops of rafters, where the strength of the fixing relies only on the resistance of the nails to pull out (Figure 3.12).

Clipped

BRE does not recommend the use of so-called clipped or flush eaves since they do not provide any protection to the wall beneath. However, they are used.

Figure 3.12: Rebuilding sprocketted eaves following deterioration of the timbers. The beam-filling above the wall plate carries the sprockets. The open eaves have no soffit

Figure 3.13: A good projection was given to this single-lap tiled verge, and the tiles were neatly clipped

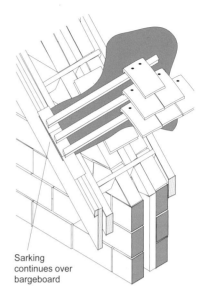

Sarking continues over bargeboard

Figure 3.15: A bargeboarded verge with gable ladder

Verges

Mortar-bedded

Verges should oversail the wall or bargeboard by 50 mm, and should be undercloaked with a durable and compatible material (Figure 3.13).

Pointing between the undercloaking and the tile above should not exceed 35 mm. Pointing in excess of this amount was, on the BRE site inspections, frequently seen to have cracked. The use of tile inserts as a form of galleting between the tile surface and the undercloaking, similar to the dentil slips used on ridge tiles, can help to minimise the appearance problems caused by cracking. A variation of this detail is to use a double undercloak (Figure 3.14).

Plain tiles should not be used for undercloaks on roof slopes of below 30° pitch.

The appearance of plain tile verges is enhanced by tilting the verge tiles inwards. This small amount of tilting or canting does not affect performance. Single-lap interlocking tiles should not be used in tilted verges.

Dry-laid

Dry-laid verges are an alternative to ordinary verges; the specially shaped tiles form a downstand similar to a bargeboard. Potentially, this form

of verge is more weatherproof than the mortared-and-pointed verge, provided the fixings are secure.

Bargeboarded

A bargeboarded verge gives more protection to the wall than a verge set on the wall head (Figure 3.15). However, the following points have been made on sites investigated by BRE and are worth noting:
• ensure that the underfelt is carried over the top of the bargeboard to meet with the undercloak,

• ensure that the inboard edge of the gable ladder is securely fixed to the rest of the roof,
• ensure that the cantilever effect of the tiles does not impose too great a load on the bargeboard.

A variation in verge detailing common on pantile roofs in Suffolk is to use a section of timber nailed to the top of the bargeboard, overlapping the verge tiles (Figure 3.16). This serves the dual function of providing a reasonably weatherproof side lap for the verge

Figure 3.14: A double undercloak on a deep verge. The pointing mortar was probably too strong leading to shrinkage cracks above the lower course

Figure 3.16: Common verge detail in a Suffolk pantile roof. The timber cappings shown are prone to decay since it is impossible to repaint the underside without removal

tile and, also, a positive fixing to resist suction in that part of the roof which is most sensitive to wind. Durability is the main problem, for the section cannot be re-painted underneath.

Abutments

Problems may be found in large numbers of terraced houses of late Victorian and Edwardian vintage where local byelaws required that separating walls were built to project through the roofs. Many houses with this feature can now be found 'flashed' with cement fillets or flaunching. This is not a realistic substitute for a metal flashing over a secret gutter, or soakers, in this situation. If, despite this, a cement flaunching is used, it should include a gritty fine aggregate with some lime in the mix.

The copings of projecting separating walls for the most part rest on kneelers on the slope or footstones at the eaves, and are bedded in mortar up the slope. Those seen on site have mostly been in good condition, though reports have been received by BRE of more recent attempts to emulate the detail, usually at gables, without the use of kneelers or footstones, where the copings have slid down the slope as a result. Damp proof course materials (eg slate) should

be capable of adhering to both coping and masonry below or it may be possible to use special fixings. Some sheet materials will provide a slip plane, with consequent risk of dislodgement of the copings in high winds. Mortars in this situation could have a bonding agent added.

Cavity trays, soakers and secret gutters, with associated flashings, are also sometimes sources of problems where steps in levels occur between adjacent buildings (Figure 3.17). The problems seem to occur not so much through design as in execution, but the pre-formed cavity tray with attached flashings is proving to be a means of achieving satisfactory performance.

Where a pitched roof abuts a separating wall in a stepped-and-staggered situation, part of the roof becomes an external gable, and the outer leaf of masonry becomes an inner leaf below roof level. Particular care is needed to ensure that rainwater is prevented from reaching the interior. Soakers for plain tiles should be a minimum width of 175 mm to give an upstand of 75 mm against the wall, and a lap of 100 mm under the tiles. A cover flashing then laps the soakers by 50 mm, or more if it can be arranged. For interlocking tiles, a stepped flashing dressed over the tiles must be used; soakers are not feasible since they would interfere with the interlock. A

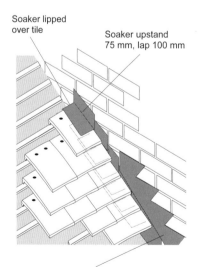

Soaker lipped over tile

Soaker upstand 75 mm, lap 100 mm

Flashing covers soakers by 50–60 mm

Figure 3.17: Soakers and stepped flashings

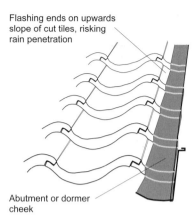

Flashing ends on upwards slope of cut tiles, risking rain penetration

Abutment or dormer cheek

Figure 3.18: An abutment of profiled tiles with a dormer window. When the flashing ends on an upward slope, as shown, the detail is vulnerable to rain penetration

secret gutter, however, is feasible. If lead is used, it should be of Code 4 thickness for soakers (Code 3 is sometimes specified, but it will be less durable) and Code 4 also for trays and flashings, but Code 5 or 6 on exposed sites, especially where tacking is not specified. Lead flashings should be protected with bitumen or patination oils where built in to any masonry. BRE Defect Action Sheet 114[19] provides further information.

Care is needed at the abutments of profiled tiles with dormer windows. Cases have been encountered where the flashing or soaker against the vertical cheek of the dormer is too short to master the roll of the tile, leaving the detail vulnerable to rain penetration (Figure 3.18).

Valleys

Valley gutters are commonly used and about two-thirds of roofs of new houses built between 1991 and 1993 had valley gutters. Most valleys seen earlier than this by BRE investigators have been formed by inserting a metal lining into the valley and cutting the tiles to shape over. Much clumsy cutting has been seen, with appearance compromised. Better solutions, though seen more rarely, involve the use of purposemade valley tiles of various configurations. Valleys formed from these tiles may be swept or laced, either to a radius or to an acute angle (Figure 3.19).

Figure 3.19: A swept valley in plain tiles. The camera viewpoint has foreshortened the tiles, and this has exaggerated their apparent irregularity. The parapet capping has not succeeded in throwing rainwater run-off clear of the wall below

A double thickness of sarking felt is needed in valleys, lapped at least 600 mm over the centreline of the valley. Also seen on site have been a number of cases of insufficient support to tiling battens where they abutted valleys.

Hips

Hips in plain tiled roofs are most often formed with ridges bedded in mortar (Figure 3.20), and held at the foot on hip irons or hooks. Hip irons should be hot dip galvanised. For roof pitches under 35°, they can be 4 mm thick; but for roof pitches over 35°, that is to say for the majority of roofs, they should be a minimum of 6 mm thick (the pitch of the hip will naturally be less than that of the roof because of its geometry). Since the 1970s it has become more common to clip each ridge tile to the hip rafter. Alternative methods of forming hips, giving a neater appearance and probably better performance, include purposemade hip tiles and bonneted hip tiles.

Mitred hips in plain tiles should be laid over soakers lapping at least 100 mm to each side of the hip. Single-lap tiles are generally not suitable for exposed mitres at hips.

Ridges

A great variety of shapes of ridge tile is available. Traditional practice is to bed them solidly on mortar (Figure 3.21a); but even when so bedded, ridges can be subjected to loss of tiles in gales, particularly the end tiles facing edge-on to the wind. There is also considerable thermal movement, and chimneys intercepting the ridge have been known to cause hogging along a run of ridge tiles. A 1:3 mortar mix, preferably air-entrained, can be used for bedding but this is a comparatively strong mix and could lead to cracking. A high durability mortar is nevertheless required, and an air-entrained 1:0.5:4.5 cement:lime:sand mortar is an alternative, though is arguably less suitable for use with concrete tiles for which the 1:3 mix is better.

Purposemade roofs do not seem to suffer much from storm damage.

Some decorative finials are still available.

Dry-laid ridge tiles are also available which may be vented

(a)

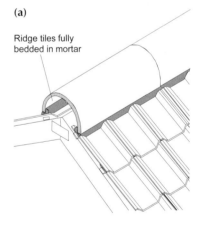

Ridge tiles fully bedded in mortar

(b)

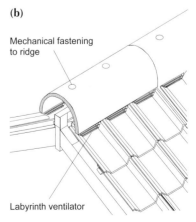

Mechanical fastening to ridge

Labyrinth ventilator

Figure 3.21: (a) A mortar-bedded non-ventilated ridge, (b) a dry-laid ventilated ridge

Figure 3.20: The mortar is already breaking up in this comparatively new hip and the galleting is becoming loose

(Figure 3.21b). Dry fixing and clipping gives a more positive fixing. Some fixings rely on there being a ridge board at the apex of the pitch, and noggings may be necessary where they are used with trussed rafters. These noggings need to be fixed particularly securely to ensure the stability of the ridge tiles.

Roof lighting

In dwellings built in the 19th and early 20th centuries, it was common in some areas of the UK to use cast iron roof lights, single glazed with a central glazing bar. Many of these are still in existence. With draughty roofs, where internal and external conditions differ little, there will be few problems of condensation; but where higher standards of thermal insulation and heating have been introduced, and ventilation within the roof is seriously affected, condensation could be a problem. Unless there are conservation reasons for their retention, it is sensible to consider replacing them with glazed units set in thermally broken metal frames or timber frames (Figure 3.22). New double-glazed proprietary copies of the old designs are available (Figure 3.23).

Dormers

Dormers are, of course, a fairly common method of introducing windows into the plane of a tiled roof. They can take many different shapes (see Figure 3.154) and are a major regional characteristic; for example, the bayed dormers of the Scottish Highlands (Figure 3.24). Although the swept, or Sussex, dormer is sometimes seen throughout the UK, it has not been encountered during BRE surveys.

Service perforations

Provided lead or other durable 'slates' are used, there seem to be few problems with perforations of roofs for vent pipes and fixings for PV systems. However, lead slates should be checked to ensure that they are dressed to a tight fit with the profiles of the tile or tiles.

Rainwater disposal

BRE inspections and surveys have shown consistently that gutter and

(a)

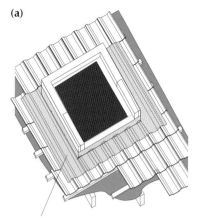

The flatter the pitch, the wider the flashing should be

(b)

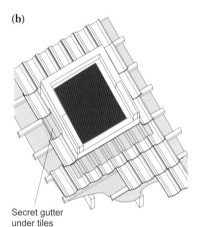

Secret gutter under tiles

Figure 3.22: Typical roof light installations: (a) the roof light sits 'on' the roof and has flashed abutments, (b) the roof light sits 'in' the roof and has secret gutters. Few problems arise with either of these types of roof light

downpipe installations on many houses were deficient. If gutters are undersized, water spillage can saturate walls and lead to dampness internally.

Other things being equal, it is preferable to position downpipes in the centres of runs of guttering rather than at ends. This will reduce the maximum water load on gutters and help to prevent surcharging and overflow. Steps-and-staggers on terraces can upset this simple rule.

Sizing of gutters and downpipes for domestic installations has become relatively straightforward, although the design of such systems for larger buildings is rather more than a simple calculation of roof area and water load. A simple design

Figure 3.23: Conservation-grade roof light

Figure 3.24: Bayed dormer common in the Scottish Highlands

procedure, taken from BRE Defect Action Sheet 55[20], is given in Box 3.1; however, this procedure can be used only with domestic roofs of short span. Other roofs of more complex design will need special calculations to the requirements of BS EN 12056-3[21].

Gutters should be fixed with a slight fall to allow for some vertical movement to assist drainage. A fall of about 1 in 350 (10 mm per 3.5 m run) is recommended in Defect Action Sheet 55[20].

The design of valley gutters is frequently defective, to judge from

Box 3.1: Calculations of rainwater run-off from short span domestic roofing and gutter size

Using Figure 3.25, the effective catchment area that will discharge to each gutter is:

- for the slope of a pitched roof, the plan area, A (m²), plus half the elevation area, B (m²) (Figure 3.25a),
- for a pitched roof abutting a wall, the plan area, A, plus half the elevation area, B, plus half the wall area, C, above the roof slope (Figure 3.25b),
- for a flat roof, the relevant plan area.

The run-off rate to each gutter is the total catchment area for the gutter divided by 48. This produces the run-off in litres per second using the recommended rainfall (thunderstorm) rate of 75 mm/hour.

The size of the guttering is shown in Table 3.5 using the flow capacity that will accommodate the run-off rate. As part of this process, the number of outlets should be considered: more outlets from a run of guttering spreads the total loading on the gutter, but the loadings will vary according to where the outlets are positioned (Figure 3.26).

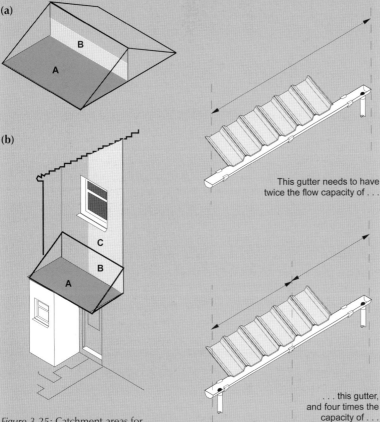

Figure 3.25: Catchment areas for calculating rainwater run-off

Table 3.5: Flow capacities of standard eaves gutters (when level)

Size of gutter (mm)	Flow capacity (litres/sec)	
	True half round	Nominal half round
75	0.38	0.27
100	0.78	0.55
115	1.11	0.78
125	1.37	0.96
150	2.16	1.52

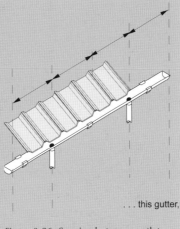

This gutter needs to have twice the flow capacity of . . .

. . . this gutter, and four times the capacity of . . .

. . . this gutter,

Figure 3.26: Spacing between outlets to reduce water load on gutters

BRE site inspections. Drawings have been seen that specifically state that the design of gutter systems should be left to the site staff to sort out! The pitch of valley gutters is less than the pitch of the roof they join, and valley gutters on pitches of less than 20° are particularly prone to leaking. Poor detailing was commonly found in the BRE quality assessments.

BRE site inspections revealed a number of cases where gutters had not been installed on porches and bay windows: decisions which may be marginal where the drips cause no inconvenience to the occupants, but less acceptable for doorways (Figure 3.27).

A room in the roof

Where a room is formed within a pitched roof void, it is sometimes difficult to achieve adequate ventilation for the roof to the outside (Figure 3.28). BRE Defect Action Sheets 118[22] and 119[23] deal with this problem, particularly where it involves the careful placing of thermal insulation to ensure an adequate ventilation gap. *Thermal insulation: avoiding risks*[24] is also helpful. The most important points to watch are:

- that a vapour control layer (eg of at least 500-gauge polyethylene) is installed in the sloping part of the roof under the insulation,
- that a vapour permeable sarking is used in new construction,
- that cupboards should be within the insulated envelope, ensuring continuity of the lining.

MAIN PERFORMANCE REQUIREMENTS AND DEFECTS

Choice of materials for structure

Timber has been used as the main structural material in the vast majority of tiled roofs: those dating from before the early 1960s for the most part designed with strutted purlins (Figure 3.29), more recent designs using trussed rafters. The structure of most of these is in good condition (only 1 in 8 of these structures being reported as faulty[5]), provided attention has been paid to routine maintenance of the covering.

Figure 3.27: The occupants of this house are not inconvenienced by drips from the unguttered raking verge of the bay window

There are probably around 100,000 dwellings in the housing stock with rolled or cold-formed steel section trusses and, in some cases, cold-formed steel sheets. Such systems include AGB, Arcon, BISF, Coventry, Crane, Cruden, Dorlonco, Hills, Howard, Modular, Nuttall, Quality, Roften, Telford (Figure 3.30) and Trusteel[25].

However, new designs of steel structure roofs have become available consisting of cold-rolled Zed purlins which carry trays ribbed at the edges to correspond with rafter positions on a conventional roof. Thermal insulation is laid over the trays to form a warm roof for which no eaves and ridge ventilation is required. This kind of roof is useful

where speed of erection is needed to achieve a watertight shell, and can also be useful where security against forced entry could be a problem. Adequate protection (eg galvanising) to the steel is needed even though it is not exposed to the weather. (For most buildings built in the late 20th century, this would have been to BS 5493[26], but for those built subsequently, BS EN ISO 12944 applies[27].

Strength and stability

The strength of the tiles themselves is covered in BS EN 538[28] and BS EN 490[16]. The strength requirements of the underlying structure are covered by the various national building regulations. It is common, now, for manufacturers and suppliers to carry out design calculations for specifiers: for example, those involving dead and live loadings on the roofs, and providing fixing requirements.

Lateral restraint strapping between roof and walls is covered in BS 8103-1[29]. Although the horizontal straps seen on sites generally complied with the requirements of the time for cross-sectional area (150 mm²), length of downturn (100 mm), and minimum straight length, it is the fixings that usually proved to be deficient, particularly with respect to numbers (not less than 4), sizes (50 mm × no. 10 wood screws, or 3 mm diameter × 65 mm nails) of fixings and their support by the necessary packings (Figure 3.31). There is more information in BRE Defect Action Sheets 27[30] and 28[31] and Good Building Guide 16[32].

Some single-pitch roofs seen by BRE investigators have given concern. The lateral stability to an external wall in some roofs was partly provided by the heel of the truss which was hooked over the head of the wall. In a number of cases where the masonry was out of true, the heel of the truss was not tightly fitted to the shoe and the shoe was not fitted tightly to the wall. The shoe bent down and became detached from its correct position; therefore, there was no effective lateral restraint.

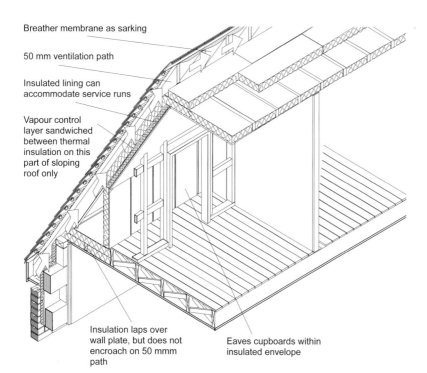

Breather membrane as sarking

50 mm ventilation path

Insulated lining can accommodate service runs

Vapour control layer sandwiched between thermal insulation on this part of sloping roof only

Insulation laps over wall plate, but does not encroach on 50 mmm path

Eaves cupboards within insulated envelope

Figure 3.28: A room in the roof. The rafters of attic trusses normally allow adequate depth for ventilation over insulation. The free area of ventilation at the eaves needs to be equivalent to a continuous 25 mm gap, and ridge ventilation equal to a continuous 5 mm gap. The vapour control layer is not easy to install, and needs a precise approach to the task

Figure 3.29: The rafters on this 1920s house were in good condition but the purlins needed replacing because they were built into wet brickwork. The packs are temporary

Figure 3.32: Damning evidence of unauthorised adjustment of trussed rafters. In this case, all the trussed rafters on one span had been shortened on site and the plates inadequately nailed back. The evidence usually only comes to light if the roof suffers structural movements in service. However, any nailing of truss plates should arouse suspicion

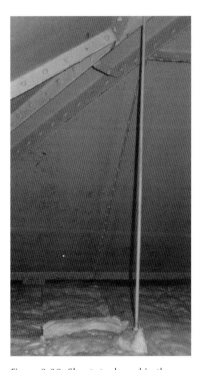

Figure 3.30: Sheet steel used in the roof of a Telford system house. This example has already passed its design life and the steel is beginning to corrode, especially on the edges of sections

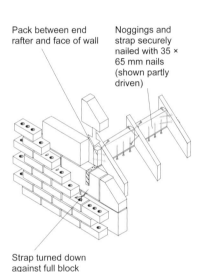

Pack between end rafter and face of wall

Noggings and strap securely nailed with 35 × 65 mm nails (shown partly driven)

Strap turned down against full block

Figure 3.31: Fixing of lateral restraint strapping

It is not normally necessary to provide anchorages against wind loadings or uplift in all domestic roofs of traditional construction, although some provision may be necessary in low-pitched roofs of lightweight construction, particularly in areas of high exposure. Again, it is not necessarily the quality of the strap that may be at fault but the manner of its fixing. In calculation

of loadings and resistance to uplift, it should be noted that, depending on design:
• clay tiles can vary between 39 and 71 kg/m² and
• concrete tiles can vary between 28 and 93 kg/m².

Boxes 3.2 and 3.3 give brief lists of problems that turned up most frequently in BRE site inspections. Box 3.4 lists additional items that have been observed on new roofs over timber-frame construction.

Cases have even been found on the BRE site inspections where bracing and binding of trussed rafters have been recklessly cut away

to provide space for water tanks and access hatches; even ad-hoc adjustment to the spans of trussed rafters have been known (Figure 3.32). No alterations to the structure of a trussed rafter roof should be made without the authority of a structural engineer.

Many early trussed rafter roofs are inadequately braced (see Figure 3.10). Before the early 1970s, for example, there was limited guidance on the bracing requirements for trussed rafter roofs. Roofs built since then may have bracing present, but it may be deficient; between a quarter and a sixth (depending on the particular survey) of all bracing faults noted by BRE investigators on site were, in their opinion, infringements of building regulations in force at the time of the inspections[33].

In a survey of house construction using trussed rafter roofs, misplaced, interrupted and inadequately fixed raking braces, and ridge and ceiling binders, were relatively common.

Inadequacy of bracing has resulted in distortion or leaning of trusses; and when movements have become appreciable, they are revealed by disruption of tiling, especially evident at verges. BRE investigators called onto sites have seen several instances of whole roofs leaning sideways until the end truss,

Box 3.2: Strength and stability problems most frequently seen in BRE site inspections of rehabilitation work on existing roofs supported by masonry construction

- Sagging of purlins caused by slipping of struts. In one case, the occupier had cut away the struts to give himself more headroom in the attic space
- Unrestrained purlins twisting
- Spread in collared roofs indicated by cracks in the walls. Although a collar will reduce the tendency to spread, it will not eliminate it entirely, and BRE investigators have seen several examples where the walls have been pushed outwards
- Settlement caused by the weight of a roof being taken on inadequate supports
- Rusting of bolted connections
- Insect and rot attack of main roof members
- Failure to allow for increased loads when tiles replace slates (see Chapter 3.2, and Figures 3.90 and 3.91)
- Collars single-nailed to rafters
- Rotation of wall plates
- No hip irons; or hip irons, clips or hooks corroded (see Case study 1)

Case study 1: Failure in a hipped roof

A development of 1930s semi-detached houses had been built with hipped roofs pitched at 45°. The roofs were covered in plain tiles, with half-round ridge tiles on the hips which were fully bedded in mortar. Hip irons, nailed to the foot of the hip rafters, terminated in a scroll.

While visiting the site 40 years after the houses were built, a BRE investigator was lucky to escape injury when most of the half-round tiles on a hip of one of the houses slid off, landing on a footpath 5 m away. Investigation of the debris showed that the hip iron and its fixing nail had corroded completely away, and the mortar bed had cracked and no longer provided sufficient adhesion.

Box 3.3: Strength and stability problems most frequently seen in BRE site inspections of work under way on new trussed rafter roofs over masonry construction (in reports to BRE in the mid-1990s, they were still occurring on a significant scale)

- No packs between trusses and separating walls
- Insufficient straps at gables, or straps provided but not fixed properly, or no packs under straps
- Binders not in contact with gables
- Joints in roof trusses not tight, trusses distorted or not plumb
- No diagonal bracing, or not in accordance with drawings, or cut too short (Figure 3.33), or only single-nailed
- No effective, or insufficient, lateral restraint
- Straps holding down wall plates inadequately fixed
- Skew nailing of trusses splitting both wall plates and trusses
- Tiling battens not fixed at every truss
- More than one in every four tiling battens joined over a single rafter (Figure 3.34)
- One side of a roof loaded with tiles before the other, causing unequal stresses in trusses
- Hip end trusses incorrectly supported
- Eaves tiles not nailed
- Side shunts in single-lap tiles
- Straps not nogged, and too short even to engage second (let alone third) ceiling joist

Figure 3.34: All these battens have been jointed over a single rafter, whereas the joints should be staggered. Of course, such deficiencies will not be exposed unless and until the tiles are stripped, but correctly fixed battens do have a bracing function

Figure 3.33: Incomplete diagonal bracing. The view is taken looking up to the apexes of two adjacent trussed rafters: the diagonal braces should continue to the ridge. Whether such deficiencies need remedial work depends on professional judgement

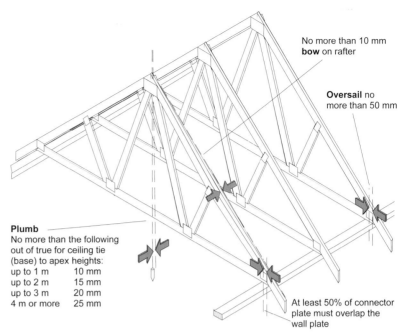

No more than 10 mm **bow** on rafter

Oversail no more than 50 mm

Plumb
No more than the following out of true for ceiling tie (base) to apex heights:
up to 1 m   10 mm
up to 2 m   15 mm
up to 3 m   20 mm
4 m or more   25 mm

At least 50% of connector plate must overlap the wall plate

Figure 3.35: Permissible deviations in the installation of trussed rafters

which has not been packed, touches the gable or separating wall.

Whether all roofs built without the benefit of bracing require remedial bracing is a matter for judgement on the part of a surveyor or engineer. Remedial bracing is possible and is dealt with in BRE Defect Action Sheets 110–112[34–36].

If lateral deviations exceed 10 mm in a trussed rafter measuring 1 m from tie to ridge (or 15 mm for a 2 m rise, or 20 mm for a 3 m rise), the advice of a structural engineer should be obtained (Figure 3.35). Where distortion does not exceed 40 mm, remedial measures applied internally may be viable. Where deviations exceed 40 mm, rebuilding will probably be necessary.

A BRE survey has revealed many instances of poorly designed support for cold water storage tanks in trussed rafter roofs. If loads from tanks are concentrated on too few trusses, or in the wrong places on trusses, truss members may deflect and trusses may be damaged. Bearers need support to prevent them from rotating, and need to be placed close to the node points of the trussed rafters (see Figure 3.9). Chipboard tank platforms have been seen which have deteriorated following saturation caused by condensation[37].

The traditional cut roof (purlin double roof) rarely gives problems, but BRE has not heard of a single whole roof collapsing, unless it has been from overloading when heavy tiles replaced slates. However, this is not to say that cut roofs are entirely problem-free. NHBC inspectors[9] reported the following in 1993:

- lack of bearing for hip rafters,
- no proper joint between hip rafters and purlins,
- wall plates set out of level,
- lack of triangulation in close-couple roofs,
- purlins inadequately strutted,
- excessive birdsmouthing of jack rafters.

Tile clips

Difficulty in inserting clips to hold tiles down to rafters means that too often they are omitted, and this practice has been observed on many sites inspected by BRE. Information available to BRE indicates that provision for fixing tiles and slates was deficient in around 1 in 5 houses built since 1990.

Extra clipping of tiles (and slates) may be needed for roofs near airports (Figures 3.36 and 3.37). Following a two-year investigation, BRE recommended methods of clipping tiles to resist damage caused by vortices from aircraft landing and taking off (BRE Digest 467[38] and Figure 3.38).

Snow guards

From a structural point of view, snowguards are not always fixed securely; guards should be firmly screwed through to the rafters, not surface fixed to the fascia alone.

See also the section on snow in Chapter 2.10.

Figure 3.36: A tiled roof damaged by vortices from landing aircraft

Figure 3.37: A video still of tiles blowing off a roof under test in one of BRE's wind tunnels

Figure 3.38: A non-standard but effective method of clipping tiles to tiling battens being tested by BRE

Dimensional stability

The type of movements to be anticipated in pitched roofs of strutted purlin design relate in the main to:

- spreading where ties or collars become defective, signified by cracks in external walls just under the wall plate and rotation of the wall plate (Figure 3.39),
- sagging of purlins giving rise to apparent hogging over separating walls.

Leaving aside the occasional catastrophic damage caused by gales (Case study 2), some of the most serious cases of movement of these roofs have occurred when natural slates have been replaced with heavy concrete slates or tiles. This is dealt with in greater detail in Chapter 3.2.

In trussed rafter roofs, any deficiencies in bracing may lead to subsequent movement and rain penetration (Case study 3).

Other faults observed on a substantial scale are soffit boards under projecting eaves which have not been nogged, leading to sagging, and the practice of painting one side only of a fibre cement board soffit. After a few years, progressive carbonation of the cement content of a fibre cement board, followed by shrinkage, leads to bowing. Cement-based boards should be painted on *both* sides before installation, or not at all. They should also be installed with slightly oversized holes for the fixings to allow for thermal and moisture-generated movements.

Exclusion and disposal of rain and snow

The relatively small units that form the coverings of most domestic roofs depend for their effectiveness on two main factors:

- the pitch of the roof,
- the amount of the end and side overlap of the units.

In general, the flatter the pitch the greater the overlap required, though there are exceptions. Pitch may also have an effect on durability since, with some porous tiles, drying out may take longer on the flatter pitches, with increased chance, therefore, of freezing under winter conditions.

Sarking felts

On a third of the building sites visited during a BRE survey, the installation of sarking felt underlays was found to be faulty in various ways. In particular, sarking felts were not fitted closely around SVPs, nor properly lapped, nor dressed out to eaves gutters and bargeboards.

Poor detailing of sarkings is seen frequently; also, many instances of tears in sarkings leading to problems with rain or melting snow. BRE Defect Action Sheet 10[39] deals with these problems.

It was noted earlier (see the section on hailstones and snow in Chapter 2.3) that if the space underneath a layer of tiles is not sealed, there is a risk that fine snow will be blown between the joints even though water may not penetrate. In this respect, the current Scottish boarded sarked roof can be expected to perform much better than the English counterpart. The unsupported underfelt used in a traditional English roof can billow in the fluctuating wind pressures of a snow storm allowing the snow to be pumped through the gaps in the covering.

The reinforced BS 747[40] class 1F felts that were commonly used in this situation as an underlay disintegrate after a few years' exposure to external conditions, leaving the detail vulnerable to rainwater blown back by the wind onto the head of the wall. The more satisfactory solutions, from the point of view of durability, for clipped eaves (as well as for overhanging eaves) are those in which the detail has the benefit of a strip of better quality material lapped under the sarking felt and dressed into the gutter. BRE Defect Action Sheet 9[41] suggests a strip of dpc-quality material is used (see Figure 3.11). BS 5534[17], the Code of practice for slating and tiling, recommends an eaves starter strip of type 5U (BS 8747[42]) polyester-reinforced bitumen roofing underlay as the base is rot-resistant, giving added durability.

At the eaves, a tilting fillet should be provided to prevent the sarking from sagging and thus providing a reservoir for rainwater to enter the

Figure 3.39: A rotated wall plate caused by spread of the roof. There are no ties at wall plate level

Case study 2: Wind damage to the gable ends of roofs

The roofs of a group of terraced and semi-detached houses of varying styles were damaged in the gales of October 1987. The slopes of the roofs facing south west suffered extensive damage and some collapsed. Both newly built houses and older houses on the site suffered.

With respect to the newly built houses, the BRE investigator on site noted that there was no packing between the walls and the end trussed rafters; and fewer than the required number of straps were in place, although those that were had been nogged. The gable ladders were not built in tight to the tops of the walls, and no fixings were in place at these points. Bracing of the trussed rafter roofs was not fully in accordance with requirements: diagonal bracing was too shallow, nails were missing at eaves, and

Figure 3.41: The more usually seen kind of wind damage to a verge

braces and binders were not always lapped over two trusses. The coverings were of single-lap interlocking tiles.

While on site, the BRE investigator also examined some of the purlin cut roofs that had suffered in the same gales (Figures 3.40 and 3.41).

Figure 3.40: Unusually severe wind damage to the gable of a double purlin roof

Case study 3: Investigation of problems in a trussed rafter roof

The roofs of 300 houses on two sites in the Home Counties were inspected at the invitation of the building owners. The roofs were 22.5° pitch carrying large concrete plain tiles that had been nailed and clipped at eaves and verges.

Inspections revealed that the trussed rafters were out of plumb and there was some evidence of buckling. The roofs had humped over the separating walls, the joints between the tiles had opened, in some cases by as much as 35 mm, leading to rain penetration. In some cases, too, the verges had an outward bow.

The reason for the movements appeared to have been that the bracing to the roofs originally consisted of longitudinal braces only. The owner was advised to rebuild the roofs that had moved, and to install diagonal braces in the remainder.

Figure 3.42: Sarkings (eg felts) should be neatly cut for soil and vent pipes

roof. The sarking should also be correctly dressed around service pipes such as soil and vent pipes (SVPs) (Figure 3.42). Poorly detailed sarking has been seen frequently.

Gutters and rainwater pipes

At the extremity of the eaves of a pitched roof, the shape characteristics of the covering will have an influence on the positioning of the rainwater gutter. As has been pointed out in *Principles of modern building*[43], the ideal profile of the

covering (from the point of view of rainwater disposal at the eaves) is for the upper edge to be rounded and the lower edge to be sharp to provide a drip from which raindrops will have less chance of being blown back up the underside of the slope. The minimum projection into the eaves gutter is of the order of 50 mm. The normal trajectory of droplets from drips of various profiles is shown in BS EN 12056-3[21]. Further protection to the eaves will be given by overlapping the

lower edge of the sarking felt (or a substitute, a more durable felt strip) into the gutter.

Gutters that are not designed for storm conditions are seen frequently. The worst cases involve valley gutters where the increased water load from very large areas of roof completely overspills the eaves gutters, soaking the wall beneath (see Figure 3.89). This is undesirable for all forms of construction, but especially so in the case of timber-frame construction.

(a)

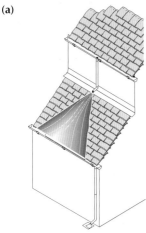

(b)

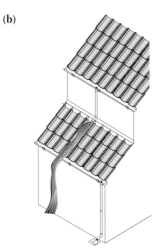

Figure 3.43: Patterns of discharge from rainwater pipes over plain (a) and profiled (b) tile roofs

BRE investigators have seen, on a number of occasions, rainwater pipes discharging over lower roof slopes. Where a roof is of plain tiles (or slates), there is usually no problem since the discharge fans out before reaching the lower gutter (Figure 3.43). However, where the tiles are heavily profiled interlocking tiles, there is no such possibility. The profile of the tiles confines the whole of the run-off to a single valley. Consequently, there is risk that the tile laps will leak, and the virtual certainty that the discharge at the foot will overshoot the lower gutter in all but the lightest rainfall. The tiles will also be preferentially stained.

Joints with neighbouring roofs
Also of concern has been the number of instances where the roof of an adjoining dwelling has been re-covered, with insufficient attention being paid to the weathertightness of the joint between new and old material. A common method of dealing with the junction between profiled and non-profiled materials is to cover the joint with ridge tiles laid on the slope, but such tiles will need dogging at the foot, as with hips. A suitable bedding mortar is described in the section on ridges earlier in this chapter.

Abutments, change of pitch, etc.
Loose flashings permitting rain to penetrate behind were seen all too frequently on BRE site inspections. Flashings may need additional clipping, especially at abutments. Flashings should be chased a minimum of 25 mm into brickwork. A particularly vulnerable flashing in pitched roofs occurs at the change of slope in a mansard. This flashing and the first row of tiles above it have been seen stripped by wind action. The heavier the weight of the flashing, the more robust it will be to resist wind uplift; the lower edges of thinner flashings in this position should preferably be held down with tacks. If a flashing is of lead, without tacks, Code 5 is more appropriate than Code 4.

In a large sample of new housing surveyed in 1992–93, approximately 1 in 6 cases had unsatisfactory weatherproofing at abutments.

A further problem observed on about 1 in 10 sites involved the bottom course of interlocking tiles which have been tilted upwards at the eaves: the tiles had not interlocked satisfactorily making the eaves more vulnerable to rain penetration.

Box 3.5 gives a brief list of those items relating to weathertightness that occurred most frequently in the site inspections of work under way on new roofs over masonry construction.

As might be expected, older roofs display many different kinds of weathertightness faults. Some of the more frequently noted items in the BRE site inspections of older pitched tile roofs are those included in Box 3.6.

Energy conservation and ventilation
Heat loss is one of the principal determinants of energy use in a building, alongside heat gains (both internal and solar). The way in which the building is used is a further factor in heat loss and gain. Realistic estimates of energy use can be obtained only if all the relevant factors, of which one of the main ones is heat loss through the roof structure (and which is more significant for non-domestic single-storey buildings), are allowed for. Chapter 3.8 gives guidance on thermal insulation in lofts.

The external envelope of a building, which includes the roof, allows both intentional and unintentional penetration of air. The unintentional component is often very difficult to predict and can vary substantially between one building and another with a nominally similar design. In some circumstances, this adventitious leakage through the building fabric is a source of excessive ventilation that can lead to energy waste and, in some cases, discomfort. In the interests of fuel economy, more controlled ventilation of living spaces and a requirement for airtightness in buildings has been introduced, eg in national building regulations:
- England & Wales, ADs F and L[44],
- Scotland, Sections 3 and 6 of the Technical Handbooks, Domestic and Non-domestic[45],
- Northern Ireland, Technical Booklets F and K[46].

This can be achieved by more airtight construction through, among other things, better draughtproofing of roofs, and ensuring adequate ventilation by installing background ventilators and mechanical ventilation at points where the need for it is greatest (eg where water vapour is being produced in quantity) in addition to opening windows. This is the best way of meeting the opposing requirements of reducing heat loss and providing sufficient ventilation to prevent condensation and to remove indoor pollutants.

Box 3.5: Weathertightness problems most frequently seen in BRE site inspections of work on new roofs over masonry construction

- Sarking missing or torn and not repaired
- Sarking stretched too tightly across the slope to allow passage of water under the tiling battens
- Sarking felts not dressed up around SVPs (Figure 3.44)
- No tilting fillet at eaves, so sarking allows ponding of rainwater
- Sarking not turned up under outer layers at dormer cheeks
- Sarking does not project to overlap into gutter or has rotted away
- Tiles broken or missing
- Flashings at steps in terraces defective
- Flush (sometimes called clipped) eaves used on exposed sites offering little protection to walls below
- Mortar flaunching used as flashing on new construction
- Lowest courses of tiles lifted too high by tilting fillets: tiles almost level will lead to rain penetration
- Lead slate (or other shaped material) round SVPs not large enough to lap adequately with tiles (Figure 3.45)
- Undercloaking mortar fillets cracked and falling out
- Tiles laid to incorrect laps (nibs not engaging); the worst case BRE has found was 50 mm where 75 mm was actually required
- Flashings with insufficient upstand (sometimes as little as 50 mm instead of 150 mm)

- Tiles laid at too low a pitch for adequate performance
- Cement fillets only covering joints between re-roofed properties and adjacent original roofs
- Sprayed treatment of undersides of roofs, completely 'wet poulticing' the battens and possibly leading to rot

Figure 3.45: An undressed (and therefore totally inadequate) lead slate round a soil and vent pipe on a newly built house. The tiles have been roughly cut, and even if an attempt had been made to dress the lead over the cuts, gaps would have been left. In this case, the sarking felt provides the only line of defence against rain penetration

Box 3.6: Weathertightness problems most frequently seen in BRE site inspections of older pitched tiled roofs

- Absence of sarking and torching
- Strips of sarking tacked to battens between rafters, leaving rafters unprotected
- Many cement fillets cracked and displaced
- Flashings working out of joints
- Clay roof tiles delaminating
- Replacement tiles overhanging eaves gutters too far so rainwater run-off overshoots
- Rainwater from long valley gutters overshooting undersized eaves guttering

Windows set vertically within the slopes of mansard roofs sometimes present problems. The vulnerable point that demands special attention is the point where protrusion changes to inset (Figure 3.46).

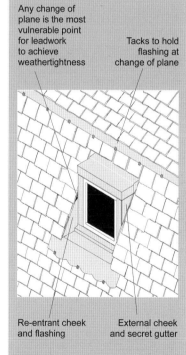

Any change of plane is the most vulnerable point for leadwork to achieve weathertightness

Tacks to hold flashing at change of plane

Re-entrant cheek and flashing

External cheek and secret gutter

Figure 3.46: Windows set in mansard roofs need careful detailing

Figure 3.44: Torn sarking around a soil and vent pipe. If this had occurred underneath the example in Figure 3.45, a leak would have ensued immediately it rained

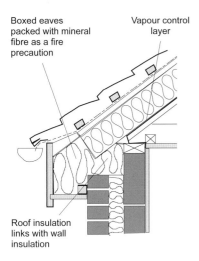

Boxed eaves packed with mineral fibre as a fire precaution

Vapour control layer

Roof insulation links with wall insulation

Figure 3.47: A warm roof eaves detail

Figure 3.49: Rainwater leakage through torching (mortar joints between pantiles)

Warm pitched roofs

In a warm pitched tiled roof, the insulation boards are placed at rafter level, usually rebated and slotted to lie between and over the rafters. The boards should be rebated to reduce the possibility of air infiltration, otherwise a membrane will be required and the boards should have a vapour resistance greater than 200 MNs/g. They usually have sufficient strength to take the loads of the counterbattens, battens and tiles (Figure 3.47). Roofs insulated in this way may be worth considering where the covering as a whole has to be replaced, or where a habitable room is to be constructed in the roof void. But, in warning, BRE investigators have seen a number of cases where lack of understanding of the principles of this form of construction has led to problems, particularly related to condensation. Counterbattening under the breather membrane can be a useful precaution against condensation.

When constructing a warm roof, the thermal insulation should be carried over the wall head into the eaves (Figure 3.48), and also over the tops of any gable walls to meet with the insulation in the cavity. All old thermal insulation at ceiling level must be removed.

Cold pitched roofs

Ventilated roofs: type 1F sarking felt (high resistance underlay)
With cold roof construction, so far as condensation is concerned, an older cold roof with torching (see Figure 3.49) (or even no sarking at all) is unlikely to present problems; there may be, though, evidence of melting of wind-driven snow or capillary attraction between the mortar and the tile (Figure 3.49). However, the lack of adequate provision for ventilation leading to condensation has been observed in many roofs of more modern construction, in spite of the provisions of national building regulations (Figure 3.50). In one particular survey, in none of the houses visited did the provision for ventilation of the cold roof void measure up to the requirements set by AD C[44] and BS 5250[47]. Unless ventilation is achieved over the whole of the roof space, water vapour leaking into it from the dwelling below is likely to condense on the cold underside of the roof covering. If excessive, this can wet the timbers. It may also wet any thermal insulation, reducing its effectiveness, and may short circuit electrical wiring.

For these reasons, it is very worthwhile restricting the entry of water vapour into a cold roof space. The most significant points of entry are gaps around loft access hatches, service pipes and cables. Remedial action includes sealing the gaps, providing hatches with

Figure 3.48: Placing thermal insulation over the eaves to provide continuity between wall and ceiling insulation. Lack of such insulation is a common cause of cold bridging and mould growth on ceilings at roof/wall junctions

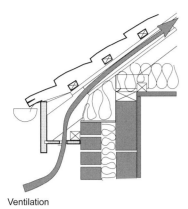

Ventilation

Figure 3.50: Requirements for roof ventilation in a cold roof. National building regulations and British Standards provide for appropriate free areas at eaves and ridges

compression seals, and, in the worst cases, a vapour control layer above bathrooms and kitchens (eg of 500-gauge polyethylene joined with double-sided adhesive tape, not welted).

Ventilation requirements for cold roofs are set out in BS 5250[47]. In summary, the Standard calls for the equivalent of a 25 mm free area at eaves for rooms in the roof and all pitches below 15°, and the equivalent of a 10 mm gap for pitches over 15°. In addition, where the pitch is more than 35°, or the span is more than 10 m, or there is a room in the roof, further ventilation is required at or near the ridge equivalent to a 5 mm gap. Where spaces exist within the roof void that are difficult to ventilate from eaves and ridge, it may be necessary to introduce yet more ventilators to remove moist air (Figure 3.51). The eaves vents should be provided on both sides of the span to give a flow of air through the roof space, and they should be screened to prevent the entry of larger insects and small birds. A 3 or 4 mm mesh should suffice.

If there are ventilation ducts within the roof void (eg for ventilation of internal bathrooms) then the ducts should be insulated with at least 25 mm of insulation material to reduce the amounts of condensation formed when the extracted warm moist air is cooled by the ducting passing through the colder air in the roof space.

Figure 3.51: Extra ventilators are often needed on the slopes of cold deck roofs to provide ventilation in dead areas round roof lights and dormers, or in clipped eaves. Although this example shows installation in new-build, ventilators may be required to remedy deficient ventilation in cold roofs, particularly where extra thermal insulation is added to an existing cold roof

Condensation in such cases cannot entirely be prevented, and provision for its removal will be needed. *Thermal insulation: avoiding risks*[24], gives more information.

Dormer windows are a particular problem, so far as achieving adequate ventilation in a cold roof is concerned, and may need special provision of vents (eg at ridges).

Some dormer designs to be seen are complex in shape, and probably impede the ventilation of the main roof in which they are incorporated. Other faults too may be found in dormers, as shown in Figure 3.52.

There may be a good case, when specifying sarkings, for requiring all sheets to be vapour permeable. It is sometimes argued that impermeable sarkings are ventilated adventitiously through the laps in the sheets, though, in practice, in most roofs seen by BRE investigators, the laps have been tight enough to prevent such ventilation. Furthermore, permeability should not be achieved by relying on an imprecise process such as poorly executed lapping of sheets.

Vapour control layers go on the warm side of the thermal insulation (Figure 3.52). There is more discussion in the section on vapour control layers in Chapter 5.1.

The most common faults relating to thermal insulation and condensation seen by the BRE investigators on ordinary cold pitched domestic roofs include those listed in Box 3.7.

Cold sealed roofs
This form of construction is similar to the traditional cold roof but with a low-resistance underlay (breather membrane). The ventilation is provided only through the

Figure 3.52: A transparent polyethylene vapour control layer (shown by the glint of the sheeting) installed incorrectly against the cheek of a dormer in a house undergoing rehabilitation. (The builder was about to place thermal insulation between the studs. He should have placed the vapour control layer on the inside of the thermal insulation quilt and not on the outside!)

Box 3.7: Thermal insulation and condensation problems most frequently seen in BRE site inspections of cold pitched domestic roofs

- Insulation quilt packed into eaves, completely restricting ventilation (Figure 3.53). This occurred on many sites
- Cold roofs in some dwellings built as late as the early 1980s have received no provision whatsoever for ventilation
- Insulation missing from parts of the roof near eaves, under sloping soffits with 100 mm deep rafters, and under hips where installation is difficult. This has even occurred in buildings that were intended to be highly insulated. Clearly, if the rest of a loft space is highly insulated, the thermal bridging could cause excessive and serious condensation on the ceilings below
- Some single pitches which seemed to be difficult to ventilate properly. Opportunities should be taken to install ventilation in the vertical walls opposite the eaves

- Different layers of insulation not laid to break joint
- Insulation placed under cold water storage tanks in the roof space
- Vapour control layers sandwiched between two layers of thermal insulation giving the risk of condensation
- Mould or fungal growth following condensation in new construction before handover; it is caused by water in the new construction being released into the internal air, and by the lack of ventilation
- Access traps uninsulated and not adequately sealed
- Insulation thickness below that required by building regulations in force at the time of construction
- Dormers unventilated
- Inadequate provision made for cross-ventilation in roof spaces that have separating walls (eg for fire requirements)

Figure 3.53: Insulation quilt packed into the eaves, blocking ventilation paths

Timber rafters, with or without underfelt, sarking, boarding, woodwool slabs, plywood, wood chipboard or fibre-insulating board

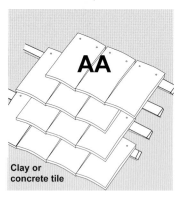

Figure 3.54: Designation of tiles for fire protection

breathable underlay. It is important for this type of construction that a vapour control layer is included. Roof construction should comply with the requirements of third-party certification for the underlay. It should be noted that the initial moisture loading from new-build construction may cause initial condensation problems and some evidence of mould on the roof timbers. This type of roof construction is not covered by BS 5250[47] so third-party

certification must be relied on to ensure regulatory requirements are satisfied. The advantage with using this construction is in the saving of energy loss because of reduced ventilation paths.

Fire precautions

Both clay and concrete tiles carried on timber rafters, with or without underfelt, sarking or boarding (whether the boarding is of ply, woodwool, chipboard or fibreboard) are listed as having notional AA

designations (Figure 3.54) under the Building Regulations 2000 Approved Documents B1 & B2 (2006 edition)[44].

There is increasing concern about fires in roofs caused by blow torches used for stripping paint or for plumbing. In a BRE survey of house design and construction[33], there was found to be a potential route for fire between adjoining dwellings at every continuous boxed eaves detail inspected (Figure 3.55a). Most also had other routes for fire past the end of the separating wall and some of these routes were wider than they might otherwise have been as a result of incomplete blockwork that had been concealed from subsequent inspection by the adjacent rafters. To comply with building regulations, these routes must be sealed (Figure 3.55b). Fire can also spread between tiling battens (Figure 3.55c) and these, too, must be sealed (Figure 3.55d). BRE Defect Action Sheets 7[48] and 8[49] give further information.

Where a pitched roof is to be built over flats or maisonettes, and there is a requirement (eg under the Building Regulations for England and Wales) to provide compartmentation within the roof space over all separating walls, it is usually impractical to build such walls between and around complicated trussed rafters. In these circumstances, it is much better to provide a simple purlin construction and also to adopt the warm roof principle. Failure to extend compartmentation into the roof void

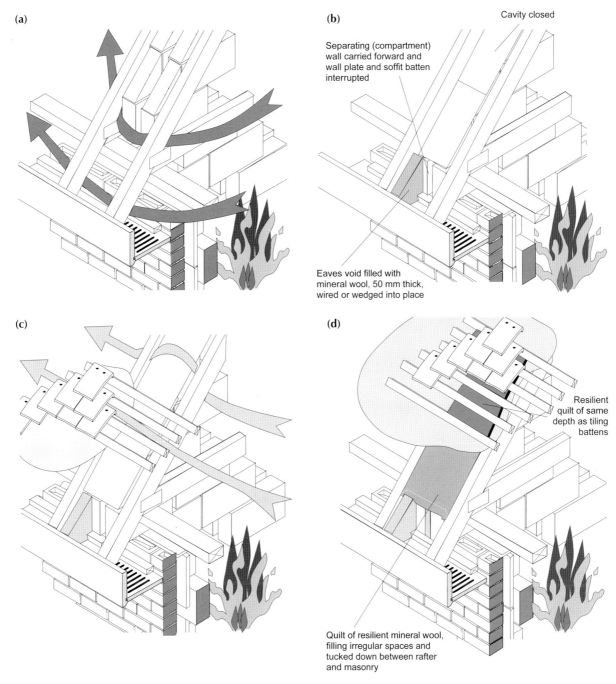

Figure 3.55: Routes by which fire can spread at boxed eaves

and the absence of an alternative fire-resisting ceiling of appropriate design were common faults found on housing sites. Some sites were seen where totally unsuccessful attempts had been made to build separating walls through the struts and ties of trussed rafter cold roofs.

BRE found the faults listed in Box 3.8 to be those most commonly occurring in relation to fire risk.

Roof lights in pitched roofs adjacent to roof escapes need to be fire resisting and an AA designation does not automatically provide this. The underside of roof lights will also need to meet the requirements for the control of surface spread of flame in the room in which they are placed. More detailed information on this topic is given in the section on fire from inside in Chapter 2.6.

Daylighting

Detailing round pitched roof dormer windows is sometimes a problem, although many roofs with dormers dating from Victorian times have performed well (Figure 3.56). One problem found during BRE investigations has been where extensions were being added to existing dwellings in which the main roofs do not have a sarking membrane. The problem relates mainly to the valley gutters. The new sarking should overlap the old at the valley, but if there is none, the structure is automatically vulnerable to rain penetration. In these circumstances, the old roof needs to be stripped back a short distance, and a batten laid along the length of the valley over which the new sarking can be lapped.

There is no British Standard for roof lights, but note that AD L of the Building Regulations in England and

Box 3.8: Fire risk problems most frequently seen in BRE site inspections of pitched tiled roofs

- Diagonal bracing, ridge and ceiling binders, strapping and wall plates crossing separating walls, making integrity against fire impossible. (Structural continuity can be given by strapping each side of the wall)
- Voids in boxed eaves bypassing separating walls
- Inadequate fire stopping (sometimes quite large gaps) over separating walls
- Gas flue terminals too close to battens and sarking felt, too close to soffits and plastics rainwater gutters at eaves, and too close to cellulosic loft insulation
- Electric cables buried in thermal insulation. It has been known for an electric cable, encased in thermal insulation in a roof void, to overheat and catch fire
- PVC-clad cables in contact with polystyrene insulation may also constitute a risk due to plasticizer migration
- Some older properties have been observed to have no separating wall in the roof space whatsoever. Although building regulations are not retrospective, surveyors encountering cases of this kind may need to recommend that a wall be provided for both fire and security reasons

Wales[44] may require roof windows or roof lights to be replaced where the building is subject to a change in energy status and the U-value of the component is worse than 3.3 W/m²·K. Guidance is given by the National Association of Rooflight Manufacturers[50]. However, most roof lights intended for installation in pitched roofs come with explicit installation instructions.

In the BRE site inspections, the increasing use of prefabricated dormers in glass-fibre-reinforced polyester was noted. These were sometimes damaged during installation on site.

Sound insulation

The most efficient domestic pitched roof from the point of view of reduction of external noise is a double-lap tiled roof with airtight (but vapour-permeable) sarking. The heaviest tiles, up to 80 or 90 kg/m², together with a heavy plastered ceiling, should give up to 40 dB, provided the sarking is reasonably airtight[43] (see also *Sound control for homes*[51]). A lighter construction, with a single layer of plasterboard on the ceiling, will give around 27–31 dB. Where a roof is required to achieve a reduction of more than 40 dB, expert advice should be sought.

Durability and ease of maintenance

Tiles

Clay tiles, provided they have been properly fired and properly fixed can last 100 years and more, though manufacturers' guarantees tend to be much less than this. The odd tile may need replacing from time to time, most probably after strong winds. However, some tiles are without nibs (peg tiles), and these, like slates, all need to be nailed. When the nails corrode, sometimes loosely referred to as 'nail sickness', the whole roof covering is at risk (Figure 3.57).

Although rarely encountered in roofs built since the 1970s, some poor quality clay tiles can still be found in older roofs (Figure 3.58).

Concrete tiles should last for periods comparable with the best of clay, though some concrete tiles may lose their outer surfacing rather earlier which may affect their appearance. It will be worth checking this risk with potential suppliers.

In the BRE site inspections, the common use of 'special' tiles at changes of pitch, and at valleys, eaves and verges was noted frequently. Some of these were already damaged, and replacements were not easily available. Consideration needs to be given to the likely future availability of replacement specials.

Pressed steel tiles giving some 20 or so years of expected life are available, and might be covered by third-party certificates.

Figure 3.56: A secret gutter was formed round this window in a plain tiled roof dating from the 1870s. Although it cannot be seen in the photograph, the gutter was still in good condition

Figure 3.57: The plain nibless or peg tiles on this 100-year-old roof are suffering from nail sickness: severely corroding nails have allowed the tiles to slip. Quite soon, the roof would need re-laying

Figure 3.58: Some of these double-lap handmade plain clay tiles are underburnt, giving rise to delamination and deterioration. They do not ring true when tapped, and some can be snapped by light hand pressure. A pale colour can also be a good indicator of poor quality

Industry research may lead to new concrete tiles which will incorporate polymers to increase strength and reduce weight without affecting durability.

Fastenings, flashings, etc.

Aluminium alloy nails to BS 1202-3[52] are satisfactory for most exposures, but polluted areas probably warrant copper or silicon bronze nails. Where there is concern about possible increases in wind loads due to global warming, consideration may be given to the use of stronger nails such as stainless steel. However, such fixings will make it much more difficult to replace defective tiles or to reclaim the tiles for use elsewhere, since any attempt to use the traditional tiler's method for stripping the roof will break the tile rather than the nail. Where nail sickness has occurred, a temporary solution may be found in spraying the underside of an unsarked roof with, for example, polyisocyanurate foam. Care should be taken to allow the battens to breathe, otherwise premature rot may ensue.

Rainwater run-off from organic growths (eg mosses and lichens) is acidic and will attack metal flashings.

Timber

Extensive or widespread decay in roof timbers is rare, but where they bear on brickwork which may be damp, particularly close inspection may be necessary (Figure 3.59). Portions of the roof with limited headroom make inspections difficult, and diagnosis may depend on cutting away plaster, plasterboard and brickwork around ceiling joists and rafters. Although roof timbers seriously affected by wet rot will need to be replaced, those with localised rot may be repairable by removal and replacement of the decayed portions. Timbers that are damp, but show no signs of decay need to be dried out to 20% moisture content or below to avoid the risk of subsequent decay. Application of wood preservatives may be appropriate if drying time cannot be reduced to less than about six weeks[53].

Limited wood-borer infestation can usually be treated effectively where necessary. The species of insect must be established[54], and also whether the attack is still active. Three damage categories are indicated:

- remedial treatment usually needed,
- treatment necessary only to control associated wood rot,
- no treatment needed.

Most infestations of common furniture beetle, even if of long standing, are of little structural significance and usually require little or no replacement of timber. House longhorn beetle (*Hylotrupes bajulus*) infestations commonly cause structural damage, and deep probing is necessary to discover the extent of attack (Figure 3.60). Other kinds of insect damage are described in *Recognising wood rot and insect damage in buildings*[54].

Steel

Corrosion of steel sections, sheets and fixings in domestic roofs can be anticipated where the original protection has disappeared, though most corrosion of steel in these

Figure 3.59: Wet rot in the bearing of a truss. Such deterioration is not easy to spot in the limited headroom at eaves, but it may show first as damp staining in the ceilings beneath

Figure 3.60: House longhorn beetle attack in a domestic pitched roof. Repair of this roof will require all affected timber units to be replaced

situations seen by BRE investigators has been superficial (Figure 3.61). The majority of examples of steel-framed roofs have already survived for their design lives, but should be capable of giving further service. The most vulnerable are those where thin steel sheet has been folded to form the loadbearing members: where the surface protection has gone, there is comparatively little steel left to corrode. Particular attention needs to be paid to those systems that made extensive use of thin metal fabricated sections (eg Trusteel and Hills[55]). Corrosion of structural steel is described further in Chapter 5.1.

So far as trussed rafters are concerned, red rust corrosion on truss connector plates (as well as the more usually found zinc oxide or so-called white rust) has sometimes been seen on the BRE site inspections; this has raised questions of future durability. Occasionally, a roof has been found to be in a totally unsatisfactory condition due to corrosion of the plates. Following a special survey on trussed rafter roofs, BRE published a report in 1983[56]. The authors concluded that corrosion of metal plate fasteners did not appear to be a significant problem generally, though the situation might be exacerbated by

preservative treatment with CCA (copper-chromium-arsenic) salts or by proximity to the coast. That situation continues. However, where the thermal insulation covers the plates, there is a risk of the insulation becoming saturated from condensation or from leaks in the outer covering; the poultice effect of the insulation will create conditions that accelerate corrosion.

Other aspects of durability

Other faults relating to durability that have been seen during BRE site inspections are listed in Box 3.9.

Maintenance

It is recommended that gutters are inspected at least once a year, especially if there are trees in the vicinity. In areas of high pollution, more frequent inspections may be necessary. The condition of the gutter supports should be checked too (Figure 3.62). One of the most serious causes of damage to gutters and fixings is the build-up of ice. Plastics rainwater goods are the most vulnerable to damage (Figure 3.64).

Nails and other fixings such as clips should preferably not be made of galvanised steel since the galvanic coating may be destroyed by hammering and their durability will suffer. Alternatives include composition and aluminium nails (BS 1202-3[52]).

As mentioned earlier, sarking felt to BS 747[40] class 1F has been seen frequently to deteriorate where exposed to ultraviolet light, particularly at gutters. The use of an eaves starter strip of Type 5U (BS 8747[42]) which has a rot-resistant base giving added durability and

Figure 3.61: Inside the roof of a Hills system house. Some of the principal rafters and ceiling ties are galvanised, others are not. Though the steel units are beginning to show corrosion, it is superficial and there is no cause for immediate concern

Figure 3.62: This gutter bracket has broken leading to rapid saturation of the wall beneath

Box 3.9: Durability problems seen in BRE site inspections of tiled pitched roofs

- Non-durable nails used to fix tiles
- The moisture content of trusses on occasion exceeding 26% (third-party certificates for metal plates require 22% maximum)
- Ordinary plywood used for soffits not sufficiently durable
- Unprotected timber in contact with potentially wet brickwork, particularly at bargeboards and clipped eaves. Fascias on several sites were not primed before fixing
- Galvanised nails driven into CCA-treated battens
- Site-cut ends of pre-treated timber were not brushed with preservative
- Access hatches under hip ends did not have sufficient headroom to allow entry to roof spaces
- Unprotected copper nails being used to fix lead flashings. Copper nails can be used, but they should be kept well away from rainwater by an upper surface of lead sheeting. Alternatively, patches of lead can be soldered over the nail heads
- Gaps large enough for birds to enter roof spaces were being left in fascia boards

- Evidence of attack (fresh exit holes) by wood-boring insects in newly installed timbers
- Clay roofing tiles disintegrating (Figure 3.63). Those seen were likely to have originated from a particular source which is no longer available

Figure 3.63: Plain double-lap tiles and half-round ridge tiles deteriorating. The causes probably included foot traffic compounded by the roof being set at far too flat a pitch, though frost action on underburnt clay was a major factor

will protect the wall below from rain being blown back under the overhang[41].

It is possible to treat lead flashings with a coat of bitumen to prevent the staining that is sometimes seen where leadwork lies above susceptible tiles and slates. Treatment with a patination oil is an alternative.

Some of the detailing on 17th and 18th century houses can be rather complicated and comparatively inaccessible, but that should not provide an excuse for delay in rectifying any faults (Figure 3.65).

Moss and lichen will grow on pitched tiled roofs, whether of clay or concrete, and might need to be brought under control. If growths become problematical, the advice given in Chapter 2.9 applies. There is evidence that moss grows less well on steeper pitches (Figure 3.66).

WORK ON SITE

Access, safety, etc.

HSE issues guidance on work in connection with pitched roofing at heights greater than 2 m above ground. The CDM regulations

Figure 3.64: Icicles hanging from a plastics gutter. When they fall, they can perforate any built-up roof below (Photograph by permission of DM Currie)

Figure 3.65: The detail at the back of this fine old parapet wall has been allowed to deteriorate : the capping to the cornice is loose or missing and the replacement plastics gutter is sagging over the painted lead hopper. Fortuitously, rainwater flowing into the gutter discharges into the hopper (the replacement guttering was designed to carry rainwater in the opposite direction)

Figure 3.66: An experimental tiled roof at BRE showing moss growing on the flatter pitch but not on the steeper pitch

---

Box 3.10: Safety advice for site inspections, maintenance and work in progress on roofs

- Use properly made roof ladders which should not be leant against gutters
- Where extensive work is to be undertaken, use correctly erected scaffolding with guard rails and toe boards
- Use a safety harness to British Standards on secure anchorages and taut lines
- Use mobile elevating work platforms
- Use crawl boards on fragile roof coverings and valley gutters

The following further points should be noted:

- Packs should be placed as near as possible to the end-use point to avoid unnecessary handling and risk of damage
- Surface biocides used to clean roofs affected by organic growths should have an HSE number
- Timber treatment operatives will need to comply with procedures required under the Control of Substances Hazardous to Health (COSHH) Regulations 1988[60]

---

introduced in 1994 (and updates in 2007)[57] were intended to put the onus of considering safety issues on roofs at the design stage. This issue was further discussed by a joint BRE/industry group and recommendations incorporated in BRE Digest 493[58]. This has coincided with work by another industry group, the Advisory Committee for Roof Work[59].

Particular attention is drawn to the points listed in Box 3.10 which apply to all site inspections as well as to maintenance and other work in progress.

Storage and handling of materials

A BRE survey[33] revealed that, on many sites, timber trussed rafters were stored in such a way that they had become damaged and wet before being fixed into place. Although wet rafters dry down during the first few months, perhaps years, of service with little long-term detrimental effect on the timber (but corrosion of the connector plates is another matter), distortion and damage can go unnoticed. BRE Defect Action Sheet 5[61] gives recommendations for storage.

Restrictions due to weather conditions

It is recommended that work on tiling roofs should not proceed when weather conditions exceed the following levels:

- rainfall greater than 0.5 mm/hr,
- any snow, sleet or hail,
- wind 3-second gust speed greater than 16 m/s.

Refurbishment of a roof will need protection from the weather whch usually involves extensive scaffolding and sheeting (Figure 3.67).

Advice on practices during poor weather is also available from the National Federation of Roofing Contractors[62].

Workmanship

Stretching of gauge, particularly on small roofs such as those over porches, has frequently been seen on BRE site inspections[33] (Figure 3.68).

Tiles to be bedded on mortar should be docked (wetted) beforehand.

Although torching a roof is not recommended these days, if a roof has already been torched, and it does not warrant stripping and remaking, it may be possible for an experienced craftsman to make a repair using a haired lime mortar.

In site studies of new-build housing[33], one of the more common defects with trussed rafter roofs was that difficulties ensued where trussed rafters had been used for buildings of non-rectilinear shapes

Figure 3.67: Partially completed protection for a house undergoing total refurbishment

Figure 3.68: The construction of this porch roof has been disastrous. Not only have tiles of different profile been used, failing to interlock, but the head lap is insufficient to keep out rainwater. Also the verge had not been pointed

Figure 3.69: Ventilation of cold deck roofs can often be impaired by thermal insulation with this form of construction, particularly at the eaves of dormers. Signs of condensation may show in dormers that are not adequately ventilated

or with hipped roofs. Binding and bracing defects were also common, as also was splitting of timber by skew nailing where the use of clips would have been preferable. There may also be potential problems in achieving adequate ventilation (Figure 3.69).

Other workmanship defects seen to occur commonly were:

- failure to construct water tank platforms correctly to BS 5268-3[63],
- trussed rafters erected out of plumb. The maximum deviation from the vertical allowed by BS 5268-3[63] for a 2 m high trussed rafter is 15 mm,
- staining of tiles by rainwater run-off from fresh mortar,
- tile perpends wandering out of true,
- patchy appearance of the finished roof (batches of tiles should be mixed to avoid this problem, though this was rarely seen to be done with repair or extension work (Figures 3.70 and 3.71)).

Supervision of critical features

The following should be given particular attention:

- clear allocation of responsibilities for the design of the building, the roof and the roof members, whether they are trussed rafters, trusses, purlins, etc. (and their interfaces),
- identification of any special features of the design such as girder trusses taking eccentric loads and explaining these in

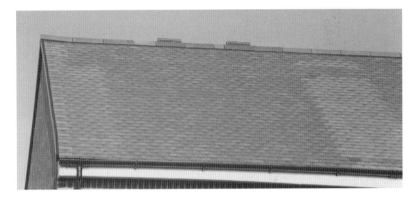

Figure 3.70: A roof completed in 1992. Batches of tiles had already been rejected because of colour mismatch. Batches should be sampled for acceptability before laying begins and may need to be randomly mixed to produce acceptable results. Tiles-and-a-half used at verges have frequently been found by BRE not to match the remainder

Figure 3.71: Replacement tiles on this roof in a conservation area leaves an unsatisfactory appearance. It is debatable whether the sagging rafters which give rise to the humping effect over an internal wall should have been packed to give a better profile

clear (written) instructions to the site staff,

- nailing and strapping of verges, eaves and ridges where called for by the design,
- prevention of tile hogging at the top of any separating wall (the wall should be kept below the level of the underside of the tiling battens and the space filled with incombustible material to accommodate any differential movement),
- binders should tightly abut separating and gable walls in order that adequate lateral restraint is provided between roof and wall (lateral restraint requirements at the gables of timber-frame dwellings though, may differ from requirements for masonry gables (TRADA[64]),
- filler units used with profiled tiles where gaps at the fascia would otherwise provide access for birds and large insects,
- roofs that are loaded out symmetrically, as called for by BS 8000-6[65]; trussed rafters in particular can be overstrained if the roof is loaded asymmetrically,
- tiling battens continuous to verges and not jointed over the last two rafters (Figure 3.72),
- projection of durable sarking felt underlay or dpc material into the rainwater gutter.

Figure 3.72: An incomplete roof, not very well constructed and showing a number of defects (eg chimney flashing not stepped, battens too short and sarking not carried to eaves)

Box 3.11: Inspection of concrete and clay tiled roofs

Checklists for surveyors have been published in the *Surveyor's checklist for rehabilitation of traditional housing*[66] and in *Quality in traditional housing*, Volume 3: an aid to site inspection[33].

The problems to look for are listed below.

Inspection of new and recently built tiled roofs

- Trussed rafters modified on site without authorisation
- Gaps at the intersections of truss timbers
- Node connecting plates in trussed rafters which have not been fully pressed home
- Trussed rafter plates corroding
- Insufficient numbers of 'teeth' of trussed rafter plates embedded in each member
- Faulty finger joints in truss members. (In the case of glued finger joints in timber, it is common practice to make these joints with gaps at the ends of the fingers in order to ensure that the faces are in close contact. These gaps do not make the joints defective)
- Quality assurance marks not shown on timber
- Stress grading marks not shown on timber
- Moisture content more than 22%
- Trusses fixed out of plumb, with bow or twist beyond that allowed
- Timbers not preservative-treated when treatment was specified
- Trusses inaccurately spaced
- No bracing and binding to trussed rafters in place (eg no web bracing where specified)
- Missing raking braces to the struts in roofs under 5 m span
- Bracing not connected to wall plates
- Binders not in contact with wall faces
- Spacing of restraint straps exceeding 2 m centres on rafters alongside walls

- Restraint straps incorrectly sized and fixed
- Wall plates not fully bedded in mortar (except in timber frames)
- Holding-down straps not in place and incorrectly fixed (if required)
- Ventilation gaps not of required size at eaves in cold roofs
- No ventilation gaps at ridges in roofs with pitches over 35°
- Holes in ceilings not sealed
- Trussed rafters not fixed to wall plates
- Trussed rafters split by nailing
- No tilting fillets fixed to rafters to support sarking and tiles
- Joints in soffits not nogged
- Structural timbers penetrating separating walls
- Boxed eaves not fire stopped at separating walls
- Eaves tiles not clipped
- Bottom edges of tiles not extended over gutters
- The pitch of the lowest courses of tiles lifted too high by fascias
- Verge tiles and intermediate tiles not clipped as required
- Verge pointing unsound
- Fire stopping not present over separating walls
- Gaps in thermal insulation
- Thermal insulation in roofs not linked with that in walls

External inspection of older tiled roofs before or during rehabilitation

- Rafter feet spreading
- Leaning or lateral movement (eg at gable ends or eaves)
- Gaps between tiles (an indication of movement)
- Undulations in the covering
- Tiles that have aged
- Tiles broken, missing or delaminating
- Tile nibs broken
- Tiles incorrectly fixed

- Nails that have corroded
- Ridge or hip tiles incorrectly fixed
- Tiles and lap unsuitable for pitch and exposure
- Interlocking tiles tilted at verges
- Flashings missing or damaged
- Metal flashings or valley gutter linings corroded
- Lead flashings too thin
- Gutters rusting, missing, damaged, leaking or blocked
- Gutters not properly supported
- Gutters not aligned to ensure fall (or are tilted back)
- Timber or concrete gutters not lined
- Junctions with abutments and neighbouring roofs unsound
- Tiles oversailing gutters
- Asbestos cement gutters and downpipes lacking durability

Internal inspection of older tiled roofs before or during rehabilitation

- Fire stops at the tops of separating walls absent
- Holes through separating walls, or walls missing
- Joints in brick and blockwork not filled (or not adequately filled)
- Wall thicknesses inadequate
- Timbers built into or through separating walls
- Walls that have cracked
- Loose brickwork in separating walls
- Tiles missing or damaged
- Sarking felt failed or incorrectly fitted
- Snow or rain penetration
- Condensation
- Insulation at eaves in cold roofs blocking ventilation pathways
- Thermal bridges at eaves or verges
- Vapour control layers installed on the cold side of thermal insulation
- Loose bolts in bolted trusses, with or without connectors

3.2 SLATE AND STONE TILES

Since many of the details of construction with natural slate and stone, and manmade slate are similar, these materials are taken together for the purposes of this book.

Natural slate and stone

Natural slates are available in a range of colours, and it will be necessary for surveyors to match originals in any replacement work. Reddish colour is given by a dominance of hematite in the composition of the slate, green by chlorite, black from carbon, and rustic from iron oxide in the planes of cleavage. Although, in the past, green colours have been mostly associated with Cumbria, mid-greys with Scotland, and blue-greys with Wales, this pattern is not infallible. There are illustrations of the colours of natural slate in the BRE book, *The building slates of the British Isles*[67] and in two books published by Historic Scotland[68, 69].

Stone slates, sometimes called tilestones, may be of limestone or sandstone, according to locality. Although the use of stone material overall in the UK is fairly sporadic in areas outside the main stone quarrying regions (the actual numbers are not identifiable separately from the national House Condition Surveys), some of the most common examples which the surveyor may encounter in particular areas include limestone slates and sandstone or gritstone slates.

Limestone slates
Limestone slates are normally riven. They include, for example, Filkins and Stonesfield (both from the Cotswolds), Collieweston (Northamptonshire) and Purbeck (Dorset). They are mostly cream or grey in colour. They are usually 12–25 mm in thickness, and up to 600 mm long. They are normally laid at pitches of around 50°.

Sandstone or gritstone slates
Sandstone or gritstone slates may be riven or sawn. Examples of these slates include Kerridge (Cheshire), Horsham (West Sussex), Mount Tabor, Grenoside and Apex (all from Yorkshire), Cilmaenllwyd (South Wales) and Caithness (Figure 3.73). They are mostly grey to brown or black in colour. Thicknesses vary. As one example, the Cilmaenllwyd tilestones were usually split to 12 mm, and holed for wooden pegs. The larger stones, known as flags, can be up to 75 mm in thickness, and 750 mm in length, and are normally laid at pitches of around 35° (Figure 3.74).

Data for dwellings in the UK in 1991 having a natural slate or stone covering for the whole or part of the roof are given in Table 3.6 as recorded in national House Condition Surveys in 1990[5–8].

Table 3.6: Reported problems with natural slate or stone roof coverings in the UK. Data from national House Condition Surveys circa 1990[5–8]		
	Number of roofs (whole or part) covered	Proportion with a problem (approx.)
England:		
Houses	3 million (20% of total stock)	1:2
Flats	565,000 (15% of total stock)	1:3
Wales:	No data	—
Scotland:		
Houses and flats	657,000	1:2
Northern Ireland:		
Houses and flats	39,500	No data

Figure 3.73: A random width slated roof. Slates of this thickness in some parts of the UK are sometimes referred to as tilestones. Most of these are in good condition, but some show signs of delamination (Photograph by permission of BT Harrison)

Table 3.7: Reported problems with manmade slate roof coverings in the UK. Data from national House Condition Surveys circa 1990[5–8]

	Number of roofs (whole or part) covered	Proportion with a problem (approx.)
England:		
Houses	689,000	1:6
Flats	305,000	1:12
Wales:	No data	—
Scotland:		
Houses and flats	17,000	1:2
Northern Ireland:		
Houses and flats	59,000	No data

Figure 3.74: A flagged roof in Caithness with head laps but no side laps on a farm building. Such roofs depend on the width and integrity of the underlay at the sides, just seen at the flag with the broken corner (Photograph by permission of BT Harrison)

If particular ages of houses in England were examined, twice as many slate or stone roof coverings built in the period from 1900 to 1918 showed faults as those not showing faults. Roofs built before 1900 had probably been repaired, whereas roofs built from 1900 to 1918 were reaching the end of their effective lives.

By 2005/6 there were 2.9 million (13.8%) dwellings in England with the whole or part of their roofs covered in natural slate, stone or shingles (categories were not identical with former years), representing a probable reduction in numbers over the intervening period. Current data specific to the repair of these roof coverings has not been identified, but experience indicates that the figures quoted in the previous paragraph are not likely to differ by more than one or two percentage points.

In Scotland in 1991, roughly the same proportion of natural slate roof coverings were in need of repair as in England (Figure 3.75).

In 2005/6 in Northern Ireland there were 56,363 (8.1%) of dwellings roofed in natural slate, stone or shingles, a substantial increase in the intervening 15 years.

Comparable data for Wales are not available, but the proportion of houses in Wales covered in slate should be substantially higher than in England given the former widespread local availability of the material and low transport costs.

Manmade slate

The majority of manmade slate roof coverings in 1991 would probably have been of time-expired asbestos cement: other materials would have been of relatively recent origin where faults usually reflected premature failure, such as bowing.

Data for dwellings in the UK in 1991 having a manmade slate tile covering for the whole or part of the roof are given in Table 3.7 as recorded in national House Condition Surveys in 1990[5–8].

By 2005/6, 1.2 million (5.7%) homes in England had manmade slates covering the whole or part of their roofs, nearly twice the number of 15 years earlier. During that time, there has been a progressive increase in the quality of manmade slates, and the proportion showing faults in service can be expected to have been reduced in consequence.

Data for housing in Scotland in 2005/6 were not available for this book.

In 2005/6 in Northern Ireland 146,483 (21.1%) houses had a manmade slate roof covering, nearly three times the number of 15 years earlier.

Figure 3.75: A slated roof in need of repair (eg sagging rafters, slipped and broken slates, racking dormers, dormer hip rolls missing and rendered flashings broken)

Comparable data for Wales were not available for 1990/1 or 2005/6.

Manmade slates are available in a wide range of materials, coatings, sizes and colours. Also commonly available is a range of matching accessories such as ridge pieces and flue terminals, depending on the material. Materials of manufacture include those listed below.

- *Unreinforced concrete*, made with Portland cement and a variety of aggregates such as granite, crushed stone or PFA (pulverised fuel ash). Colouring is either by a pigment incorporated into the mix, or by a surface layer of coloured sand and cement. Polymer modifiers may be added. Slates may be extruded or moulded.
- *Portland cement reinforced with manmade fibres* including glass, PVAL (polyvinyl alcohol) and polypropylene.
- *Resins* (thermosetting, polyester or acrylic) with a crushed slate or other inert filler.

Asbestos cement slates were formerly used on a considerable scale in the UK. It is almost universally accepted that asbestos fibres pose serious hazards to health[70]), and therefore asbestos cement has been phased out of production in many countries. The problem with asbestos relates to the small diameter of the fibres, 3 micrometres (µm) or less, and to their insolubility when taken into the lungs. In contrast, the alternative fibres for reinforcement are mainly of the order of 10 µm diameter, which present less risk to health because they are not easily inhaled. Specialist contractors will need to be employed when such roof coverings are being replaced.

Boards and sheets reinforced with manmade fibres are dealt with in Chapter 5.1.

CHARACTERISTIC DETAILS

Basic structure

Most of the existing stock of small dwellings roofed with slate or stone will have a structure of rafters carried on purlins, with very few differences in character from the roofs that

carry tiles; there may be differences, though, in the smaller sizes of timbers made possible by the lighter weights of the covering materials. However, some of the natural limestone and sandstone slates can be heavy, and will necessitate an increase in scantling sizes well above those normal for other natural and manmade slates. Calculations of loadings should be carried out. Slates can range in weight from about 25 kg/m² for best Welsh to about 78 kg/m² for Westmorland thick slates. Stone flags are heavier than slates because of their greater thickness.

Main roof areas

Natural roofing slates are available in many sizes and thicknesses. In years past, sizes were given names by the trade, eg:

'smalls'	12 in × 6 in,
'doubles'	13 in × 7 in,
'ladies'	16 in × 10 in,
'countesses'	20 in × 10 in,
'duchesses'	24 in × 12 in.

These names are no longer in universal use.

Natural slates are now specified by their metric dimensions. Shaped bottom edges are obtainable, though they will be cut more roughly than their manmade counterparts.

The thinnest natural slates are known as 'bests' and the thickest known as 'strongs' with 'mediums' in between.

Slates should conform to BS EN 12326-1[71]. Slating practice is covered in BS 5534-1[72], with workmanship aspects in BS 8000-6[65].

Surveyors may encounter both centre-nailed and head-nailed natural slated roofs, with the centre-nailed occurring in general on the larger slates and the flatter pitches. Head-nailed slates should have a larger lap than centre-nailed slates of the same size. All natural slates should have their 'grain' running parallel to their length for maximum durability (Figure 3.76).

Stone slates may occasionally still be encountered laid in a mortar bed on a boarded low-pitched roof. These roofs are prone to decay, with moisture trapped in contact with

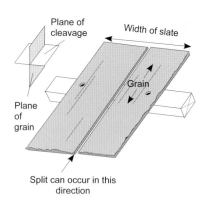

Figure 3.76: A split centre-nailed natural slate showing the direction of 'grain'

the boarding; this form of roof is not used nowadays.

Sawn stone slates of regular dimension can be laid using details similar to those suitable for tiles. On the other hand, riven stone slates are random in size and should be sorted before slating begins so that the largest stones are near the eaves and the smallest near the ridge. A high degree of skill is needed to achieve adequate side and head laps with such stones.

Manmade slates are normally centre-nailed, with lower edges clipped or riveted (Figure 3.77). Clips or rivets should be shown by the manufacturer to be suitable for the conditions of service. In all cases, manufacturers' recommendations should be followed. Normally, fixing is by copper nails and rivets, or by stainless steel hooks.

Nails for slates must be chosen to suit the size and thickness of the slate. The length and thickness of

Figure 3.77: Centre-nailed manmade slates. Rivets on the fixed slates are all in place, though on the top row a few of the rivets have slipped down the slope. These will be slid up again to fix the tails when the row above is laid

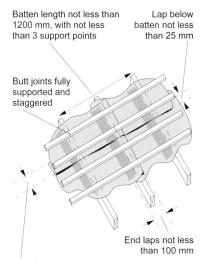

Batten length not less than 1200 mm, with not less than 3 support points

Lap below batten not less than 25 mm

Butt joints fully supported and staggered

End laps not less than 100 mm

Edge laps:
pitch >35°, not less than 100 mm
pitch 15–34°, not less than 150 mm
pitch 12.5–15°, not less than 225 mm

Figure 3.78: Sizing of battens, fixings and laps for slate

Figure 3.80: Carnass Bay Coast Guard Station, Isle of Anglesey, being roofed in ordinary natural slate in 1961. This is an unusual method of laying slates in a very exposed situation

Figure 3.82: The absence of tilting fillets at these clipped eaves leads to the possibility of future rainwater penetration to the interior of the building when the underfelt perishes

the nail should also be checked against the batten size to ensure that the battens will not split and that no penetration of the underlay can occur. BRE recommendations for sizing of battens, fixings and laps for slates are given in Figure 3.78 and BRE Defect Action Sheet 142[73].

Occasional regional and other variations may be encountered. For example, in Renaissance times stone vaults were occasionally roofed over with stone flags, the first layer spaced apart but overlapping in a shallow pitch (say 10°) down the slope, and the second layer laid to lap both adjacent flags (Figure 3.79). Occasional examples of similar techniques used in more recent times may be found, as in Figure 3.80.

It is customary in some areas of the UK to fix slates with a single head nail, but slates fixed in this way

will sometimes be pivoted sideways by strong winds.

Eaves

Overhanging
The first course at the eaves is normally a row of shorter slates laid over the tilting fillet or fascia in a similar fashion to tiling practice (Figure 3.81). The overhang of the slates over the gutter should be 50 mm, with a strip of dpc material laid over the underfelting and dressed into the gutter, or a

purposemade unit. Ventilation must be provided for a cold roof.

Clipped
Clipped eaves are not recommended by BRE since no protection is offered to the wall below (Figure 3.82).

Verges
Since wind suction on pitched roofs is usually greatest at the verges, and since slates are comparatively light in weight, detailing at the verges must be carefully specified to avoid disruption. In particular, slates-and-a-half, which provide a greater area for fixings, should be used instead of half slates to provide the side lap. There is, in any case, the requirement in BS 5534[17] for all edge tiles and slates to be mechanically fixed, and BRE recommends that the last three slates are fixed in this manner. The overlap of the slates over the bargeboard or brickwork needs to be 50 mm. In the case of the bargeboard, the slates should be nailed to the top of the board, and where the verge is laid over brickwork, the undercloak should be as wide as possible to overlap the head of the wall and be bedded in mortar.

Abutments
Stone slated roofs often have the walls carried above the roof lines at gables to form low-coped parapets. In these cases, secret gutters provide the most weatherproof detailing at junctions. Mortar fillets

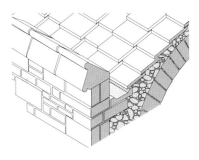

Figure 3.79: A stone vault roofed with overlapping flags

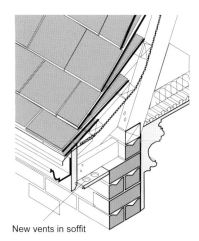

New vents in soffit

Figure 3.81: Slated eaves. If extra thermal insulation is put into a boarded or sarked roof, vents will be needed in the soffit. If the roof does not have an underlay, adventitious ventilation will be sufficient to remove condensation

Figure 3.83: Lead abutment flashing

Figure 3.84: Mitreing a valley in manmade slate. A mortar-bedded detail for natural slate will require a wider lining to the gutter than a comparable dry fix detail. In wet fixing, the natural slate should not bear on the timber fillet. Some of the batten ends are shown unsupported

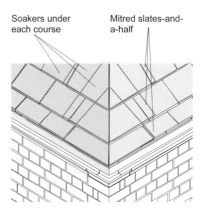

Figure 3.86: A mitred and soakered hip. Mitred slates-and-a-half are most vulnerable because of the limited area for holing and fixing. Mitred hips should not be used on shallow pitches (ie not on pitches of less than 40°)

or flaunchings at abutments of slated roofs are traditional solutions in some parts of the UK, but, as with fillets over tiles, are prone to deterioration, especially if the mix is too rich. Cement mortar fillets should be avoided wherever possible since, in time, they crack away from walls. The alternative of a lead flashing may not always prove successful, however, Figure 3.83 illustrates sound practice.

Flashings at abutments at the heads of slopes and changes of pitch should be held with tacks at about 400–500 mm centres depending on exposure, especially if the flashing is of Code 4 lead. Thicker materials need less tacking.

Readers should also see the section on Abutments in Chapter 3.1.

Valleys

Valleys in slate and stone-covered intersecting sloping roofs are normally formed by mitred slates laid over Code 4 lead soakers or Code 4 or 5 lead or zinc valley gutters (Figure 3.84). However, it is possible for a skilled operative using random or hand-dressed stones to sweep or lace a valley using techniques similar to those for such details in tiles. It must be remembered that the rainwater run-off load in a valley can be an order of magnitude greater than the rainwater load from a normal slope. Faults in the coverings therefore lead to greater risks of rain penetration. Wherever possible, inspection of valleys should be made from inside the roof void as well as from outside.

Some old slate-covered butterfly roofs encountered by BRE investigators in site inspections had valleys running from front to back of the dwelling: a valley profile must be carefully detailed to avoid water overtopping the gutter lining and seeping into the building below (Figure 3.85).

Manmade slates in valleys should not be bedded in cement mortar, though it is common to see natural slates so bedded. Weathertightness

does not seem to be affected; indeed, mortar adheres quite strongly and assists in resisting wind uplift on slates where mechanical fixings might be awkward.

Hips

Although hip ridges in natural slated roofs are commonly covered with clay ridge tiles on a mortar bed with a hip iron at the eaves, this is a relatively clumsy detail. Neater methods, that tend at the same time to give more durable results, include mitred slate-and-a-halfs overlying Code 4 lead soakers of adequate width (Figure 3.86), or lead and zinc rolls. Care needs to be taken in considering mitred hips in

Figure 3.85: The rafters in this old butterfly roof bear from a separating wall to an internal partition which also carries the valley gutter. Sagging can occur if the partition deforms. Water discharges to a hopper head behind the parapet, but the sides of the gutter are shallow and could be overtopped

exposed locations to make sure that workmanship skills are available to complete a satisfactory job.

Ridges

Ridges can be found in many materials, with practice usually varying according to region. In many parts of England, it is common practice to cap the ridge of a slated roof with fired clay ridge tiles bedded in mortar; a suitable mortar is described in Chapter 3.1. In parts of Scotland, by contrast, lead or zinc ridge cappings are common. While fired clay ridges are usually satisfactory, lead is not suitable for use with some kinds of natural slate since staining may occur. Specifiers should check with the supplier of the slate.

Sawn natural slate ridge cappings are occasionally seen with the courses of slate nearest the ridge bedded in mortar. Manmade slates are normally provided with their own ridge cappings which can be laid dry. Displacement can and does occur if manufacturers' instructions for fixings are not followed.

Roof lighting

Thinner materials will affect detailing around roof lights set into in the plane of the roof. Dormers present no particular problems (Figure 3.87), except that swept dormers in slate are unusual.

Figure 3.88: An industrial building with a large roof area that has led to problems with rainwater run-off at eaves which overtops the gutter

Readers should also see the section on Roof lighting in Chapter 3.1.

Service perforations

Service perforations, such as soil and vent pipes and heating terminals in roofs covered in the largest natural slates and flags, are probably best avoided wherever possible because of the size of the lead slate required and the consequent difficulty of support. The smaller slates do not present these problems.

Rainwater disposal

Slate roofs encountered by BRE investigators of rehabilitation in progress have provided some interesting examples of old practices with rainwater disposal. On more than one site the rainwater gutters were made of softwood timber boards. If there had ever been any lining to these, or even paint, it had long since disappeared. On other examples with parapets at the front elevation, and perhaps the need to avoid rainwater downpipes on the fronts of elegant dwellings, the rainwater was taken in open troughs through the roof void to discharge at the rear of the building.

Water run-off from valleys and large-span roofs (Figure 3.88) may occasionally prove difficult to handle and result in overflow of gutters and staining of wall surfaces (Figure 3.89).

Readers should also see the section on Rainwater disposal in Chapter 3.1.

MAIN PERFORMANCE REQUIREMENTS AND DEFECTS

Choice of materials for structure

So far as the structure is concerned, most of the faults in tiled roofs occur also in slated roofs.

Readers should also see the section on Choice of materials for structure in Chapter 3.1 and the inspection list at the end of that chapter.

Strength and stability

Natural slates chosen and fixed in accordance with BS 5534[17] should withstand most wind loads,

Figure 3.87: Natural slate roofing on an Inverness-shire crofting cottage. The cheeks to the dormers are in single-lapped slate, and the ridges and valleys were originally of zinc but some are now in lead. The roof was originally covered in heather thatch, being re-covered in the early 1930s; since then it has needed only two or three slates replacing, mainly at the ridge

Figure 3.89: Rainwater run-off from a valley gutter has spilled over the stop end of the eaves gutter. Soakers are needed to divert the flow away from the stop end

Figure 3.91: In the same house as shown in Figure 3.90, the purlin has been displaced carrying with it part of its unbonded half-brick supporting wall

especially uplift, experienced in the UK. However, cases have occurred, particularly in Scotland and the north of England, where the slates themselves rather than the fixings have been damaged by wind forces. The slates have proved to be too thin and have snapped across the grain. In these parts of the UK, 'bests' in lower strength materials should not be specified and Spanish light grey slates need to be thicker.

Even the strongest slates, however, will not provide a satisfactory cover if the structure is defective.

A problem that has been brought to the attention of BRE on several occasions has been distress in roofs where the original lightweight covering of slates has been replaced with plain or lapped tiles of much heavier weight without suitable strengthening of the structure (BRE Defect Action Sheets 124[74] and 125[75], and Figures 3.90 and 3.91). The increase in dead load when, for example, Welsh bests are replaced with concrete tiles, can be as much as 100%. Roofs showing signs of distress are not uncommon, and cases of actual collapse have been seen by BRE investigators. Consequently, the Building Regulations (England &

Wales) 2000 AD A[44] now provides for suitable calculations to be made for replacement coverings of more than 15% heavier than the original (or significantly lighter to ensure anchorages against wind uplift are adequate).

In certain parts of the UK, it was customary practice to hang stone slates on riven oak pegs, with the pegs relating to the size and weight of the slate. While there is no question that the pegs are sufficiently strong if they remain in good condition, inevitably, over time, some will suffer decay and slippage will occur. It has become customary, therefore, to fix these roofs with non-ferrous nails. Also, it is not possible to use sarking felts on a pegged stone slate roof, unless thicker battens are used, since the pegs are longer than the normal batten thickness. Such roofs were in the past normally torched with lime mortar gauged with cow hair. Torching is not now recommended.

Concrete slates have been tested in accordance with the now superseded BS 473, 550[76] (replaced by BS EN 490[16] and BS EN 491[77]), although it is not strictly applicable to concrete slates. Nevertheless, some indication may be obtained of the relative strengths of these slates.

Again, although not strictly relevant to the material, fibre-reinforced slates have been tested to BS 690-4[78] and BS 4624[79]; glass-fibre-reinforced slates have been

Figure 3.90: New, heavy, single-lap interlocking concrete tiles (replacing slates) have pushed out the wall plate and eaves. The temporary green cover has been pushed aside to reveal the extent of the damage

tested against BS 6432[80] (replaced by BS EN 1170-3[81]).

Bending strengths of manmade fibre-reinforced slates, generally speaking, are lower than for otherwise comparable former asbestos cement products. They should therefore be handled and installed with due care.

Dimensional stability

While natural slate is dimensionally stable, there can be problems due to impurities that cause some disruption within the material. Tests described in the standards listed above should reveal, in most circumstances, any risks.

Manmade slates can bow or curl after laying, even if recommended slating practice has been followed. Specifiers should seek assurances from suppliers and should preferably use products that have performed well over a period of time.

In the early stages of bowing (Figure 3.92), performance with respect to weathertightness is unlikely to be affected, and the decision on whether to replace depends entirely on visual acceptability. After all, handmade clay tiles are frequently bowed in moderation during manufacture, and this is sometimes considered a desirable feature. However, manmade slate roofs have been seen where the bowing has been enough to straighten rivets, which is quite unacceptable. Manufacturers are well aware of the problem, and hopefully the incidence of bowing has fallen in recent years.

Exclusion and disposal of rain and snow

Natural slate is relatively non-porous, and weathertightness depends primarily on adequate lap and gauge rather than material thickness. Reference should be made to BS 5534[17] and to BS 8000-6[65]. An underlay of suitable material is needed, and boarded roofs should be counterbattened to allow drainage under the slates.

Natural tilestones can be relatively porous and therefore pitch chosen in relation to their thickness is crucial to their satisfactory performance. In those areas of the UK where heavy, thick tilestones are used, pitch may be as low as 25–30°. Where light, thin tilestones are used, pitches should be steeper, ie at least 45° and preferably even steeper. Since tilestones are normally produced in random sizes, no recommended dimensions for width and gauge are given, but local custom is the preferred guide. Lap will depend on the pitch with the greatest lap on the lower pitches.

For those dwellings with slate roofs that were seen in the BRE site inspections (see *Assessing traditional housing for rehabilitation*[1]), the most frequently occurring faults relating to weathertightness are listed in Box 3.12.

A number of concrete slate roofs were seen, several of which had a high proportion of cracked slates; these were found on the shallower pitches particularly, presumably as a result of work on chimneys and television aerials, leading to greater risk of slippage and rain penetration.

Fire precautions

Some examples of generic roof coverings and their likely performance in the BS 476-3[82] external fire exposure test are given in Appendix A to AD B1 and B2 of the Building Regulations 2000 (England & Wales)[44]. For example, natural or fibre-reinforced cement slates carried on sarking, boarding, woodwool slabs, plywood, wood chipboard or fibre-insulating board carried, in turn, on timber rafters,

Figure 3.92: Manmade slates just beginning to bow

> **Box 3.12:** **Weathertightness problems most frequently seen in BRE site inspections of pitched slate roofs**
>
> - Missing slates, caused in the main by corroding fixings
> - Cracked or split slates
> - Hard mortar fillets (used instead of flashings) cracking and disintegrating
>
> - A range of faults associated with sarking, including its absence (Figure 3.93), as described in the section on Exclusion and disposal of rain and snow in Chapter 3.1

Figure 3.93: A slated roof with no sarking showing signs of water penetration, particularly on the tightly built-in wall plate. The ceiling has been stripped and plaster removed from the one-brick-thick wall below prior to rehabilitation

Timber rafters, with or without underfelt, sarking, boarding, woodwool slabs, plywood, wood chipboard or fibre-insulating board

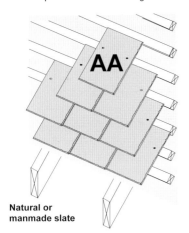

Natural or manmade slate

Figure 3.94: Designation of slates for fire precautions

with or without underfelt, would be expected to give an AA rating (Figure 3.94).

As with tiled roofs (see section on Fire precautions in Chapter 3.1), slated roofs may also be encountered with deficiencies in separating walls (Figure 3.95).

Sound insulation

A pitched slated roof with flat plastered ceiling underneath can give values in the range 30–40 dB(A) depending on the airtightness of the surface and, to some extent, on the weight of the slates. The heaviest tilestones or flags will give the best sound insulation[43].

Energy conservation and ventilation

Cross-ventilation is less easy to achieve in single pitches, and extra ventilation at both high and low level may be necessary to promote a through-flow of air (Figure 3.96). Many older semi-detached and terraced houses had rear extensions where a dual-pitch roof covers two rear extensions. However, with the separating wall between the two extensions built up to, and running the length of, the ridge, for ventilation purposes the effect is the same as for separate single pitches but without the facility to ventilate via the external wall.

Readers should also see the section on Energy conservation and ventilation in Chapter 3.1.

Durability and ease of maintenance

Most roofs using natural slates of UK origin should last at least 100 years (Figure 3.97). Indeed, some roofs using Welsh slate have lasted several hundreds of years. On the other hand, roofs of inferior slate can give problems in a few months. Although relatively infrequent, cases of natural slates becoming friable and disintegrating into powder have been seen by BRE investigators.

With older slates, it is the fixings that usually give problems before the slates themselves deteriorate. Since there are no nibs on slates to provide extra purchase on the

Figure 3.96: A ventilator inserted to achieve cross-ventilation in a cold roof having a separating wall at the ridge. The effect should be equivalent to a continuous 5 mm gap the length of the eaves (Photograph by permission of FCD Harrison)

battens, corrosion of the nails will lead directly to slippage of the slates. This nail sickness can be quite widespread in roofs over about 50 years of age, depending naturally on the specification of the original nails and the pollution conditions that the roof has experienced.

Spraying the underside of unsarked roofs to fix slipping slates should be considered only as a temporary measure before re-roofing. Dampness problems ensue from completely encapsulating the battens, leading to rot.

Figure 3.97: This natural slate, removed from a house after many decades of service, shows little sign of deterioration. There is some discoloration where the part of the slate not covered by another slate is exposed to weathering, but the slate is fit for many more years of service

Figure 3.95: A short-span domestic roof in natural slate. A significant part of the separating wall was missing and an attempt to rectify the deficiency is shown to have been under way

In replacement work, it could be advantageous to use nails of aluminium, austenitic stainless steel, copper or silicon bronze (BS 1202-2 and -3[52]). Nails of steel (whether galvanised or not), aluminium or copper alloys should not be used where battens have been treated with CCA (copper-chromium-arsenic) preservatives.

Disintegration of natural slate begins with powdering, flaking or blistering of the surface. There are two main reasons for this:
- oxidation of iron pyrites,
- sulfur dioxide in the atmosphere combining with rainwater to form sulfuric acid which converts calcium carbonate in slates to calcium sulfate.

Delamination may also occur where clay minerals are present.

Assessments of the durability of natural slate can be made by using the tests of BS EN 12326-1[71].

The water absorption test measures the resistance of the slate to the uptake of water and, hence, its probable frost resistance. Slates should have a water absorption of not greater than 0.3% and the best slates have an absorption of less than 0.2%[67].

The test for pollution resistance by immersion in a bath of sulfuric acid is sometimes dispensed with in the case of roofing slates where extra thickness has been used to compensate for a less durable material, and the slates have a performance record proved in the UK environment. The less durable slates contain impurities of expanding clays, calcite or pyrite, and these are identified by the acid immersion test; they are suitable only for use in areas where pollution from atmospheric sources is low and the performance of the slate has been proved in the UK environment (Figure 3.98).

Since the 1980s there has been a considerable increase in the use of imported slate, mostly from Spain, and more recently from China, at about half the cost (size for size) of the comparable UK product. In a series of tests carried out at BRE, only about one-third of the imported slates tested passed the

Figure 3.98: A roof in imported natural slate showing the initial signs of decay after about five years due to the presence of iron pyrites and other impurities

then current BS 680-2[83] tests. This is not necessarily a bar to their use, however, provided they are carefully selected for the actual situation of use and the degree of pollution to be encountered. As the product quality of even a single quarry can vary considerably, it is most important that the specifier insists on random testing of slate delivered to site, and a check should be made that the material in the thickness chosen is sufficient to withstand the maximum wind-loading characteristic of the area[84].

The mechanisms of deterioration of natural sandstones and limestones is described in *The weathering of natural building stones*[85] (Figure

3.99). In particular, there is a section on the growth of lichens, which may be considered to be a problem in some circumstances. Growths of lichen and moss are best removed by scraping. Fungicides can be used but might alter the appearance of a slated roof considerably.

Some slated roofs seen on BRE site inspections have proved remarkably durable in the face of obvious deficiencies. The moral here is that action may not always be imperative (Figure 3.100).

Manmade slates are available in a wide range of materials (eg fibre-reinforced cements and fillers such as natural slate bound with resins) and it is therefore not possible to give universally applicable guidance on durability to be expected. There is no single standard that covers all of them, though there are individual standards associated with individual materials and third-party certificates cover many. It should be noted, however, that manufacturers' guarantees are usually for 20 years. Although shorter lifespans can be expected than are normally achieved from the best natural slates, premature deterioration may occur with all kinds of slating materials.

Asbestos cement products were first introduced at about the end of the 19th century, but probably very few roofs of more than about 50 years of age are expected to have survived.

Figure 3.99: One or two slates (or tilestones) are beginning to delaminate in this fine old roof over an outbuilding in Caithness. Moss grows in the joints but should not affect performance. The roof-light frame is of cast iron with an integral cast iron apron flashing (Photograph by permission of BT Harrison)

Figure 3.100: This slated porch roof is about 15 years old. The ridge has moved, the laps have been stretched and there is no projection at the verge, yet it is still watertight

Figure 3.101: Loss of surface coating on a manmade slate

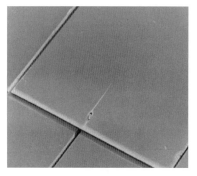

Figure 3.102: The colour of this slate has faded at the edges. There is also a crack originating from the rivet hole in the centre; whether this was induced when laying or fixing rather than being a fault in the component is open to question

Figure 3.103: These slates have discoloured, probably as a result of water seeping into the material at the sides and lower edges, and lying in contact for long periods just under the head lap. There is also fading in the centres

When manmade fibre-reinforced slates were first substituted for asbestos cement slates, a number of failures occurred that gave these manmade slates a bad reputation. While many of these have been resolved there are, however, points to watch for in manmade slate products that have remained on the market. Stability of colour is a problem, although not thought to be widespread; distortion and cracking are more widespread but not in all products. Orientation of the roof does not appear to play a significant role in deterioration, although there is a slight tendency for cracks to occur sooner on south-facing slopes than on other slopes.

Some slates may lose some of their surface coatings but this is relatively rare in BRE's experience (Figure 3.101).

Discoloration can take one of two forms: either pale or dark edges (Figures 3.102 and 3.103). These phenomena are unlikely to affect performance, at least initially, and decisions on replacement are usually based on visual acceptability.

Since slates have no nibs to retain them in place, they rely entirely on the integrity of the nails holding them to the battens. Replacement of the occasional slate is possible by using a tingle inserted under the slate above, and turning up its tail to retain the inserted slate. Proprietary

clips that can be nailed in the gap between the upper slates are also available. Modern manmade slates use a proprietary clip at the bottom edge held in the gap between the slates below.

It is the practice in west Wales coastal areas to render a deteriorating roof to improve weathertightness, but BRE has no detailed information on its effectiveness (Figure 3.104). Durability will naturally depend on the specification for the render as well as how it is applied. The effect of any significant increase in the loading on the rafters should be checked.

WORK ON SITE

Access, safety, etc.
The cutting of fibre-reinforced slates is controlled by health and safety legislation.

Many old roofs in fibre-reinforced slates, particularly asbestos cement slate roofs, will not withstand foot traffic. Even bridging with roof ladders is difficult, and access may demand the use of hydraulic platforms. Removal of old asbestos cement sheets is covered in HSE leaflet HSG 189/2[70].

Readers should also see the section on Access, safety, etc. in Chapter 3.1.

Storage and handling of materials

Some manmade slates are prone to breakages if handled roughly, and care must be taken in unloading and moving packs around sites.

Restrictions due to weather conditions

Readers should refer to this section in Chapter 3.1.

Workmanship

Roofing using natural slates calls for considerable skill. The thickness of natural slates, since they are still cleft by hand, varies considerably. They therefore need to be sorted so that the thickness of slates for use together varies only fractionally; otherwise, adjacent slates of differing thicknesses will leave a slight gap against the slate in the course above, and a path for wind-blown rain and snow to penetrate. BS 8000-6[65], recommends sorting into three or four groups, and that slates narrower than 150 mm should not be used.

If manmade slates are not clipped or riveted down sufficiently, they may rattle in the wind.

Figure 3.104: A rendered roof. Such sand–cement rendering is meant to improve weathertightness on a temporary basis and has been seen in varying thicknesses, from barely a wash to a coat or coats of substantial thickness (Photograph by permission of BT Harrison)

Box 3.13: Inspection of slate and stone tiled roofs

In addition to most of the items listed in Box 3.11 at the end of Chapter 3.1, the problems to look for are listed below.

- Natural slates not obtained from known and reliable sources
- Faults in slates (tapping before use may reveal faults)
- Slates delaminated on existing roofs
- Slate surfaces eroded where lapped
- Slates slipped and nails corroded
- Slates, lap and pitch unsuitable for the exposure likely to be experienced
- Ridges and hips incorrectly fixed
- Asbestos cement slates which are time expired (Figure 3.105)

Figure 3.105: A bungalow, dating from the 1920s, roofed in diamond pattern asbestos cement slates that are now in a fragile condition from carbonation of the cement content

3.3 FULLY SUPPORTED METAL

Data for dwellings in the UK in 1991 having a metal covering for the whole or part of the roof are given in Table 3.8 as recorded in national House Condition Surveys in 1990[5–8]. The data collected for the English House Condition Survey 1991[5] did not differentiate between metal and glass, nor between fully supported and self-supporting metal.

By 2005/6 there were 138,991 (0.7%) dwellings in England that had pitched roofs covered in glass, metal or laminate (again, the data do not differentiate between materials nor between fully supported and self-supporting metal). This is a significant reduction from 15 years earlier. The situation may be further complicated by the recent increase in the number of conservatories with glazed roofs.

By 2005/6 there were 5255 (0.8%) dwellings in Northern Ireland that had pitched roofs covered in glass, metal or laminate (the data do not differentiate between materials nor between fully supported and self-supporting metal), or roughly the same number as 15 years earlier.

CHOICE OF MATERIAL FOR A METAL ROOF

The colour of the weathered surface of a metal roof normally plays a significant role in the choice of metal. The following notes give a brief description of the main properties, together with relevant standards.

Lead
Lead normally weathers to a medium grey following the formation of a coating of lead carbonate; the coating then protects the metal from further corrosion. In

Table 3.8: Reported problems with metal pitched roof coverings in the UK. Data from national House Condition Surveys circa 1990[5–8]

	Number of roofs (whole or part) covered	Proportion with a problem (approx.)
England (metal/glass roofs):		
Houses	341,000	1:6
Flats	70,000	1:10
Wales:	No data	—
Scotland:		
Houses and flats	25,000	1:7
Northern Ireland:		
Houses and flats	6000	No data

polluted atmospheres the carbonate may be associated with sulfates but this does not affect the protection afforded by the carbonate layer. The main Standards for lead are BS 12588[86] and BS 6915[87].

Thicknesses of lead sheet used in roofing are mainly in the range Code 4 to Code 8, with Code 8 more than half as thick again as Code 5; Code 5 should provide adequate durability for most situations.

Occasionally, largely for repairing old cast lead sheet roofs, there is a demand for lead cast in the traditional way instead of milled; more recently, though, industrially produced cast lead sheet has become available.

Copper
Copper sheet used in traditional small bay-size roofing (Figure 3.106) is normally soft tempered whereas sheet used in long-strip roofing, with sheets running the whole length of the slope, is normally one-quarter hard tempered. Thickness of sheet should be not less than 0.6 mm for normal uses in both traditional and long-strip applications. Thicker sheet is available for more onerous service conditions. Lighter gauge sheet has been available in the past, factory-bonded to various substrates. The thickness of lighter gauge sheets has meant, in some cases, that durability has been less than for the standard thickness sheets.

Figure 3.106: Copper-covered roofs on houses in open country. These roofs were laid in 1947 and have given excellent service

Figure 3.107: The protection given to the copper-covered cheeks of this swept valley dormer has inhibited the formation of the green patina

Copper weathers over the years to a blue–green, though in some polluted climates the colour of the patina may be brown or black. This patina, as in the case of lead, forms a protective coating to the metal. Sometimes, the formation of the typical green patina may be delayed for as many as 15 years from laying. Any form of protection to the covering will also delay, and even inhibit, the formation of the patina (Figure 3.107).

The main Standards for construction in copper are contained in BS EN 1653[88] and BS EN 1654[89] and for the material itself, BS EN 1172[90] and BS EN 1652[91].

Zinc

Zinc usually forms a grey-coloured carbonate when exposed to the atmosphere. This carbonate affords protection to the bare metal. The rate of carbonation may be predicted reasonably accurately and the life of the covering therefore also predicted. On the other hand, zinc oxide or hydroxide ('white rust') affords no protection. Zinc is often alloyed with small amounts of titanium.

The main Standards for zinc are CP 143-5[92] and BS EN 988[93]. Type A is used for normal roofing and type B for flashings.

Aluminium

Sheet aluminium is available in various degrees of purity and various degrees of temper (see BS EN 485-1–4[94]. Generally speaking the softest tempers are used for flashings, though the highest purity metal (99.99 %) needs to be used with a slightly harder temper (H2). The harder tempers (eg H4) are used only for machine-formed seams. Thickness should be not less than 0.8 mm. Various finishes of the sheet are available and are dealt with in the section on durability and maintenance later in this chapter.

Aluminium, if unprotected, will form white corrosion products in unpolluted atmospheres. Pre-coated aluminium has become popular and is available in a wide range of colours.

Stainless steel

Stainless steel has for some time been available for roofing purposes in the UK but, so far as is known, it has virtually no use in domestic roofs. It is further described in Chapter 5.1.

CHARACTERISTIC DETAILS

Basic structure

Most metal roofs are laid to comparatively low pitches and any basic roof structure appropriate to loading, span and pitch may be used. The continuous decking required to support the metals in this category will contribute substantially to the overall stability of the roof structure.

The traditional decking material for most fully supported metal pitched roofs is tongued-and-grooved timber boarding. This is superior to plain edge boarding as it is less likely to curl. Other sheet materials are also suitable though they should be thick enough to take the shanks of nails used for fixing the cleats. The deck should be overlaid with a barrier (eg polyethylene rather than felt) to prevent leaching of damaging materials which would attack the metal, then a slip membrane (eg a geotextile permeable membrane) placed between the barrier material and the metal covering.

Main roof areas

All the four metals in common use can be laid to a variety of types of detail, though the standing seam

Figure 3.108: A standing seam roof in pre-coated aluminium. The dark colour of the coating is not in accordance with best practice because of its inability to reflect solar heat and probably contributes to the characteristic slight billowing in the sheets. The gutter needs clearing

and roll cap forms are perhaps the most common and well tried (Figures 3.108 and 3.109). Suitable literature is available from the trade associations, and their comprehensive advice is not given here.

Overhanging eaves

The longer the sheet of metal on the roof slope, the greater the amount of thermal movement to be expected. With large spans this can be considerable and should be calculated to ensure that, for instance, gutters are not overshot in summer rainstorms. Coefficients of thermal expansion are given in Table 3.9. Buckling of sheets along the length of the slope is normally avoided by means of a sliding joint cleat fixing (Figure 3.110).

Gutters for metal-covered roofs are often to be found formed integrally with the roof covering itself and covered in the same material. With copper such gutters are formed by folding and welting the sheet material, though, usually with aluminium, site welding is used. Movement joints will be needed in long gutters of all metals.

Joints along slopes

Metals are laid to standing seam or to batten roll as appropriate.

(a)

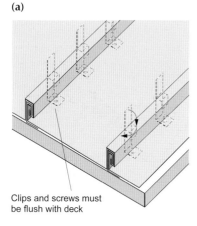

Clips and screws must
be flush with deck

(b)

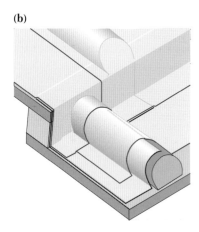

Figure 3.109: Typical metal roof
details: (a) standing seam, (b) roll cap

Figure 3.110: The eaves of a long-strip
aluminium roof before the verge cap is
positioned

Standing seams for pitched roofs
have been seen as low as 25 mm,
though batten rolls tend to be higher
for ease of forming the joints.

Hand forming can be used on all
metals though it is becoming more
common for a degree of machine
forming to be used, particularly
where the material is less malleable.
The seam height with machine-
formed seams is somewhat higher

than that for hand-formed seams.
For all the materials, the seams will
occur at between 500 and 750 mm
centres, depending on the type
of seam adopted, the clip fixing
centres required and the width of
the original material. Manufacturers'
recommendations should be
followed since any stretching of
seam centres with the thinner
materials may leave the panels
vulnerable to wind suction.

Joints across slopes

It is considered good practice
to stagger the positions of all
joints formed across the slope,
so simplifying the forming of
the intersection of longitudinal
and transverse joints. Four-way
intersections are difficult to make.
Ridges in all metals are normally
provided by means of a roll formed
over a timber section, though it
is possible in some materials to
form a standing seam without that
support. Unevenness in the ridge is
more likely to be apparent with this
method.

Lead
Joints will occur at intervals of
1.5–2.5 m, depending on the pitch
of the roof and the thickness of the
lead.

Copper
Double-welted joints on slopes
between 10° and 45°, and single-
welted joints on steeper pitches, are
usual. Long strip sheets are available,
reducing the need for cross-joints.

Zinc
Single-welted joints are usually used.

Aluminium
Although traditionally, lengths
up to 7 m were possible without
joints, practice in more recent years
has allowed far greater lengths
provided thermal expansion can be
accommodated.

Abutments

Upstands should be at least 150 mm
at abutments and flashed over with
a compatible metal (Figure 3.111).
Zinc–lead alloys are used with zinc
roof coverings because of their
better workability.

Figure 3.111: The flashing on this
chimney is less than 50 mm above the
roof slope. A deeper flashing would
have kept the chimney drier from
rainwater splash-up from the shallow
pitch of the roof

Roof lighting

Readers should refer to this section
in Chapter 3.1.

Rainwater disposal

Readers should refer to this section
in Chapter 3.1.

MAIN PERFORMANCE
REQUIREMENTS AND DEFECTS

Choice of materials for
structure

Care should be taken with the
choice of underlays. Under some
conditions it will be necessary to
provide a barrier to contaminants
leaching upwards from, for example,
timber decks. Permeable sarking
felts are ineffective for this role
and a barrier such as 500-grade
polyethylene underneath the paper
or felt underlay may be necessary.
Ventilation to prevent condensation
forming on the underside of the
covering may also be required.

Strength and stability

The provision of anchorages will
need to be considered to hold
down the roof within this category
against wind forces causing uplift.
It is, however, relatively unusual to
encounter fully supported metal
roofs that have been sucked off by
wind action, though damaged roofs
of sheets that are not fully supported
have been seen more frequently.

Since wind suctions are greater
at the ridge, eaves and verges than
over the remainder of the roof slope

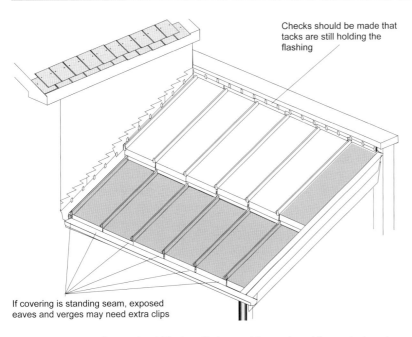

If covering is standing seam, exposed eaves and verges may need extra clips

Checks should be made that tacks are still holding the flashing

Figure 3.112: Extra fixings should be installed at positions vulnerable to wind suction

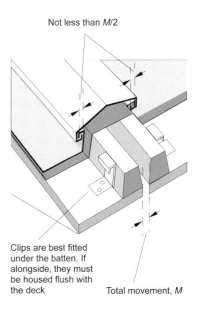

Not less than *M*/2

Clips are best fitted under the batten. If alongside, they must be housed flush with the deck

Total movement, *M*

Figure 3.113: A movement joint cap. Caps must have provision for movement at the clips, otherwise fatigue cracking may ensue

Table 3.9: Thermal movements of metal			
	Coefficient of linear thermal expansion $\alpha = °C \times 10^{-6}$	Movement in mm for 1 m length for temperature range (winter or summer) of:	
		75 °C	90 °C
Lead	30	2.2	2.6
Copper	17	1.4	1.6
Zinc	23–33	1.7–2.6	2.3–3.3
Aluminium	24	1.7	2.3

(Figure 2.50), it is good practice to provide more fixings in appropriate positions (Figure 3.112). This may necessitate reducing the width of sheets at verges and providing clips at more frequent centres in the last bay or two adjacent to the verges.

A deck that is of comparatively soft material may not provide sufficient support to the metal covering to resist the loads imposed by maintenance foot traffic. Also, when a plumber is on a roof to lay the metal and form the seams, his movements demand a reasonable degree of stability in the deck material.

Dimensional stability

Long lengths of roofing in one piece will move by significant amounts at the eaves, as calculations from Table 3.9 above demonstrate. The clips holding the seams in place will need to provide for the covering to slide,

in some cases by several millimetres. Furthermore, roof coverings will creep down the slope if fixings are insecure. One of the most important factors in controlling thermal movement is to keep the bay sizes small. Another is to make provision for movement at movement joints which does not involve continuous flexing of the metal (Figure 3.113).

The harder tempers of some of the metals, when inadequately fixed, may suffer from metal fatigue. One such example is aluminium which can suffer from star cracking.

Exclusion and disposal of rain and snow

No particular problems have been reported to BRE with any of the metals in common use provided that the relevant code of practice and the recommendations of trade organisations have been followed.

Entrapped water

Metal-covered roofs are prone to problems from entrapped water, and from moisture being pumped through the joints. The section on pressure differentials causing water 'leaks' in Chapter 2.4 describes the phenomenon. Lead is particularly susceptible to deterioration from condensation forming on its underside, though other metals are less susceptible (see the section on Ventilation later).

Control of solar heat gain

Particular care should be taken with dark-coloured finishes (eg on aluminium sheet).

Fire precautions

For an AA rating, the Building Regulations 2000 (England & Wales) ADs B1 & B2[44] recommend fully supported sheet metals carried on timber rafters, with or without underfelt, on:
- decks of timber joists and tongued-and-grooved or plain edged boarding, or
- steel or timber joists with decks of woodwool slabs, wood chipboard, fibre-insulating board or 9.5 mm plywood, or
- concrete or clay pot slab (in situ or precast) or non-combustible

deck of steel, aluminium or fibre-reinforced cement (with or without insulation).

These specifications can all be expected to achieve an AA rating when tested to BS EN 13501-1[95] or B_{ROOF}(t4) (European Class)[96, 97] (Figure 3.114).

Lead sheet on plain edge boarding may give only a BA rating. To achieve AA, tongued-and-grooved boarding must be used.

Sound insulation

Sound insulation of low-pitched metal-covered roofs depends mainly on the type of material used for the deck[43]. Approximate values are given in Table 3.10.

Drumming caused by rain or hail falling on sheet metal roofs may be a problem in some circumstances. The conventional solution is to sandwich an underlay and a layer of fibreboard between the metal roof covering and the structural deck, the fibreboard being next to the deck to provide a degree of damping of the noise. This gives only a marginal improvement.

Noise may also be caused by continuous thermal expansion and contraction of the harder tempered sheets, particularly if they intermittently stick on the cleats instead of sliding freely.

Durability and ease of maintenance

Metals should be selected for their compatibility with other materials as well as with other metals, and for their inherent resistance to corrosion. From the corrosion standpoint, metals perform best in a clean, dry environment. It is never

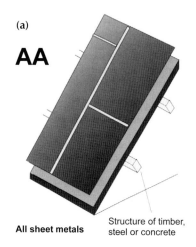

(a)

AA

All sheet metals — Structure of timber, steel or concrete

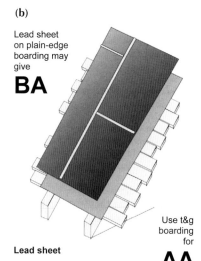

(b)

Lead sheet on plain-edge boarding may give

BA

Lead sheet — Use t&g boarding for

AA

Figure 3.114: Designation of metal roof coverings for fire protection. All sheet metals on timber joists with tongued-and-grooved (t&g) boarding, steel or timber joists with decks of woodwool slabs, wood chipboard, fibre-insulating board or 9.5 mm plywood give AA rating

possible in roofing to achieve these ideal conditions but the design should be such as to prevent, as far as possible, the lodgement of dirt, dust and moisture on the surface. While it is impossible to avoid exposed surfaces getting both wet

and dirty, designs should be free-draining to reduce the duration of wetness.

Metal roof sheets may be subjected to attack from certain kinds of preservative in treated timber decks and a sheet of polyethylene laid over the deck before fixing the covering will prevent most attacks from this source. A sheathing felt of class 4 to BS 747[40] does not give the same degree of protection from deleterious substances migrating from below, though it has been seen in common use. This type of felt is not included in BS 8747[42] with the intention that its description will be transferred to a mastic asphalt standard (see also BS EN 13707[98]).

Other things being equal, unprotected metals have a life related to their thickness, though other factors, including pollutant levels, will play a significant role.

Lead

Cast lead sheet has been known to last for 400 years, though the thinner milled sheets will have a shorter life, probably upwards of 100 years depending on thicknesses and pollution levels. Lead is also susceptible to condensation. It is sensitive to acid waters, and run-off from areas of roofing covered with algae; certain kinds of wood such as cedar or oak may cause local deterioration of the surface. Lead is also attacked by alkalis and should therefore be protected with a coat (eg of bitumen) where it is embedded in mortars or renders; for instance, with flashings (Figure 3.115). Other points to note include:

- fixings for lead sheet should be of protected copper or stainless steel,
- lead sheet can be patched, for example to lead-burn in a patch, but the fire risk needs to be considered. However, patching with bitumen is not recommended.

Copper

Copper generally forms a durable roof and examples in the UK have endured for more than 100 years. Points to note include:

Table 3.10: Approximate values for sound insulation for low-pitched metal-covered roofs

Type of structure/deck	Sound insulation (dB)
Timber structure supporting timber boards:	
without ceiling	20–25
with ceiling	30–35
Concrete deck:	
of weight not less than 200 kg/m²	45
of weight not less than 200 kg/m² with roof lights covering approximately 10% of the roof area	25

Figure 3.115: This lead abutment upstand, which has been rendered over to reduce its visual impact, is likely to give long and trouble-free service. Although hidden in the photograph, the lead under the render was coated before the rendering was completed

- the metal is unaffected by alkalis and is attacked only slowly by acids,
- salts, such as those carried in the winds around coasts, can accelerate corrosion,
- the run-off from copper roofs can cause problems for other metals, particularly zinc and steel. The run-off may also disfigure stone or brickwork,
- nails used for fixing copper cleats should be of copper or brass,
- attempts artificially to introduce green patina on copper roofs are unlikely to be entirely successful,
- if bays are too large, fluctuations in wind suction can cause fatigue and consequent cracking of the sheets.

Zinc

Zinc sheeting, provided it is used in substantial thicknesses (eg 14 gauge), can be expected to last around 50 years if it is not exposed to industrial pollution and if the roof pitch is around 45° or more. The metal is attacked by both acids and alkalis, and protection of the metal by means of a coat of bitumen is necessary where the metal is embedded in mortar. Zinc can be soldered, so, in theory at least, can be patched, though this is unlikely to be economic in the long term. Cleats for zinc roofing should be of the same material as the covering and fixed using galvanised steel nails.

Aluminium

Aluminium is a suitable material for the external finish of a roof as it has a low rate of corrosion, but it must be expected that its appearance will deteriorate with time. As the white corrosion product forms, the surface will become rough, and will entrap dirt and become unsightly. Pollutants and contaminants will also collect on the surface, and there will be a risk of accelerated corrosion.

Aluminium is normally available in mill finish and in various grades of anodising or paint finish. Mill finish has the shorter life with the protected surfaces lasting longer. The durability of mill finish depends on the amount of washing the metal receives. Examples of mill finish metal inspected by BRE which have been regularly washed only by rainwater, though in a relatively unpolluted urban environment, have lasted practically unmarked for 12 years, and were expected to give satisfactory service for at least another equivalent period.

A protected aluminium-covered roof can be normally expected to last upwards of 50 years depending on alloy, finish and position.

The surface of aluminium can be protected by anodising. The coating is brittle, and cannot be used therefore when post-forming the metal. These anodic coatings can be coloured. Anodising produces a layer of oxide on the surface of the aluminium which in practical terms delays the onset of corrosion. The corrosion product of aluminium is white; therefore, if dark-coloured anodic coatings are employed, when deterioration occurs the corrosion is readily seen. If seen it is probably an indication that the anodised surfaces have not been regularly washed.

Aluminium sheeting, pre-coated with organic coatings, is now available. The visually acceptable life of the material is essentially the life of its organic coating, but it depends on the type of coating, environment, thickness and bond. In aggressive environments it will be necessary to consider eventual painting or repainting to restore appearance. The use of organic dyes should be avoided if medium- to long-term colour retention is

required. Aluminium extrusions as well as sheet can be coated with organic coatings, the more successful of these being the fluorocarbon PVDF (polyvinylidene-fluoride or sometimes abbreviated to PVF2), and polyester-based materials.

Other points to note in relation to aluminium include:

- the metal is attacked slowly by most acids; it does not last very long in the vicinity of flue terminals,
- it is also sensitive to alkalis, and needs protection (eg using a coating of bitumen) where it is to be embedded in mortar,
- aluminium must also be isolated from contact with other metals (except zinc) or corrosion will ensue,
- fixings for aluminium-covered roofs should be of aluminium or one of its alloys.

Ventilation

Care may need to be taken, in the specification of fully supported metal roof coverings, to ensure that there is some ventilation over the top of the thermal insulation and underneath the outer sheeting on its deck. Otherwise, the sheet may be at risk of deterioration through inadvertent rainwater penetration to its underside. The mechanism of failure is that solar heat expands and expels the trapped air; on cooling during rainfall, a partial vacuum is created that can suck in rainwater over the top of any vulnerable standing seams. The popular name for this effect is 'thermal pumping'. Prevention is by means of small shielded ventilators penetrating the roof covering.

Although it may be thought that stainless steel and copper sheets do not need this protection, being inherently more durable than, for instance zinc, there is always the risk of rainwater deterioration of any timber supporting structure.

WORK ON SITE

Access, safety, etc.

Readers should refer to this section in Chapter 3.1.

Workmanship

All nails and screws used for fixing decks, cleats and sheets should be punched or screwed flush to prevent any raised contact with the underside of the sheet and, consequently, perforation of the metal as it expands and contracts.

When lead-burning on site, care must be taken to prevent fire damage to underlying timbers.

Supervision of critical features

When repairing copper roofs on site, it should be remembered that the metal must be annealed since it will have work-hardened during the original process of fabrication.

Zinc can be soldered on site for repairs, although care needs to be taken to avoid antimony in the solder since this will cause embrittlement of the surrounding metal.

Further information on suitable techniques for repairing old metal-covered roofs is given in *Practical building conservation*[99].

Box 3.14: Inspection of metal roofs

In addition to most of the items listed in Box 3.11 at the end of Chapter 3.1, the problems to look for are listed below.

- Seams unfolding, particularly at verges
- Creep down the slope if bay sizes are too large or fixings insecure
- Fatigue cracking of movement joint caps
- Condensation and corrosion on the undersides of metal sheets
- Fixings proud of decks wearing through outer coverings
- Aggressive run-off from adjacent surfaces
- Incompatibility of metals in lightning protection, and inadequate sheathing

3.4 FULLY SUPPORTED BUILT-UP BITUMEN FELT AND FELT STRIP SLATES

Data for dwellings in the UK in 1991 having a built-up felt covering for the whole or part of the roof are given in Table 3.11 as recorded in national House Condition Surveys in 1990[5–8].

The data from the English House Condition Survey 1991[5] did not distinguish between pitched and flat felted roofs.

By 2005/6 the numbers of dwellings in England whose pitched roofs were entirely covered in built-up felt numbered 49,196, which, even allowing for the fact that the former figure combined flat and pitched, probably represents a considerable reduction from 15 years previously (Figure 3.116).

By 2005/6 the figure for roofs in Northern Ireland wholly covered in felt had dropped to 871, or about

Figure 3.116: A steeply pitched felt roof on a house. Tails were not properly stuck down leaving them vulnerable to wind damage

half of the number obtained 15 years earlier.

BRE investigators have seen, in the main, two kinds of felt roofing used as the cap layer on pitched roofs.

Granule-surfaced, bitumen-fibre-based felt (BS 747[40] class 1B)
This felt was often also used until the 1970s as a single layer on short-life buildings or on the porches of permanent dwellings. This material is to be distinguished from bitumen-mineral-fibre laminate (class 1E) which also has been used in a single layer, or as a cap sheet to multi-layer, but proved to be more durable.

Glass-fibre-reinforced bitumen strip slates (or tiles)
These slates are normally made from felts manufactured to BS 747[40] class 3B or 3E. They are surfaced with mineral granules in a wide range of subdued colours and shapes. These slates are, confusingly, commonly referred to as 'shingles' in the trade. BS EN 544[100] is the new European standard covering these products. They are more widely used on mainland Europe.

Pitched roofs built since the 1980s may also be found covered with built-up felts containing materials defined and included in a revised version of BS 747[40], ie BS 8747[42], or BS EN 13707[98]. These materials are dealt with in Chapter 4.1 since they more often than not seem to have been encountered by BRE on nominally flat roofs rather than pitched.

Also noted for convenience in this chapter is the use of concrete gutters to pitched roofs, normally lined with bitumen sheet.

CHARACTERISTIC DETAILS

Basic structure
Decks for pitched roofs which are covered in felt are normally of plain-edged or tongued-and-grooved (t&g) timber boarding, nailed to the rafters. Tongued-and-grooved is more durable than plain-edged boards since the edges of the boards are held in the same plane, and ridging does not occur to split the felt. Decks of 9.5 mm plywood have also been used. BRE investigators on site inspections have not seen woodwool slabs used on the roofs of dwellings for a considerable number of years, though, no doubt, examples remain in service.

Occasionally, the surveyor may come across a domestic roof of parabolic shape, constructed of plywood suspended from a ring beam; the single layer coverings (dealt with in Chapter 5.2) are suited better, perhaps, to this kind of roof because of methods of laying. Roofs of this design and specification are rare.

So far as pitch is concerned, the steeper slopes usually provide better durability though some of

	Number of roofs (whole or part) covered	Proportion with a problem (approx.)
Table 3.11: Reported problems with built-up felt roof coverings in the UK. Data from national House Condition Surveys circa 1990[5–8]		
England (pitched and flat roofs):		
Houses	2.85 million (18% of total stock)	1:6 or 1:7
Flats	442,000 (23% of total stock)	1:10
Wales:	No data	—
Scotland (pitched roofs):		
Houses and flats	19,000	1:3
Northern Ireland (wholly covered roofs):		
Houses and flats	1600	No data

the heavier cap sheets have shown evidence of movement and rucking (eg at service penetrations). This problem has been chiefly evident on south-facing slopes.

Overhanging eaves
Since bitumen tiles are not sufficiently strong to cantilever over gutters, the eaves need to be constructed in sheet material dressed over the fascia to form a welt. Eaves details for sheeted pitched roofs are similar (Figure 3.117a).

Verges
Because of the difficulty of forming a welt on the slope so that it remains true to line, it has been common practice to finish the verge with an aluminium trim of suitable shape (Figure 3.117b). Any joints in the trim must be fixed down securely immediately adjacent to the join in order to avoid differential movements tearing the felt underneath. The trim should be kept to lengths of less than 1.2 m to avoid major thermal movements and be fixed at 300 mm centres.

Roof lighting
Roof lights for pitched roofs are available with prefabricated trim suitable for installation into built-up felt (Figure 3.118). Those seen by BRE have not shown problems.

Figure 3.118: Roof lights in the plane of the roof can be installed with little difficulty, but the detail that is normally required does rely on sticking the felt slates to the flange of the trim

Rainwater disposal
Although in the past they could be found on pitched roofs covered in different materials, proprietary precast concrete gutters fitted at the head of the wall often have been used on roofs of fully supported felt construction. These gutters have been used also as permanent shuttering for lintels carrying the eaves and wall plates over first floor windows. They were lined with bitumen felt and, in the experience of BRE investigators, many leaked. The cause of the leaks was splitting of the linings which were fully bonded to the concrete, the joints of which had opened.

MAIN PERFORMANCE REQUIREMENTS AND DEFECTS

Choice of materials for structure
The fire performance of decks that have been used under pitched felted roofs are listed in Tables 3.12 and 3.13. Provided the fire requirements can be met, there is no doubt that the most popular choice in the past has been tongued-and-grooved timber boarding.

Some of the main problems occurring with this category of roofing relate to movements in the deck and the inability of fully bonded built-up felts to cope with movement. However, provided partial bonding is adopted, there should be relatively few problems, especially with polyester felts given their greater inherent extensibility. Those products that have SBS (styrene-butadiene-styrene) polymer-modified bitumen can cope with usual movement induced by substrates and could theoretically be fully bonded. However, the possibility of molten bitumen falling between the timber joints means that partial bonding by nailing would be preferred. BS 8747[42] covers the specification and selection of reinforced bitumen membranes and should be consulted. Older roofs, however, often show bonding problems (Figure 3.119).

Exclusion and disposal of rain and snow
The materials in built-up roofing in theory provide a continuous and impervious barrier to rainwater. Provided the surfaces are not perforated there should be no problems. However, damage to many felted roof surfaces of comparatively low pitch from maintenance foot traffic (eg when fixing aerials on chimneys) has been a source of problems in the past.

Energy conservation and ventilation
Condensation has actually been seen by BRE investigators to occur on the underside of built-up felt roofing over a sloping chipboard deck in the cold deck roof configuration. The chipboard would not last long in

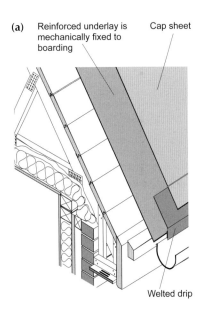

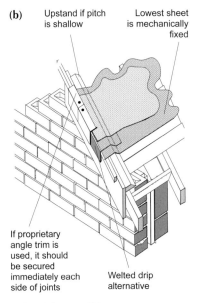

(a) Reinforced underlay is mechanically fixed to boarding Cap sheet

Welted drip

(b) Upstand if pitch is shallow   Lowest sheet is mechanically fixed

If proprietary angle trim is used, it should be secured immediately each side of joints   Welted drip alternative

Figure 3.117: Typical details for a built-up covering: (a) eaves, (b) verge

Figure 3.119: Ridging and billowing of built-up felt roofing over a comparatively short-span roof caused by unsatisfactory bonding

these conditions. The main reason is a lack of ventilation at the eaves, and is most serious when external temperatures are at their lowest.

The construction of a warm deck roof, as described in Chapter 2.11 in relation to flat roofs, would be a satisfactory remedy.

Control of solar heat

The provision of a reflective upper surface to the built-up roof is one of the most effective ways of ensuring a reasonably trouble-free life for the covering; Chapter 1.5 broadly describes a traditional mix. However, alternative and more durable finishes of equivalent high reflectivity have been developed, some integral with the material, and high reflectivity should be specified.

Fire precautions

Most BS 747[40] class 1 and 2 bitumen felts have generally ceased to be available from UK sources. These materials have been used on many roofs in the past (some of which still exist) and reference to them remains in the building regulations for England and Wales; for completeness they are therefore included in Tables 3.12 and 3.13. These Tables indicate what the materials were expected to achieve when tested to BS 476-3[82].

Asbestos-based bitumen felt strip slates, to BS 747[40] class 2E, with an underlayer of class 2B, carried on timber rafters and timber boarding,

plywood, woodwool slabs, wood chipboard or fibre-insulating board were expected to achieve an AA rating when tested to BS 476-3[82].

Strip slates of class 1 or 2 carried on timber rafters and timber boarding, plywood, woodwool slabs, wood chipboard or fibre-insulating board were expected to achieve a CC rating.

Built-up felt coverings on pitched roofs with a class 1E upper layer (now obsolete) and underlayers of class 1B with a minimum mass of 13 kg/10 m^2, were expected to achieve the performance shown in Table 3.12. Substitution by class 3 (glass-fibre) is intended to give equal or better performance.

Built-up felt coverings on pitched roofs with a class 2E upper layer and underlayers of class 1B with a minimum mass of 13 kg/10 m^2 were expected to achieve the performances shown in Table 3.13.

Glass-fibre-based built-up felt coverings on pitched roofs with a class 2E upper layer and underlayer of class 2B, and laid on any of the decks quoted in Table 3.13 is intended to achieve AB.

Built-up felt coverings on pitched roofs with a class 3E upper layer and underlayer of class 3B or 3G, and laid on 6 mm ply, 12.5 mm chipboard, 16 mm tongued-and-grooved softwood, or 19 mm plain-edged softwood boarding is intended to achieve BC.

Table 3.12: Expected fire performance of built-up felt coverings in pitched roofs: BS 747[40] class 1E (obsolete) upper layer

Structure on which the covering is laid	Expected performance
Plywood, 6 mm	CC
Wood chipboard, 12.5 mm	CC
(Finished) tongued-and-grooved softwood, 16 mm	CC
(Finished) plain-edged softwood, 19 mm	CC
Screeded woodwool slab	AC
Profiled fibre-reinforced cement deck	AB
Profiled steel with or without insulating board	AC
Profiled aluminium deck with or without insulating board	AC
In-situ concrete or clay pot	AB
Precast concrete	AB

Table 3.13: Expected fire performance of built-up felt coverings in pitched roofs: BS 747[40] class 2E (obsolete) upper layer

Structure on which the covering is laid	Expected performance
Plywood, 6 mm	BB
Wood chipboard, 12.5 mm	BB
(Finished) tongued-and-grooved softwood, 16 mm	BB
(Finished) plain-edged softwood, 19 mm	BB
Screeded woodwool slab	AB
Profiled fibre-reinforced cement deck	AB
Profiled steel with or without insulating board	AB
Profiled aluminium deck with or without insulating board	AB
In-situ concrete or clay pot	AB
Precast concrete	AB

Durability and ease of maintenance

Unfortunately, roofing felts previously included in BS 747[40] (meant to be used in single layers only for short-life buildings) have also been seen by BRE investigators in use for coverings to porches and canopies in permanent housing, with consequent poor performance. An indication of durability can be obtained by fatigue testing of samples. BRE tests show, typically, endurance of 150 cycles for BS 747[40] class 3B and 6200 cycles for class 5B polyester-based felts. In comparison, polymer-modified felts will give in excess of 140,000 cycles, and PVC and EPDM (ethylene-propylene-diene monomer) in excess of 1 million cycles.

With bitumen felts on a pitched roof it is wise always mechanically to fix the first layer. Faults seen on site include cupping of the underlying timber boards causing ridging in the covering. Where, in the past, sheets have been fully bonded to the deck, thermal and moisture movement of the boards has caused splitting of the covering. Bitumen felt tiles are normally not fully bonded and tend not to show this particular problem, though they do curl upwards in winds if the tails are not well stuck down.

Since there could be a risk of condensation occurring on the underside of the weatherproof layer, it is advisable for the deck to be durable. In particular, consideration should be given to specifying plywood decks to the following standards.
- Canadian Douglas fir and softwood plywoods, Finnish conifer plywood and Swedish softwood as specified in BS 5268-2[63] and having a water and boil proof (WBP) bond.
- Plywood treated with a 10-minute dip in an organic solvent preservative to Type F/N of BS 5707[101] or treated with any of the preservatives

complying with Section 8, Clause 63 of BS 5589[102] so that the outer veneers are completely penetrated. This ply should also have a WBP bond as above.

Existing precast concrete eaves gutters built into the head of the wall can be relined. A rubberised plastics sheet, which can be hot-air welded on site to form a lining in one sheet the whole length of the gutter, including the forming of rainwater outlets, should provide a suitable repair. The material is flexible and not fully bonded to the gutter so risk of splitting is reduced.

Repairs to built-up sheets are dealt with in Chapter 4.1.

WORK ON SITE

Access, safety, etc.

Readers should refer to this section in Chapter 3.1.

Storage and handling of materials

The risk of cracking membranes as they are unrolled increases with falling temperatures. The European Union of Agrément Method of Assessment and Test No 27[103] requires that a material shall be capable of being unrolled without

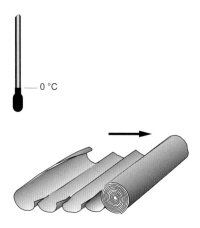
— 0 °C

Figure 3.120: The risk of cracking of oxidised bitumen felts and some PVC materials increases with falling temperatures

damage at the lowest anticipated temperature of use (Figure 3.120).

Shelf life of products should be rigidly observed.

Forming of upstands will often entail gentle heating of the materials before bending into shape, otherwise cracking could occur.

Surfaces should be dry, otherwise hot bitumen used for fusing successive sheets together will bubble. The bitumen should be hot at the time of laying (200–220 °C).

Restrictions due to weather conditions

Flexible roof membranes will not perform well if laid on a wet or frosted deck: blisters will result. High winds also increase the risk of damage to membranes during laying, and low ambient temperatures may cause lack of adhesion when adhesives are hot-applied.

It is recommended that work on felting roofs should not proceed during the following weather conditions:
- air temperature is less than 5 °C (heavy sheets) and 0 °C (light sheets),
- any rain, snow, sleet, hail or dew,
- wind 3-second gust speed is greater than 16 m/s.

Primers should be dry within 24 hours at the temperatures at which they are to be used.

Workmanship

Many of the roofing systems described in this chapter depend on a trained work force for their successful installation. Particularly, care must be taken to ensure that the overlap between successive sheets is adequate and in accordance with the manufacturer's recommendations. In practically all circumstances a minimum side lap of 50 mm is necessary[103]. The sheets should not be contaminated with soil from, for example, inadequate storage arrangements.

Box 3.15: Inspection of built-up bitumen felt and felt strip slate roofs

In addition to most of the items listed in Box 3.11 at the end of Chapter 3.1, the problems to look for are listed below.

- Surface damage from maintenance traffic
- Surface damage from movement of decks
- Blistering, splitting and rucking of the coverings
- Lack of extra courses of nails where pitches are above 60°
- In cold weather, cracking of sheets during laying
- Condensation under coverings
- Loss of solar protection
- Poor adhesion of covering to decks
- Flashings absent or damaged at abutments
- Poor detailing of fillets
- Inadequate flashing to upstands

3.5 RIGID SHEETS

Corrugated iron was invented and patented in 1829 and soon built up a large market for both roofing sheets and for whole buildings for the export market. The sheets are rolled in a variety of profiles to provide stiffness in their length. The sheets overlap at the sides and the length and are fixed to the purlin or roof structure by nails, screws or spikes. Protection of the metal was commonly a red lead paint, a bitumastic paint or coated by a galvanising process. They were often used for over-roofing thatch. Figure 3.121 shows sheets that are well over 100 years old and the ingenious stretching of the metal over the eyebrow. For comprehensive information see *Corrugated iron and other ferrous cladding*[104] and BS 5427-1[105].

Metal-covered roofs spanning directly between purlins have been used in housing since the 1914–18 war. Roofs of corrugated galvanised steel have generally been regarded with disfavour in the UK. However, since the late 1950s, there has been a gradual fall in prejudice against sheet metals and many of these roofs have been constructed using alternative materials at relatively low pitches, notably sheet aluminium. Roofs covered with these materials were insulated at ceiling level, as with conventional roofs, in a cold-deck situation. Insulated profiled coated steel or coated aluminium have not often been used in domestic construction and are therefore described in Chapter 6.

Another use of uninsulated and not-fully-supported sheets in the domestic context is in over-roofing of troublesome flat roofs (*Overroofing: especially for large panel system dwellings*[106]). Over-roofing is defined as the covering of a former flat roof by a completely new pitched roof. Although the number of dwellings where any over-roofing has been done is small, there seems to be considerable interest in the technique by owners of large stocks of dwellings; the main considerations are therefore described in this chapter.

Over-roofing is only one of many options for remedying faults in flat roofing and there are many techniques from which to choose. The solution has usually been adopted where the building is to remain in use for a period exceeding 20 years. Such over-roofing can dramatically improve the appearance of a run-down block (Figure 3.122).

CHARACTERISTIC DETAILS
Most of the solutions seen by BRE involve shallow pitched sheet steel or aluminium with a variety of finishes, although a variant, a steel tile made to look like concrete, has become available. Tiles are also available with integral battens, also in steel, spanning between rafters. This type of material has the great advantage of being light in weight and can be used in replacement works where the original building may be insufficiently robust to carry a heavyweight roof.

Basic structure
Timber purlins spanning directly between masonry cross-walls have formed the basic structure of most examples seen by BRE, the form being single or dual pitch. Hips are not easy to construct with this type of support.

Most applications of the over-roofing technique have used either a timber or a lightweight steel dual-pitch roof structure, though some single pitches have been seen. Some original structures may provide the opportunity to carry a new roof structure off intermediate

Figure 3.121: Old thatch over-roofed with corrugated iron

Figure 3.122: Over-roofing completed on a six-storey block of flats

loadbearing walls which will reduce the spans and so permit lighter members (Figure 3.123).

Rainwater disposal

On system-built blocks of flats, parapets and gutters have often been prone to leaks (Figure 3.124). Although there were existing rainwater disposal systems on the original roofs, in most cases of this type seen by BRE it has proved necessary to design a completely new system.

Readers should also see the section on Rainwater disposal in Chapter 3.1.

MAIN PERFORMANCE REQUIREMENTS AND DEFECTS

Strength and stability

The rigid sheet form of roofing, including over-roofing, was of most concern to BRE following the gales of 1990. The vulnerability of these roofs, if they were insufficiently strapped down, has been known since the 1960s, and it is clear that the lessons of earlier years were not learned (Figure 2.4). It is to be hoped that they have been learned by now.

The addition of a pitched roof to what was originally a flat roofed building will alter the wind-load

characteristics and will necessitate a new assessment. New roof structures may also add considerably to the dead load on the existing structure.

If the over-roofing is being applied to an existing concrete flat roof that has deteriorated, it will be important to check that adequate fixings can be made into sound material. There is discussion of the durability of concrete roofs and of assessment procedures in Chapter 5.2.

Energy conservation and ventilation

Practically all older roofs modified by over-roofing will be of the cold-deck variety with minimal thermal insulation at ceiling level.

In site surveys carried out by BRE investigators, cases of condensation occurring on new construction have been seen on pitched uninsulated metal decking, pointing to leakage of water vapour into the cold roof spaces which were not always ventilated. Sheet metal roofs are even more airtight than tiled or slated roofs, so require adequate provision for ventilation at eaves.

In over-roofing, where additional thermal insulation is added over the top of the existing flat roof covering, the existing covering will usually act as a vapour control layer on the correct side of the insulation (ie the warm side); this will reduce the

Figure 3.123: Steel framework for over-roofing supported off loadbearing cross-walls in a previously flat-roofed building

Figure 3.124: Rain penetration at deteriorating parapets is a major source of water penetration in system-built flats

amount of water vapour percolating into the roof void and needing to be removed by ventilation. Ventilation will be required by building regulations and, in any case, will be a wise precaution, especially if the existing roof covering has been perforated. Adding thermal insulation at the eaves will avoid thermal bridging.

Fire precautions

The Building Regulations 2000 (England & Wales) AD B2[44] states that single-skin profiled sheets of galvanised steel, aluminium, fibre-reinforced cement sheets or pre-painted (coil-coated) steel or aluminium with PVC or PVDF (polyvinylidene-fluoride) — either without underlay or with an underlay of plasterboard, fibre-insulating board or woodwool slab, and carried on a structure of timber, steel or concrete — can be expected to achieve AA rating (Figure 3.125).

Double-skin profiled sheets without an interlayer or with an interlayer of resin-bonded glass-fibre, mineral wool slab, polystyrene or polyurethane, and carried on a structure of timber, steel or concrete, will also achieve AA rating.

Sound insulation

Single-skin, thin sheet metal over-roofs are likely to make only marginal improvements to the sound-insulation value of an existing heavy roof. Where the original roof is light in weight, the noise of heavy rain and hail may still be noticeable to occupants of dwellings below. Remedial possibilities are confined in practice to an extra layer of plasterboard on ceilings. If the whole roof needs to be replaced, it may be possible to incorporate some form of damping.

Readers should also see the section on Sound insulation in Chapter 3.3.

Durability and ease of maintenance

Plastics-coated sheet steels used in domestic situations should give up to 25 years life, though there have been examples that have deteriorated badly before this time. There is further discussion of this topic in Chapter 6.1.

In cases where metal sheets had been used in domestic pitched roofs, it has been noted that cut edges where required were not always treated as manufacturers recommended, with consequences for durability of the sheeting (Figure 3.126).

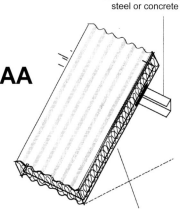

Figure 3.126: Failure of the polymeric coating over the surface and at the edges of profiled metal sheeting. (There is also mechanical damage to the sheeting)

Modifications for climate change

Metal roofing can be considered either as fully supported, such as lead or copper sheets, or as self-supported, such as profiled metal sandwich panels. The durability of some of these products is related to the type of surface coatings applied to them in the factory. When a factory-applied coating fails due to weathering it can be replaced or repaired by a product applied on site. If the life of the coating is affected by climate change, the effect is probably increased maintenance costs due to more frequent applications.

The durability of metal roofs has not been easy to predict in the past. The expectation is that corrosion would be predominantly due to moisture and heat. However, there are other factors and inter-reactions present that can affect durability both positively and negatively. Because these influences are not well understood, it is not possible to make predictions about how climate change scenarios might affect the durability of metal roofs. Simplistically, drier summers may reduce corrosion rates while the wetter winters may increase them with a net yearly rate showing little change.

Profiled sheet of galvanised steel, fibre-reinforced cement, or pre-painted (coil-coated) steel or aluminium (PVC or PVDF)

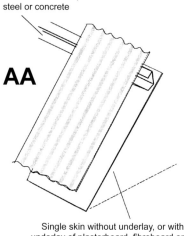

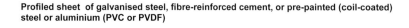

Structure of timber, steel or concrete

Structure of timber, steel or concrete

AA

AA

Single skin without underlay, or with underlay of plasterboard, fibreboard or woodwool

Double skin without interlayer, or with interlayer, or with interlayer of resin-bonded glass fibre, mineral wool slab, polystyrene or polyurethane

Figure 3.125: Specification for self-supporting sheets to achieve AA fire designation

WORK ON SITE

Access, safety, etc.
Readers should refer to this section in Chapter 3.1.

Restrictions due to weather conditions
It is recommended that work on over-roofing using large sheets should not proceed when wind conditions exceed a wind 3-second gust speed greater than 12 m/s.

If it is necessary to breach the existing weatherproof covering to provide support for the new structure, precautions against rain penetrating the building, especially when it continues in occupation during progress of the work, will be required.

Workmanship
The sequence of laying sheets and the application of seals is important for excluding the weather. It is recommended that side laps be made facing away from the direction of the prevailing driving rain (Figure 3.127). BRE investigators called onto sites to examine problems have noted a number of cases where this had not been done.

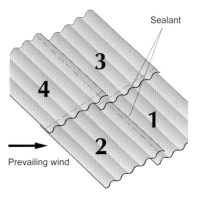

Figure 3.127: The suggested sequence of lapping sheets and positioning of seals

---

Box 3.16: Inspection of rigid sheet roofs

In addition to most of the items listed in Box 3.11 at the end of Chapter 3.1, the problems to look for are listed below.
- Lack of strapping down to resist wind suction
- Condensation on the undersides of sheeting
- Lack of durability, especially of cut ends of sheets
- Incorrect sequence of lapping of sheets

3.6 SHINGLES

Cedar as a roofing material goes back to Biblical times: 'So he built the house, and finished it; and covered the house with beams and boards of cedar.' (I Kings 6:9), and there is evidence that shingles were in use in Rome in 600 BC (*A history of building materials*[107]).

The main advantages of shingles are lightness of weight and their attractive appearance.

Oak shingles have been used in the UK for 2000 years though they are not often seen now. Other species from which shingles have been made include larch and pine. The Norwegian stave churches give impressive evidence of durability, though perhaps the shingles used on their roofs cannot quite match the 600 years or so of life of the basic structures.

Imported thin shingles or thicker shakes have been available in the UK made from western red cedar. In the USA, metal shingles have been used, though they have not been imported into the UK in any significant quantity. Shakes have become available in the USA made from wood particles bound with Portland cement. These are somewhat thicker than cedar shingles, but have a similar appearance. In addition they have improved performance against external fire exposure and an expected longer life but are heavier than the product they are designed to replace.

Shingled roofs are confined, in the main, to timber-built dwellings but a number of timber rural public transport shelters have also been roofed in shingles.

Although most timber-covered roofs are made of shingles, occasionally boarded roofs will be found with the boards laid either across the slope (Figure 3.128) or down the slope.

CHARACTERISTIC DETAILS

Basic structure
Shingles are relatively lightweight, and some economy in structural roof timbers (when compared with, say, a tiled roof), may result from their use. However, there will be a corresponding increase in the need for strapping of wall plates and jack rafters to resist wind suction.

Apart from requirements for steepness of pitch (described in Main roof areas below), the structural requirements are much the same as for other kinds of domestic roofs.

Main roof areas
Shingles are normally relatively long in relation to their width: of the order of 300–600 mm to 100–325 mm. Thickness will vary:
- oak shingles are the thickest, ie up to 15–20 mm at the butt,
- cedar are thinner, ie 8–10-mm at the butt, tapering to about 2 mm, though cedar shakes are thicker.

Oak shingles need to be laid at a pitch of not less than 45°, though cedar shingles may be laid at a pitch of 30° or even lower. Both seem, though, to perform better on roofs of pitch steeper than 45°. They have in the past normally been laid directly on boarded sarkings without sarking felt, twice-nailed in the centres and fixed to a double-lap configuration, but consideration could be given to the use of a felt underlay counterbattened to provide ventilation.

Eaves
The principles for detailing of a shingle roof at the eaves are similar to those for ordinary tiled or slated roofs dealt with in Chapters 3.1 and 3.2, though due allowance needs to be made for counterbattening (Figure 3.129). With the very light weight of most shingles, however,

Figure 3.128: Ship lap section, timber-boarded roofs on the south coast of England. The lichen growing on this roof is yellow-coloured and normally grows on south-facing slopes

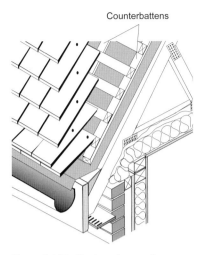

Figure 3.129: Design of eaves for a typical shingled roof

close attention must be given to soundness of nailing to resist wind uplift.

Verges

Bargeboarded verges are preferred for shingled roofs and they need to be nailed securely.

Ridges

Ridges may be found site-assembled from standard shingles laid horizontally to single lap, close-butted at the ridge. This detail does not provide absolute assurance of rain resistance, since edge nailing the thin shingle to permanently close the overlap in the joint is not feasible. Nailing to battens placed on each side of the ridge piece is necessary, otherwise any warping will tend to open the joint (Figure 3.130).

Roof lighting and service perforations

Roof lights and service perforations are best avoided, if at all possible,

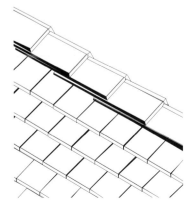

Figure 3.130: A ridge of a typical shingled roof

in view of the risk of corrosion of metal flashings from run-off water. Non-metallic flashings should be considered where perforation of a shingled roof is unavoidable.

Dormers are practicable, but conventional roof windows laid on the slope may be difficult because of the thinness of the coverings. Again, consideration of the durability of flashings is required.

Rainwater disposal

The corrosive characteristics of rainwater draining from any form of wood used on external surfaces of roofs should be taken into account in choosing rainwater gutters and downpipes. Plastics rainwater goods probably provide the most appropriate answer. Also, water may be stained by tannins in wood and consequently made unsuitable for rainwater recycling.

MAIN PERFORMANCE REQUIREMENTS AND DEFECTS

Choice of materials for structure

Shingled roofs are laid on battens and rafters which can be carried on any of the roof configurations described in Chapter 3.1.

Exclusion and disposal of rain and snow

Shingles depend to a large extent on their tapering thickness to reduce capillary attraction of rainwater between the units.

Traditionally, shingled roofs have been laid without felt underlays on the assumption that this practice could well impede the free flow of air essential for removal of moisture. However, if counterbattening is adopted, this risk will be obviated and underfelting would be beneficial. If closeboarding is used, the sarking should be laid on the closeboarding, then the counterbattens laid before battening. If sarking felt alone is used without closeboarding, the felt should be laid over the rafters and then the counterbattens immediately over the top of the rafters.

Shingles laid to pitches over 30° should have a gauge of at least 125 mm. With the use of

conventionally sized shingles of around 300–400 mm length, this will give a covering of not less than three thicknesses of the material. Flatter pitches should be laid to a gauge of 90 mm or so which should give a thickness of not less than four layers over the roof.

Energy conservation and ventilation

Although the wood of the shingle is a fair insulator against heat loss, the fact that ventilation rates need to be high to assist durability means that roofs covered in shingles will not usually be expected to perform any better than conventionally covered pitched roofs for control of heat losses. A shingled roof therefore, in practice, usually means a cold deck roof design.

Fire precautions

The Building Regulations 2000 (England & Wales) AD B1[44] sets the same limitations on shingles as AD/BD/CD or E_{ROOF}(t4) if performance under BS 476–3[82] or DD ENV 1187[97] cannot be established. However, pre-treatment with a fire retardant, which should not leach out through the action of rainwater, may give an improved designation. In practice, wood shingles that have not been treated have been restricted for the most part to residential buildings or buildings not subject to building regulations. When following AD B1, untested shingles cannot be used, for example, on roofs within 6 m of a relevant boundary and then only in small areas. There are fewer conditions on roofs at least 12 m from the relevant boundary. The Building Standards (Scotland) Regulations 2004[45] and the Building Regulations (Northern Ireland) 2000[46] impose slightly different requirements.

Durability and ease of maintenance

Principles of modern building[43] indicates that heartwood oak shingles might have a life, in normal conditions, of around 50 years, and cedar shingles could well last longer. Current views are that these lives are rather optimistic.

Figure 3.132: Extreme moss growth on shingles

Figure 3.131: Shingles made from western red cedar, *Thuja plicata*, taken from a 30° pitch roof after about 20 years' service. The wood is beginning to twist, and there has been erosion of approximately 1–2 mm where the surface is unprotected by the layer above. There has also been erosion under the join of the overlying shingles. The growth on the surfaces of the shingles is lichen

Shingles made from other species of wood (eg larch) will have shorter lives (probably around 20 years in good circumstances); longevity will depend also on the efficacy of preservative treatments, if applied.

When the use of western red cedar rose in the UK in the early 1920s, because of the recognised natural durability of the timber and the outstanding reputation it enjoyed in Canada and the USA, a life of at least 50 years was expected for cedar shingled roofs. The cedar obtains its natural durability from phenolic-type compounds that are toxic to wood-rotting fungi (Figure 3.131).

By the 1950s, some instances of premature decay were observed in cedar shingle roofs, sometimes less than 10 years old. This decay occurred more frequently in the western parts of the UK where the climate is wetter. What had happened was that the natural protection compounds were being leached out by heavy rain. Since about 1963, therefore, it has become standard practice to treat cedar shingles with a preservative giving them a life of the order of 50 years. Assurances will need to be sought from manufacturers or importers of shingles on the likely

efficacy and side-effects of any chemical treatment proposed.

Shingles and shakes in all species of woods ideally should be riven on at least one of the flat faces in manufacture, and not sawn, although sawing of the edges is of less consequence for durability.

As has been said, providing an air space underneath the shingles by means of counterbattening encourages the free circulation of air with its beneficial effects on drying and, ultimately, on durability. The unevenness of any riven face contributes to the provision of gaps, and should be sufficient to prevent capillary action between the overlapping faces of adjacent courses.

For maximum durability, both oak and cedar shingles should be twice-nailed with stainless steel, phosphor bronze or silicon bronze nails into drilled holes; some staining might occur with copper or brass nails. Galvanised iron nails are attacked by

acids in the timbers and therefore ought not to be used. Ring-shanking the nails will afford better resistance to wind suction.

Durability is very much dependent also on roof pitch and how well the shingles are made and laid.

In a few cases investigated in the early 1990s, where comparatively large, unexplained holes were found in shingles, BRE concluded that they were the work of woodpeckers.

See also Chapter 2.9 for methods of dealing with growths of lichens and algae on shingles (Figure 3.132).

WORK ON SITE

Access, safety, etc.

Since shingles are relatively thin, maintenance traffic could damage them and consideration should therefore be given to the use of crawl boards or roof ladders.

Readers should also refer to this section in Chapter 3.1.

Workmanship

Shingles should not be tightly butted to each other, but a gap of 2–3 mm left between them to allow for moisture movement, depending on moisture content measured at the time of laying; the drier the shingle at time of laying, the more spacing is needed. Although, in theory, a shingle could split if nailed at each side, in practice this does not seem to happen on a significant scale. If for some reason shingles need to be laid very wet, specifiers might consider pre-drilling holes, slightly oversized to allow for such movements.

Box 3.17: Inspection of shingled roofs

In addition to most of the items listed in Box 3.11 at the end of Chapter 3.1, the problems to look for are listed below.

- Rafters not clipped or strapped
- Lack of counterbattening
- Splitting of shingles
- Incorrect nailing patterns
- Wood rot
- Growth of lichens and algae
- Relatively large holes in shingles, possibly made by woodpeckers
- Lack of preservative treatment

3.7 THATCH

Thatch as a material for covering roofs has been used in the UK since the Iron Age. It is mainly used in domestic construction (Figure 3.133), although there are a few examples of comparatively large buildings roofed in thatch: parish churches in east Suffolk (Figure 3.134) and many farm buildings such as barns.

Hastings, writing in 1985, estimated that there were at least 50,000 thatched buildings in Britain, of these, 24,000 were listed buildings[108]. Nevertheless, it is a matter of common observation that thatched roofs tend to occur in particular localities, normally in the more sheltered and drier areas of the UK such as East Anglia, and certain parts of southern England and the Midlands. But, paradoxically, they are also used extensively in some of the wetter areas, for example in Northern Ireland and western Scotland, west Wales and the south west of England.

It is not known how many thatched roofs there are currently in the UK, although the English House Condition Survey for 1996[109] recorded 23,000 dwellings roofed

Figure 3.134: A church nave roof in east Suffolk, rethatched in 1971. In spite of covering with wire netting the ridge will weaken with time, releasing pressure on the thatch beneath which is then either blown or washed down the slope. Birds may also contribute to the damage

with thatch, while the 2005/6 data[3] recorded 34,505 which represents a growth of approximately half as many again as 10 years earlier. In 2005/6 the Northern Ireland House Condition Survey[4] recorded only 81 thatched roofs on houses. Contrary to common belief, thatch is not a material only of historic significance: it is being applied to many new buildings.

CHARACTERISTIC DETAILS

Techniques, and to some extent materials, vary according to locality or region, and it is not possible in this section to provide more than an outline description. Most examples in England are of water reed or special varieties of straw grown for the purpose. Occasional examples of heather, bent, bracken, iris, juniper, dock, rush, turf or broom may be found, occasionally in conjunction with clay, especially on farm buildings in Scotland and

Northern Ireland. Concealed fixings may normally be of hazel or willow; mild steel has increasingly replaced iron hooks. Roped roofs may use tarred twine, coir, heather, manila and straw, sometimes in conjunction with nets or chicken wire. Scallops of hazel, willow, or rope can also be found. Chicken wire weighted with stones is a common form of roof restraint in Scotland.

Further advice on thatching can be obtained from the Society for the Protection of Ancient Buildings (SPAB) and the Rural Development Commission (who have jointly published *The care and repair of thatched roofs*[110]), the Historic Scotland publications, *Thatch and thatching techniques – a guide to conserving Scottish thatching techniques*[111] and *Scottish turf construction*[112], and from specialist consultancies.

Local planning authorities should be consulted before work

Figure 3.133: A thatched roof held down by twine and batten

on thatched roofs is undertaken, particularly where stripping of the old thatch is proposed or where covering or style is to be changed. Some old roofs may be protected.

Basic structure

Pitch varies according to material and locality. Within reason, the pitch should be as steep as possible to promote quick run-off; in England more than 45°, though shallower pitches are often recorded in Scotland and Northern Ireland. Thatchers in different parts of the UK will be used to slopes of particular angles more than others, and may find it strange to work on slopes with which they are unfamiliar. The proposed thatcher should therefore be consulted at the design stage of a new roof.

A wide variety of basic structures has been used in the past: coppice pole rafters were popular in England in medieval times and may sometimes still be found. Scottish and Irish roofs tend to have an undercloak of scraws (turves) but basic structures of wattle, straw mat, straw rope or a combination of any two are common in specific areas.

Dormers and other projecting features

Thatch performs best on uninterrupted roofs. In new work (eg extensions) where thatch is intended, chimneys placed in the middle of roof slopes should be avoided wherever possible. All changes of plane of the roof surface (eg in valleys and over dormers) should be as gentle as possible (Figure 3.135).

Overhanging eaves

Eaves projections for thatch should be as wide as possible, especially where the walls below are of a vulnerable material such as rammed earth, clay or chalk. It is normal practice in some localities to leave a sharply pointed eaves projection at the extremity of the overhang (see Figure 3.134) and in other areas to reshape or round it (Figure 3.136), either by driving up the stems with a leggett (a flat-bladed implement) or by shaving. Because of the nature of the material and the natural drainage pattern, most of the rainwater will drip from a zone some 50 mm wide at the extremity of the eaves from where it can be blown inboard by the wind. Eaves projections using water reed, particularly on larger span roofs, can sometimes be greater than 450 mm with considerable advantages from both drainage and appearance points of view, but there is a practical limit on cantilevering the thatch.

Soffits at eaves should be protected with netting only, to encourage drying out of the thatch. A tilting fillet of around 75 mm height will usually be found at the wall head. Traditional Scottish thatched roofs of turf and heather do not have overhanging eaves (Figure 3.137); in some cases the run-off is directed over the hearting (infilling between the leaves) of the stone external wall by means of a puddled

(kneaded) clay course where the wide wallheads are exposed.

Verges

The stems in verges abutting parapeted gables, or in Scottish roofs with bargeboards, can be laid in line with the slope; but the last few bundles (sometimes called nitches or sheaves) in overhanging verges which need to be trimmed back over bargeboards are best laid to an angle so that the water is directed away from the wall surface. In this detail, no lengths of stem are shown on the verge: only trimmed stem butts dressed back with the leggett to a square surface. The overhang is normally less than that at the eaves even though the wall below needs just as much protection (Figure 3.136).

Ridges

English ridges are normally of straw or sedge, laced with hazel or willow. Timber, turf and concrete are more common in Scotland. In straw or sedge ridges, the stems forming the ridge may either be wrapped over the ridge in an inverted V shape or the bundles may be jointed at the apex, depending on materials and local practice. Many Scottish roofs have no defined ridge, the thatch being rounded in cross-section over an apex roof or over curved yokes with the thatch continuing uninterrupted.

Figure 3.135: Water reed thatch round a dormer in Suffolk. When completed, only the butts of stems will show (Photograph by permission of BSH Muckley)

Figure 3.136: A water reed thatched roof with a crow-stepped gabled verge. This roof is not typical of the area in which it is situated, Ross and Cromarty, and surveyors should be aware that thatchers sometimes travel long distances for work

Figure 3.137: An Isle of Skye black house roof (rethatched only a short time before the photograph was taken), roped and weighted with stones (Photograph by permission of BT Harrison)

Figure 3.139: Gale damage to a thatched roof

Rainwater disposal

Rainwater gutters for thatched roofs are rare though a number of examples have been examined. Those seen depended on rise-and-fall brackets extended the full projection of the eaves: in one particular case by as much as 600 mm for which the brackets appear to have been purposemade by a local blacksmith. Although the brackets were sagging significantly, there was nearly 300 mm of rise and fall still available in the threaded upstand to enable drainage falls to be maintained. In other cases examined, the thatch appeared to have been trimmed back to a less deep overhang in order to use standard brackets with a smaller extension (Figure 3.138).

Thatch is normally laid in layers to total thicknesses of around 300–600 mm with rainwater run-off occurring in a zone some 50 mm deep from the edge of the eaves. While it might be feasible, no doubt,

Figure 3.138: Eaves trimmed back to permit rainwater guttering to be installed on a house in Norfolk

to design rainwater gutters to catch most of the run-off from thatch, the gutters would need to be much wider than those normally used on other kinds of pitched roofs in order to catch all the water running within the thickness of the thatch. Guttering might also be considered to spoil the appearance of the roof, and presumably that is why gutters are normally dispensed with for the majority of thatched roofs.

For these reasons, a generous overhang at the eaves becomes a necessity if rainwater blowing back onto the external walls is to be reduced to a minimum. A further need is to provide generous drainage gutters on the surface of the ground or paving underneath the eaves, and channelling or broken surfaces that will minimise splashing of rainwater onto the lower surfaces of walls. See also Chapter 2.3.

MAIN PERFORMANCE REQUIREMENTS AND DEFECTS

Choice of materials for structure

Old roofs built to be thatched may not be as robust as modern roofs; for example, spacing of rafters may be at wider centres than would now be required under building regulations.

Readers should also refer to the section on Choice of materials for structure in Chapter 3.1 and the recommendations made earlier in this chapter regarding steepness of pitch.

Strength and stability

Few problems seem to exist because of the ability of most kinds of thatched roofs to resist wind uplift. This is probably due to their surface permeability, although wind damage has been known to occur (Figure 3.139). Roofs covered with fine mesh wire netting will fare better in exposed areas of the country than those not covered, but there may be other considerations which militate against wire netting.

In exposed parts of Scotland and Northern Ireland, thatch used to be held down with ropes pegged to the stone walls or weighted with large stones (see Figure 3.137).

Exclusion and disposal of rain and snow

Whatever pitch is determined for a thatched roof (for watertightness and durability the steeper the better), the stalks or stems that form the covering lie at shallower angles than the roof pitch because of the method of laying them (Figure 3.140). The principles on which thatch operates to keep out the rain are:

- first, that there is a great number of stalks in the path of any droplet penetrating the outer layers, and
- secondly, droplets penetrating the outermost layers of the stalks will track a greater distance horizontally than they will vertically.

So, unless wind blows the droplets back up the slope, drainage is assured. A check that drainage is still satisfactory is best made under

(a)

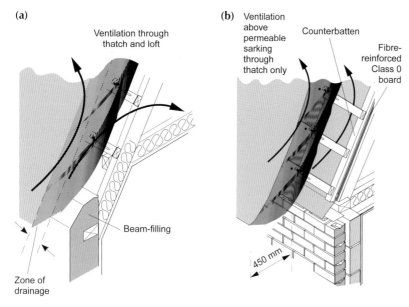

Ventilation through thatch and loft

Beam-filling

Zone of drainage

(b)

Ventilation above permeable sarking through thatch only

Counterbatten

Fibre-reinforced Class 0 board

450 mm

Figure 3.140: Eaves details: (a) traditional English and lowland Scots thatched roof, no sarking, (b) underdrawn thatched roof with fire-resistant boarding and permeable felt underlay

the **leeward** eaves after a rainstorm: the thatch at the wall head should be reasonably dry. If any underfelt is present, and is wet, it may be a sign that the rain has penetrated that far and the covering needs attention, although the wetness might be caused by summer condensation (described in the section on Energy conservation and ventilation, below).

Abutments and chimney stacks provide the greatest problems with weathertightness. Stacks should preferably be on the ridge so that complicated back gutters and their associated 'flashings' or drips can be avoided. Where the stack is on the slope, it will be necessary to provide a slightly wider back gutter to the stack than is normal with tiled roofs because of the need for a larger overhang of the thatch.

Energy conservation and ventilation

There is very little thermal insulation value in most kinds of thatch since air currents can pass through freely. In a traditional roof without sarking, ventilation through the thatch and across the loft space is probably more than sufficient to remove condensation in the roof void, though no measurements have yet been made by BRE to check this aspect of performance. With a boarded roof, ventilation paths

to the interior of the roof space will need to comply with national building regulation requirements.

The major perceived risk with thatch is fire, so limited or non-combustible board materials are often placed directly below thatch to reduce the risk of fire spread. Although some of these boards are vapour permeable, they may reduce the natural ventilation or air movement through the thatch. This can reduce the rate of drying, resulting in increased deterioration of the thatch. In addition to this, recommendations for increasingly higher levels of insulation can result in insulation being placed below the thatch. This again can restrict air movement behind the thatch; compounded by lower temperatures, there is a risk of early deterioration.

Impermeable sarking membranes, such as polyethylene and bituminous felt, are often introduced over the rafters to provide initial weather protection before and during thatching. This membrane is often left in position after thatching in the belief that it will reduce the risk of rain penetration. Impermeable membranes interfere with the natural transfer of water vapour through the thatch. Water vapour condensing as moisture either on this membrane or within the depth

of the thatch will result in fungal decay and early deterioration of the thatch.

Where a membrane is needed below the thatch, for example to prevent detritus and thatching materials dropping into the roofspace, this membrane should be permeable. Impermeable membranes, such as polyethylene and bituminous felt, are discouraged.

The membrane is draped over the rafters to allow a clear cavity of nominally 50 mm between the membrane and the underside of the thatch (Figure 3.141). Where the risk of condensation is high, this void or cavity should be ventilated at the eaves by a continuous air gap at least equal to a continuous strip 25 mm wide on both elevations (Figure 3.142).

Early deterioration of relatively new thatch has been observed

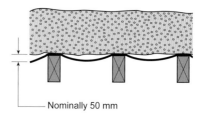

Nominally 50 mm

Figure 3.141: Flexible membrane draped over rafters

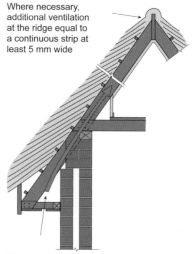

Where necessary, additional ventilation at the ridge equal to a continuous strip at least 5 mm wide

Eaves ventilation equal to a continuous strip at least 25 mm wide on both elevations

Figure 3.142: Ventilation to thatched roof with rigid boarding

in roofs opened up for repair or inspection. It is likely that this has been compounded by the issues highlighted here, for example the use of impermeable underlays or non-combustible boarding[113].

It has become more common for newly thatched roofs to be felted and boarded before thatching, which does at least afford temporary protection to the building. Any such protection should be vapour permeable — there have been problems with the bitumen-loaded impermeable felts and impermeable proprietary plastics membranes. However, there are other drawbacks; in particular, there is less ventilation than in a traditional unfelted roof to enable the drying out of both thatch and the underlying roof timbers, so monitoring of the performance of such roofs is advisable. Where underdrawing is used, counterbattening the roof to increase the amount of ventilation over the underlay will help the drying process. Traditional Scottish roofs should not be underdrawn.

So far as ventilation of the roof void is concerned, there may be conflicting requirements. It is important to avoid excessive water vapour within the roof void for the sake of both the thatch and the roof timbers; the void under a thatched roof could be at a slightly higher temperature, under certain conditions, than voids in slated or tiled roofs, and condensation, therefore, is less likely to occur. Ventilation under the eaves might exacerbate fires generated internally, and is best avoided unless the thatch is underdrawn with a Class 0 board (see *Thatched buildings. New properties and extensions (The 'Dorset' Model)*[114]. Special dispensation may be needed for new thatched cold roofs without sarkings to reduce the amount of additional ventilation which otherwise would be required by building regulations at eaves (Figure 3.143).

Rigid board or lining materials fixed along the line of the rafters should be at least of limited combustibility and can be fixed on top, below or between the rafters. A clear cavity of at least 50 mm between the top of the board

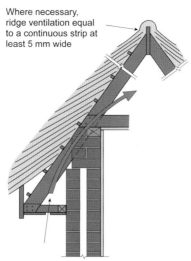

Where necessary, ridge ventilation equal to a continuous strip at least 5 mm wide

Eaves ventilation equal to a continuous strip at least 10 mm wide on both elevations

Figure 3.143: Ventilation in a conventional thatched roof space

and the underside of the thatch is recommended (Figure 3.144). If there is an additional risk of condensation, or if the cavity could become restricted by the thatch slumping, ventilation should be provided equivalent to a 25 mm continuous strip at the eaves. Additional ventilation should be considered at ridge level (see Figure 3.142). If the rigid board is fixed to the top of the rafters, counterbattens and battens may be needed to achieve the recommended depth below the thatch.

The horizontal ceiling below a thatched roof space should incorporate a board having at least a limited combustibility rating. Where practicable, a vapour control layer should be introduced to reduce vapour diffusion into the colder roof space. This may be particularly important where

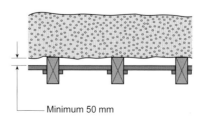

Minimum 50 mm

Figure 3.144: Rigid boarding below thatch

additional thermal insulation has been laid along the horizontal ceiling and there are showers or kitchen areas directly below. Vapour control layers are not an alternative to the recommendations for ventilation, rather an addition to them.

A vapour control layer should be incorporated directly behind the internal lining. This can reduce the amount of moisture reaching the back of the thatch. It should not be relied on as an alternative to the recommendation for a cavity and ventilation, but as an addition. Mechanical extract ventilation should be installed to remove moist air at source.

Where thermal insulation is required along the line of the rafters, it is recommended that a clear cavity at least 50 mm between the insulation and the underside of the thatch (Figure 3.144). In addition to ventilation at the eaves, where it is practicable, additional ventilation should be considered at the ridge at least equivalent to a 5 mm continuous strip (see Figure 3.142).

There is some indication that summer condensation can occur under thatch that has been felted underneath. Strong sun on south-facing thatch can drive water vapour from the thatch inwards, where it will condense on a cool membrane. This is the most likely explanation of wetness under the thatch when no rain has fallen recently.

During periods when warm moist air is being driven to the outside, there may also be an increased risk of condensation. This may be a particular problem where the thatch is laid tight, as with reed.

Fire precautions

Since medieval times, thatch has been whitewashed as a fire-preventive measure or plastered with a clay slurry gauged with lime. It is possible to treat thatch with chemical fire retardants though the treatment can affect durability and is said not to be liked by craftsmen. The protection may also be washed out in a relatively short period of time.

There is no evidence that fires occur more frequently in dwellings having thatched roofs than in dwellings having roofs of any other kind of covering. The risks from fires started internally must be approximately the same as for other roofs, though it is a wise precaution, in the case of unboarded roofs, to warn occupiers that roof spaces should be kept clear of detritus falling from the covering; regular vacuum cleaning of the loft interior, thermal insulation permitting, is good practice. It is also a wise precaution to check electrical installations periodically, and to make sure that chimney stacks and flues are maintained in good condition: sparks have ignited thatch. In fact, sparks and incandescent material from bonfires can set fire to thatch. Consequently, occupiers of thatched buildings need to be aware of these risks and should be warned to be watchful.

Spark arrestors fitted to chimneys may give a measure of protection from sparks falling on the thatch, indeed may be required, but they need constant maintenance if they are to function as intended. In the absence of maintenance, they become clogged and cease to function effectively, and the risk of fire increases. The highest risk occurs with wood-burning appliances.

While there may be a perceived higher risk externally from radiation from fires in adjoining property, this is taken care of (so far as new-build is concerned) by building regulation requirements on proximity to boundaries. In England and Wales, guidance on minimum distance from boundaries is given in Table 5 of Approved Document B1[44], clause B4. No new building of more than 1500 m³ can have a thatched roof which would rule out a building, such as that shown in Figure 3.136, unless it can be shown that the proposed thatch satisfies the requirements of Approved Document B. In the absence of test evidence, thatch is treated the same as other materials of AD, BD, or CD designation. Repair or identical replacement of existing roofs is not subject to building regulations provided that thermal performance is not improved.

For new thatched roofs, the ceiling under the loft space can

provide sufficient protection for the occupants to escape from the building if the thatch catches fire. However, a fire-protecting material such as a rigid fibre-reinforced Class 0 sarking board will improve the behaviour of the roof. The board is laid over the rafters, felted, counterbattened and battened in the usual way to hold the bundles of thatch, as illustrated in Figure 3.140. It may also be both possible and beneficial to underdraw an existing roof with board, either under or between the rafters, depending on the roof configuration and whether or not the building is listed. A fixed sparge pipe at or near the ridge has been suggested in order to maintain appropriate moisture levels in hot weather, though this practice has not been proved. Such a system may assist in protecting the roof from sparks. Generally, no special precautions are necessary for lightning protection, but if protection is required it must be fixed to chimneys.

Preliminary results from full-scale fire testing of thatch roofs with a non-combustible board directly below the thatch suggest that there is no additional fire-spread risk arising from this ventilated cavity. On the contrary, results indicate that the fire spread and resultant damage is significantly less.

Advice concerning fire precautions for thatched roofs can be obtained from the Rural Development Commission, the Society for the Protection of Ancient Buildings and the fire safety departments of local fire brigades.

Durability and ease of maintenance

In the UK, thatch can be expected to last a minimum of 20 years in the wetter parts of the UK if of straw and around 40 years if of reed. In drier areas, life expectation could be twice as long. Accurate figures cannot be given for the west Highlands and Western Isles where top dressing tends to be undertaken annually or biannually, though there is some indication that turfed and heather-thatched roofs in Scotland can last as long as 40 years or more. There is also a suspicion that some

thatched roofs over black houses (thatched houses, usually without chimneys, where the smoke from peat fires found its way through the roofs) have lasted much longer than that. This may not altogether be unexpected since creosote is a distillation of wood tar.

Thatch seems to deteriorate first on southward-facing slopes, though the reasons usually put forward to explain this tend to be speculative rather than definitive.

Inspection and repair intervals for thatched roofs should not exceed seven years, especially if bird nuisance is not controlled by means of protective wire netting. Thatch normally erodes gradually in service, but new material can be inserted by a skilled craftsman.

It is possible to repair straw thatch piecemeal rather than renewing the whole roof, although in such repairs the newer material inserted into the old is visually obtrusive for a time (Figure 3.145). Adela Wright describes, in *Craft techniques for traditional building*[115], the process of repair of straw thatch known as 'stobbing': the insertion of bundles of new stems into the old layers and shaving afterwards to conform to the slope.

It is considered to be good design in new work that dormer windows

Figure 3.145: Inner layers of thatch on an ancient Carmarthenshire clom-walled (cob-walled) house. The haphazard arrangement of bundles of straw is evidence of past clumsy attempts at repair

are not placed too close together, so that the intervening thatch is more exposed to air currents and given a better chance to dry out between rainstorms. However, no thatch should dry out completely; it has been suggested, even, that moisture content may need restoring by sparge pipe or hosepipe in periods of exceptionally dry weather.

Timber for use under thatched roofs can be treated with preservative: the situation is little different from roofs covered in other materials. Advice concerning timbers in all situations will be found in BS 5589[101].

Unprotected mild steel rod used for fixings could, in the damp environment of thatch, easily be less durable than brotches (staples) of hazel or willow, particularly if the roof is situated near the coast. (Figure 2.38 can be used to assess the risk.) Marram grass and manila twines used for fixings have proved to be insufficiently durable. Plastics string stretches, leaving the thatch bundles slack.

Ridges will deteriorate quicker than the main slopes, with expected lives of 10 to 15 years, and it is usual therefore to renew them several times in the life of a covering.

Lichens, algae and moss growth will often be found on thatched roofs, particularly on a roof that is overhung by trees (Figure 3.146). Whether these growths are detrimental to the performance of roofs is perhaps debatable, though they do have a profound effect on appearance. Moss growth is probably instrumental in holding water against the top surface of the thatch, preventing its drying out and hence accelerating deterioration rates. Lichens produce a less dense growth and do not have the same 'poulticing' effect.

If moss growths become excessive, they are best removed

Figure 3.146: In the foreground, moss growth on a thatched roof under trees. In the background the roof, which is not overhung by trees, is quite clear

by careful scraping, though this task is not easy if the roof is covered with wire netting which should be removed before tackling the moss. Chemical treatments may be an appropriate alternative (BRE Digest 370[116]) but should be used with great care, and specialist advice should be sought (they are not liked by the thatcher). The traditional method of inhibiting organic growth was by fixing a copper rod on each side of the ridge but BRE has no experimental evidence as to the effectiveness of this practice on thatch.

WORK ON SITE

Access, safety, etc.
Thatched roof coverings should not be climbed over. Access to chimneys for maintenance of spark arrestors or television aerials can damage the thatch unless access is specifically provided for or roof ladders or platforms are available.

Readers should also refer to this section in Chapter 3.1.

Restrictions due to weather conditions
Since most thatch needs to have a moderate moisture content in service, laying in slight rainfall (eg not exceeding 0.5 mm/hr) should be satisfactory. However, water reed should always be laid in totally dry conditions.

Workmanship
Durable thatching is very heavily dependent on the quality of workmanship as well as the quality of materials, and great care should therefore be taken over selection of a thatcher. The thatching trade associations which began in 1947 were formed mostly from the existing major thatching families, who were predominant in the trade. Today's members are trade descendants from these origins and have learnt their skills through a combination of intense training and on-site experience. The National Council of Master Thatchers Associations represents the county-based Master Thatchers Associations[117].

Box 3.18: Inspection of thatched roofs

In addition to most of the items listed in Box 3.11 at the end of Chapter 3.1, the problems to look for are listed below.

- Thatching material, styles and methods not typical to localities
- Thinning of total depth of thatch
- Deterioration of thatching material, especially in ridges, and on southerly and westerly aspects
- Detritus falling inside unlined roofs and not cleared away
- Lichens, algae and moss growth on the upper layers
- Wetness under **leeward** eaves after rain
- Condensation on underdrawn roofs
- Holes in wire mesh

Box 3.19: Moisture in thatched roofs — results of BRE research

BRE undertook research in the mid-1990s to assess and monitor a number of thatched roofs, some of which incorporated either an impermeable sarking membrane or a non-combustible board. The results showed that the outer layers of thatch with a southerly or westerly aspect were at risk from higher moisture levels, and that this risk was increased if the thatched roof incorporated an impermeable membrane or rigid board. This suggested that the natural movement of moisture through the thatch was being impeded, encouraging higher moisture levels towards the outer layers of thatch. These findings were confirmed by the increased level of deterioration seen on these aspects.

As a result of this monitoring work, a BRE Good Building Guide was published to encourage the introduction of cavities and ventilation within new or refurbished thatched roofs, both with and without rigid boarding[113].

To ensure that the recommendations contained in the new guide did not increase the fire risk, full-scale fire tests were carried out. These showed that a ventilated cavity between the underside of the thatch and a board does not increase the fire risk.

3.8 THERMAL INSULATION IN LOFTS

In England, by 2005, there were 2,375,000 (10.9%) homes without lofts. Of the remainder with lofts, 6,332,000 (29.1%) had less than 100 mm of insulation, 7,295,000 (33.5%) 100–150 mm and 5,778,000 (26.5%) 150 mm or more. There is therefore still considerable scope for bringing these dwellings up to current standards (see Table 4.1, English House Condition Survey 2005[5]).

There is no directly comparable information for houses in Scotland in the published Survey[118]. Dwellings in Scotland should meet the Scottish Housing Quality Standard introduced in 2004[119], which includes energy efficiency. The majority of dwellings failing to meet this standard did so because they failed to meet the energy efficiency criteria (paragraph 44, Scottish House Condition Survey 2004/5[118]). Although there is no indication that loft insulation was deficient in those cases, it is one of the criteria.

Up-to-date figures for Wales and Northern Ireland were not available at the time of writing.

A project has recently been carried out by BRE for The Cavity Insulation Guarantee Agency (CIGA) to assess the effectiveness of existing levels of thermal insulation in lofts. A survey of 193 lofts in Great Britain was undertaken, with a representative geographical distribution, and range of dwelling types. The year of installation of

the insulation was assessed, and the areas of disturbance, and modified depths were recorded, as well as subjective descriptions of the loft insulation. Established U-value calculation methods were used to assess the modification to the U-value of the affected areas, and to the overall U-value of each loft, caused by each 'disturbance mode'. A number of lofts had areas of missing insulation as a result of lighting installed in the floor of the loft requiring ventilation to dissipate the heat. Such areas cannot be insulated during a top-up, and are effectively lost as regards energy savings. A number of lofts were found that had been insulated in the 1960s and 1970s, and one individual loft insulated in 1957. The state of insulation in these lofts varied from poor to very good, and varied in depth from 25 mm to 100 mm. In general, there seems to be no evident reason why these lofts should not still be performing reasonably in reducing heat loss.

In national energy efficiency schemes, such as the Carbon Emissions Reduction Target (CERT) and its forerunner, the Energy Efficiency Commitment (EEC), energy savings and associated carbon dioxide emissions are calculated to give appropriate credits and hence priority in terms of cost effectiveness, to individual measures. Schemes such as CERT apply a lifetime calculation of the savings. For loft

insulation, the calculated savings at time of installation are used, with a 40-year lifetime (the same as for cavity wall insulation). This lifetime is an estimated parameter, and there is a lack of knowledge about how thermal properties, and hence savings, change over the years.

It is thought that such a change may be caused by natural effects such as:
• wind scour,
• biological process such as vermin attack,
• accumulation of dust and debris,
• ageing of the material,

and more significantly, as shown by the survey of lofts, householder action such as:
• works (eg lighting installed in the loft floor, re-wiring, alterations to water tanks, with movement/ folding of insulation),
• boarding out the loft,
• use of the loft for storage, with possible compression of insulation,
• installation defects (insulation missing near eaves, narrow or wide compared with the width between joists).

Some of these may be corrected when topping up insulation (moved/ folded insulation), others may not (lighting installed in the loft).

3.9 LOFT CONVERSIONS

Converting the space in the loft under a pitched roof into habitable rooms has become increasingly popular. However, loft conversions call for careful consideration of structural design, layout, fire performance, means of escape, ventilation, insulation and staircase provision. Not all houses are suitable. Most of the work will involve normal building practices and be subject to the usual national building regulations and standards.

This chapter gives advice on loft conversions in houses of not more than two storeys high. Building Regulations (England & Wales) impose different requirements for three or more storeys, particularly for means of escape, and Scotland and Northern Ireland regulations also differ from these in some respects. The advice given here applies to loft conversions with no more than two habitable rooms and a maximum floor area of 50 m², and only to dwellings in single occupancy. More detailed guidance is contained in BRE Good Building Guide 69 (2 Parts)[120].

FEASIBILITY

Preliminary inspections

Before beginning detailed work on a proposed loft conversion, it is essential to assess the feasibility of the project. This means noting the character of the surrounding buildings and inspecting the loft to assess:
- the internal space,
- the type of roof structure,
- the general condition of the existing structure.

Box 3.20 gives more detailed advice on assessing a property for its suitability for a loft conversion.

Some shapes and forms of roof lend themselves more readily to conversion than others. The key factors to assess are the roof shape and its internal height, width and pitch (Figure 3.147). Gable-ended roofs are generally easier to modify than hipped roofs since the separating or gable-end walls, together with the internal loadbearing walls, can usually support any new beams or trussed

purlins. Advice should be sought from qualified engineers.

The conversion of hipped roofs can be more complicated unless internal loadbearing cross-walls are available or extra beams are inserted at floor level to support secondary beams or purlins.

An important consideration is the need for adequate ceiling height: there is no statutory minimum requirement, but the preferred dimension for habitable rooms is 2.3 m although this height need not be maintained over the whole floor area. Eaves space can be used for low furniture and for storage. Lower ceiling heights may be acceptable in bathrooms and corridors, although a minimum ceiling height is required at the head of stairs to comply with building regulations.

Generally speaking, roofs of 45° pitch or greater are the easiest to convert. Lower pitches may possibly require dormers that extend above the level of the existing roof (Figure 3.148).

In assessing the available headroom, the thickness and depth

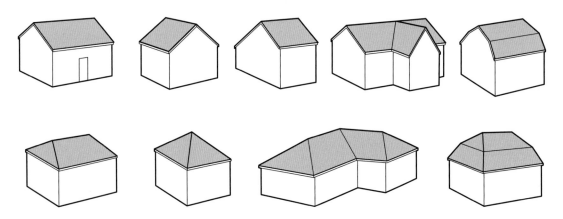

Figure 3.147: A wide variety of roof shapes may be suitable for loft conversion

Box 3.20: Checklist for assessing the condition of a property and its suitability for a room-in-the-roof

The following observations should be made:

General

- The effect of fitting dormers or rooflights on the appearance of the property. Are there any dormers on nearby properties?
- Whether or not the property is in a conservation area and whether there are listed buildings nearby?
- The age of the property and neighbouring buildings
- The legal status of the proposed loft conversion. An agreement with the owners of the neighbouring property will be needed if a separating wall in a semi-detached or terraced property is to be altered.

Defects

- Any common defects in the property and nearby similar properties (eg settlement, distortion of windows, cracking, weaknesses in chimney stacks, roof vents, soil pipes)

Roof characteristics

- Roof structure, eg:
  - dimensions of the available space
  - pitch and type of roof (eg is it gable, terraced, hipped?)
  - trussed rafters or strutted purlins
  - steel or timber rafters
- covering, eg:
  - changed or renewed
  - battened or boarded
  - the presence of an underlay and its condition
- the presence of separating and gable walls:
  - their construction (eg cavity or solid?), materials, thickness, perforations, fire stops
  - their condition and the condition of the mortar
- the details at the eaves and how they could be adapted or replaced
- the provision of ventilation and the ease of doing so

- the position and size of water storage and header tanks, pipework, ducts, chimneys, flues and possible re-routing
- wiring (eg lighting, alarm systems, TV aerials)?

Internal characteristics

- the structural condition of floors, eg any sag or spring and the likely cause
- loadbearing walls and the possibility of movement and strengthening
- lintels: their type, materials and condition
- evidence that walls or supporting features have been moved or removed
- the space and headroom required for another staircase
- fire regulation requirements for the floor and whether this will affect the headroom in the loft

Figure 3.148: Dormers and roof lights inserted during loft conversion

Planning and design

It is important to establish the legal framework for the conversion (eg complying with the requirements of the Party Wall, etc. Act 1996[121]) and the process for planning consent and building control approval, including submission of structural calculations and technical drawings.

Boxes 3.21 and 3.22 give guidance on building regulation and planning approval. Depending on the planned use of the loft space, there are a number of design considerations (Box 3.23).

of the new floor, and that of the thermal insulation to be fitted to the underside of the roof slope, should be taken into account.

Natural lighting and ventilation are usually achieved by fitting dormers or roof slope windows, although there is no building regulation requirement to provide daylighting. Nevertheless, it is advisable to provide daylighting to staircases and access routes.

The conversion must include provision for safe access and exit. This requires a fixed staircase that complies with building regulations in respect of its pitch, headroom, stair width and balustrading. For loft staircases, there are reduced requirements for headroom (see section on Access). Escape windows must also be fitted to each habitable room (see section on Fire protection and escape).

Box 3.21: Building regulation

All loft conversions must comply with the statutory requirements of building regulations. These are supported by examples of constructions that meet the requirements in England & Wales of the Approved Documents. Other methods may be acceptable but compliance must be demonstrated to the satisfaction of Building Control who should be consulted at an early stage.

Planning permission from the local authority is not needed for all loft conversions, but will be necessary if:

- the new windows face a highway,
- the new space adds more than 50 m³ to a detached house or 40 m³ to a terraced house,
- any part of the new structure will be above the existing roof line,
- materials to be used are not compatible with the existing property.

If the property is listed or in a conservation area or National Park, both listed building consent and planning permission will be needed.

Structural calculations

For almost all loft conversions, it is advisable to employ a structural engineer to check proposals for structural changes to, or adaptation of, existing members. The Building Control Body may request formal structural calculations to confirm the stability of both the existing and the new structure.

Box 3.23: Design considerations

The following may be considered when designing a loft conversion:

- the relationship between the property and its neighbours (party wall awards),
- the effect on the value of the property,
- the dimensions of the space: one large room or two small ones?
- the amount of light required (if the space is to be used as a studio, would large roof slope windows or full height dormers give the most light?),
- the lighting of the stairs (install a dormer or roof slope window?),
- the outlook from the new windows to try to ensure a pleasant view,
- the ceilings (rooms without horizontal ceilings have a more spacious feel),
- the location of the stairway (is there space on the first-floor landing or must a room be used?),
- the design of the staircase (conventional or non-standard design?),
- the requirement for storage space.

Preliminary structural questions

Before contemplating any conversion, it is important to understand the functions of existing structural members and the possible consequences of their removal. If their function is uncertain, it may be advisable to provide temporary support until they can be replaced. Trussed rafters should not be modified without consulting a structural engineer. Structural design must also take into account the need for headroom in the roof space and above the staircase, and fire regulation requirements (Box 3.24).

STRUCTURAL SOLUTIONS

Common types of roof structure

Before the middle of 20th century most roof carpentry was cut and fixed on site. The size of timbers used depended on the weight of roof covering, but, even so, rafter sizes rarely exceeded 2″ × 4″. They can be even smaller in late Victorian houses, which often have quite low-pitched slated roofs. Problems can arise with this type of roof when the slates are replaced with heavier concrete tiles (see Chapter 3.2).

Post-1945 system building introduced a variety of roof constructions, most of which are low-pitched and are unlikely candidates for room-in-the-roof conversions, although some post-1918 houses may have roofs that are more suitable. If the supporting structure is strong enough, an extra usable storey could be added using attic trusses (see Figure 1.10).

Trussed rafter roofs present particular difficulties. Although modifying timber trussed rafter roofs to form acceptable spaces may be possible, it should only be attempted using specialist advice. The members in these trusses are more highly stressed than in a traditional roof and trusses are generally wider spaced than members in traditional roofs. Some low-pitch roof-space conversions may require removing existing trusses to improve headroom or to enable roof slope or dormer windows to be installed.

Box 3.24: Checklist for structural considerations

The existing structure will need to be assessed to decide how it can be adapted and what will need to be relocated, eg:

- accommodation of the new floor and roof loadings
- the capacity and condition of the supporting walls
- support for the roof structure.

Specifically, careful consideration should be given to:

- the sizes and condition of all roof members and fixings (eg are there any signs of distress or any weakening due to decay and/or insect attack?)
- the adequacy of the foundations, gable and supporting walls and floors for the additional loads, particularly around openings, and the ease of providing additional support
- possible sound transmission through a new beam supported by a party wall
- alternative gable support when cutting binders and purlins
- the need for new collars
- the loading and structural implications of the new staircase.

Most importantly, consideration should be given to the sequence of work (eg providing supports before any structural member is removed, access for inserting new structural members and timing of the wiring, plumbing and heating installations).

For adaptation of trussed rafters, the advice of a structural engineer should be sought.

Some types of roof are not suitable for adaptation, but whole or partial replacement with special attic trusses could offer a practical solution. Support above window openings should be considered carefully.

If changes are to be made to existing members, note the advice given in Box 3.22 concerning structural calculations.

CHARACTERISTIC DETAILS

Rafters

Most loft conversions in pitched roofs require existing rafters, purlins, ceiling joists and supporting structures to be strengthened in order to give support, a clear floor area, lighting and access.

Where rafter supports, such as purlins and struts, are removed to provide space, alternative supports must be provided. Rafters may be stiffened by fixing new timbers alongside. A structural check will be needed to ascertain the condition of the rafters and to determine the degree of strengthening and stiffening required. Where rafters are cut to provide space for windows, adjacent members may need to be strengthened. The provision of a ridge beam may be useful in providing additional support, particularly in the case of a dormer (Figure 3.149), but detailing the insulation to ensure continuity can be difficult.

When deep purlins or steel joists support both the floor and

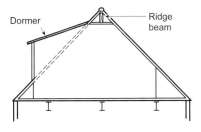

Figure 3.149: New ridge beam supporting rafters

roof loads, they impose point loads on the walls at their ends, thereby requiring a check on the capacity of the wall to cope. Beams to spread the load may be required.

Additional loads on rafters can increase the outward thrust at the eaves and may require additional ties to counteract this. Hipped ends are particularly vulnerable, and in some cases it would be a wise precaution to fit intermediate straps and dragon ties (Box 3.25).

Valleys

Pitched roofs with intersecting valleys can be more difficult to convert unless existing cross-walls are available to support new timber beams or purlins. New columns or beams may be required at these intersections to support joists or purlins forming the new roof space.

Floors

In a loft conversion it is usual for the new floor to be formed above the existing ceiling joists, ideally without disturbing the house below until the new staircase is installed. However, if there is insufficient headroom in the loft, and plenty of headroom at first-floor level, an option would be to insert additional support below the ceiling before taking out any binders.

The new floor will change the loading on the existing supporting structure, which must accommodate the new dead and imposed loads. Usually, it will not be possible to use the existing ceiling joists to support new floor loads, and these will need

additional support members when used as floor joists.

Figure 3.151 illustrates two of the many ways of providing loft floors above the existing ceiling rafters. The method chosen will depend on the spans and the loads to which they are subjected, which may include the need to support the roof structure. Again, sound-transmission and fire-resistance factors will need to be addressed. New loft floors will form part of the roof structure as a whole, for example new joists could well provide additional lateral stability to supporting walls.

It is also possible to support the new floor by hanging it off purlin beams spanning the gable walls. Deep timber or lattice beams could perform similar functions, but can restrict access to the eaves. The advantages of this system are that disturbance to the household is reduced. However, some intermediate support from below is likely to be needed, which will inevitably affect the headroom of the room below.

Existing walls

In all cases, the additional loads of the loft conversion will be carried by the outside and loadbearing inner walls of the house and transmitted directly to their foundations. The foundations may need to be exposed to check they are adequate.

In terraced and semi-detached houses, structural solutions may involve imposing loads on separating walls. Before proceeding it is necessary to discuss the proposals with the owner/occupier of the adjoining property. Guidance can be found in the Party Wall, etc. Act 1996[121] and an explanatory leaflet is available.

In older terraced houses or in buildings that have been divided, separating walls may not continue into the loft space and will have to be extended up to the underside of the roof covering. Newly built or existing separating walls must of course comply with national building regulation requirements for strength, sound transmission and fire resistance, or be modified so that this is achieved.

Box 3.25: Special note about binders

Most cut roofs incorporate binders which link ceiling joists and maintain their alignment and lateral stability, and in hipped roofs resist the outward thrust of the rafters. If the binders are removed, alternative support must be provided. Figure 3.150 shows a possible solution for a hipped roof using dragon ties and strapping.

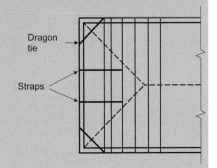

Figure 3.150: Plan of hipped roof showing possible position of added ties and straps

New floor having joists fixed to existing celing joists with bolts or screws

Diagonal strutting or strapping

Lightweight, perforated galvanised steel strapping to stabilise joists and rafters

New floor having joists independent of existing ceiling joists, which are stabilised by strapping

Figure 3.151: Common methods of providing loft floors above existing ceiling rafters

Some separating walls in the roof space are not directly above the wall below. These 'flying freeholds' give rise to issues from which structural, fire-resistance and legal problems can arise.

BUILDING SERVICES

Electrical installations

The requirements and positions for lights, socket outlets, telephone, aerials, alarms and smoke detectors should be determined at an early stage so that fixings, ducts, backboards, etc. can be provided and cables run within stud walls, partitions and floors. It will then be possible to estimate the electrical loading, the adequacy of the existing supply and earthing and bonding arrangements, and the position and accessibility of any consumer unit. The need for any new circuits should be considered and any necessary changes made. For England and Wales, requirements for electrical safety are set out in AD P[44] and for Scotland in Section 4[45].

To prevent overheating, cables should not pass through insulating material. If this cannot be done, it may be necessary to install a heavier-duty cable. Avoid contact between PVC-sheathed cables and polystyrene insulation.

When a loft is converted to living space, building regulations require that mains-wired smoke detectors are fitted in circulation spaces such as the hall, landing and stairs of all storeys. There should be at least one detector on each storey and the detectors interlinked so that detection of any one triggers all. Each must be mounted within 7.5 m of the door of any habitable room and at least 300 mm from walls

and light fittings. For the power supply, AD B1[44] suggests using a single independent circuit from the consumer unit, or, if the alarm system has battery back-up, making a connection to a regularly used local lighting circuit.

Heating

Most rooms-in-the-roof have low heat losses because of the high level of newly installed thermal insulation surrounding them. If the rooms are for occasional use, it may only be necessary to provide electric heaters or oil-filled radiators. For rooms in continuous use, it may be possible to extend the existing central heating system, provided the boiler has sufficient spare capacity and the feed and expansion cistern is suitably positioned. Otherwise, a change to the central heating system may be needed, such as adoption of a pressurised system that requires no feed and expansion cistern. The new radiators should be fitted with thermostatic radiator valves (TRVs) or similar valves so that they can be controlled independently.

Plumbing

Loft conversions often require cold water storage tanks which have been positioned in the loft space to be relocated at an early stage. If the house is occupied during the work, temporary arrangements for water supply will be needed. Re-siting tanks in the eaves using linked lidded cold water storage tanks is a common solution. Water Regulations require 350 mm space above the tank for access to the float-operated water inlet valve. If space is limited, more than one shallow tank may be needed.

MAIN PERFORMANCE REQUIREMENTS

Access

When the loft is converted into habitable rooms, a fixed stair must be provided. Non-fixed stairs, such as retractable loft ladders, are permitted only when the space is to be used solely for storage.

Wherever possible, for maximum safety and ease of use, stairs should comply with the requirements set out in the relevant building regulations. In England and Wales AD K[44] stipulates a maximum pitch of 42° and a minimum headroom of 2 m. However, if this is not possible in a loft conversion, a minimum headroom of 1.9 m at the centre line is permitted, with a minimum of 1.8 m at the edge of the stair (Figure 3.152). There must be a minimum clearance of 1.5 m between the pitch line and any bulkhead slope. The rise should be no more than

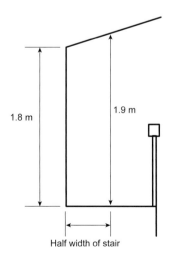

1.8 m   1.9 m

Half width of stair

Figure 3.152: Minimum clearances required above stairs in loft conversions. Modified from Approved Document K[44]

220 mm and the going (tread less nosing) no less than 220 mm. A good guide is that twice the rise plus the going (2R + G) should be between 550 mm and 700 mm.

There is no minimum requirement for width, but where there will be two new rooms, a width of 800 mm is recommended for ease of moving furniture. If there is only one room, a minimum width of 600 mm is acceptable.

If space is limited, provided they give access to only one habitable room, alternative designs of stairs are permitted.

Daylighting

Windows are needed in a loft conversion to provide daylight, ventilation, a view out and a means of escape. In some cases, windows can be installed in the gable wall, but more usually they will have to be fitted into the sloping roof as either dormer windows or roof slope windows. Sloping windows in a 45° pitch generally provide up to 40% more light than dormers with a similar glazed area, but they do not provide the increased headroom and consequent usable floor space which is available with dormers.

Maintenance and cleaning of windows in second-floor loft rooms is a crucial consideration for the designer.

There is a wide variety of styles of dormer (Figure 3.153 and 3.154) and also of finishes and claddings. Side-glazing can be incorporated into dormers to increase the light admitted.

Figure 3.154: Bayed dormers which are common in Scotland

Ventilation

Ventilating living spaces

Building regulation requirements for ventilation in rooms-in-the-roof are the same as for living spaces generally. To meet the requirements, part of the window opening should be 1.75 mm above floor level; if the opening is too low it may not be effective in letting warm air escape and this could lead to overheating. If possible, provide cross-ventilation across the loft rooms. Background ventilation should be provided and both dormers and roof slope windows should normally be fitted with trickle ventilators.

Normal requirements also apply to bathrooms, which must be fitted with an extract system, either mechanical or possibly passive stack ventilation.

Insulation and ventilation of the roof

The roof space around a loft conversion needs insulation, to prevent both heat loss and too much solar gain.

Note that insulation can be installed on the underside of the whole roof, or only around the new rooms. The advantage of the former is that any tanks and pipework are protected from frost, and they can be accommodated at the apex and in the eaves spaces, which are also warm for storage. If the depth of the rafters needs to be increased to accommodate the insulation, battens can be added to their underside. There is a wide choice of suitable materials, with insulation boards giving the best thermal performance.

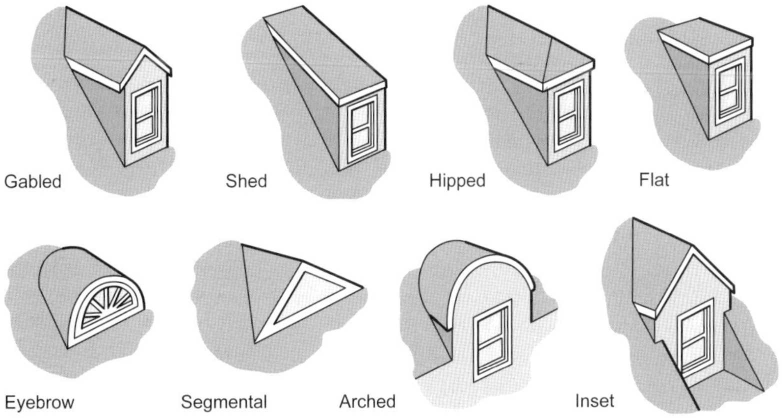

Table 3.14: U-values for elements in a loft space

Element	U-value (W/m²K)
Roof with a flat ceiling	0.16
Roof with a sloping ceiling under	0.3
External walls, including side walls between the room and the unheated roof void	0.35
Windows and rooflights: Timber or PVC-U frames	2.0
Metal frames	2.2

Note: the U values in the above table may have to be enhanced to match SAP calculations for the whole dwelling (see Parts L1A and L2A in the England and Wales Building Regulations)[44].

Gabled   Shed   Hipped   Flat

Eyebrow   Segmental   Arched   Inset

Figure 3.153: A wide range of dormer designs exists

The various elements of the new loft space will need to achieve the U-values given in Table 3.14.

With all methods of insulation, there must be cross-ventilation on the cold side of the insulation to prevent moisture condensing in the space. There should be an air path at least 50 mm deep between the insulation and the roof or wall covering and the underlay. It is important to ensure that ventilation paths are not blocked by trimmers or window frames.

Ventilation openings should be provided at ridge level, equivalent to a continuous opening of 5 mm. The ridge tiles can be replaced or altered with a vented dry ridge. If the insulation is between the rafters, there must be a ventilation slot at eaves level equivalent to a 25 mm continuous gap. For insulation at ceiling level, the eaves gap should be equivalent to 10 mm (Figure 3.155). Alternative methods for achieving these gaps are described in BRE Good Repair Guide 30[122] and in Figure 3.156.

Vapour control layers should be installed on the inside surfaces, using well-sealed 500-gauge polyethylene sheet between the lining and the insulation. This is particularly important if sufficient ventilation is difficult to achieve.

Any gaps between the floor joists at the eaves should be blocked to prevent heat loss due to air flow from the eaves passing under the floor. Insulating and ventilating the cheeks of dormers needs particular care (Figure 3.157). The risk of thermal bridging should also be considered.

If the roof covering is being replaced, insulation could be added immediately below the covering, creating a warm roof, in which case ventilation is not needed.

SOUND INSULATION

Separating walls

In a terraced or semi-detached house, the separating wall must provide adequate sound insulation, irrespective of whether or not the next door house has a habitable room in the loft. This is because sound can travel through the wall into the neighbouring roof void and through its ceiling.

If the separating wall is one brick thick or is of 225 mm coursed stone, or is a cavity wall, any gaps between the top of the wall and the roof should be made good, all gaps and holes should be filled with mineral wool, and the whole wall should be rendered with a sand and cement coating before finishing with a coat of plaster. If the wall is only half brick (100 mm thick), there must also be an independent insulated stud partition, not attached to the wall. This should comprise two layers of plasterboard, with staggered joints, with minimum mass per unit area of 20 kg/m². A layer of mineral wool with minimum density 10 kg/m² and minimum thickness 25 mm should be included in the cavity between the panel and the existing wall, making sure it is not compressed (Figure 3.158). The partition only should be fixed to the floor and roof structure, and any gaps filled with acoustic sealant. The frames will slightly reduce the available space in the rooms, but there is no viable alternative. See also the BRE companion book, *Walls, windows and doors,* Chapters 6.2 and 10.2[123].

In the loft of a timber-frame house, additional plasterboard will be needed to improve sound insulation. The minimum thickness should be at least 30 mm, with staggered joints between the layers. Internal walls should be one of the four generic constructions shown in AD E (2003) sections 5.18–5.21[44], or comply with Scotland and Northern Ireland regulations[45,46], or have been tested in an accredited laboratory and have a minimum airborne sound insulation of 40 *R*w dB.

Floors

There is no building regulation requirement to improve the sound insulation between floors in a house in single occupancy, but it is often desirable to do so.

Some improvement in structure-borne sound insulation can be achieved by placing a layer of mineral wool, minimum thickness 100 mm and minimum density 10 kg/m², over the ceiling (Figure 3.159). The floor should be timber

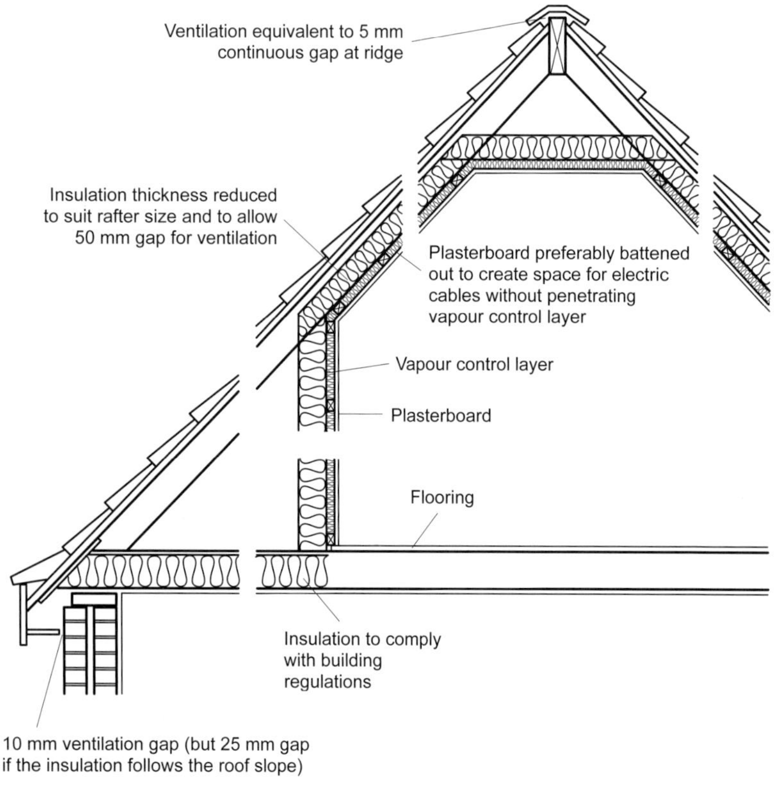

Ventilation equivalent to 5 mm continuous gap at ridge

Insulation thickness reduced to suit rafter size and to allow 50 mm gap for ventilation

Plasterboard preferably battened out to create space for electric cables without penetrating vapour control layer

Vapour control layer

Plasterboard

Flooring

Insulation to comply with building regulations

10 mm ventilation gap (but 25 mm gap if the insulation follows the roof slope)

Figure 3.155: Provision of ventilation paths above loft insulation

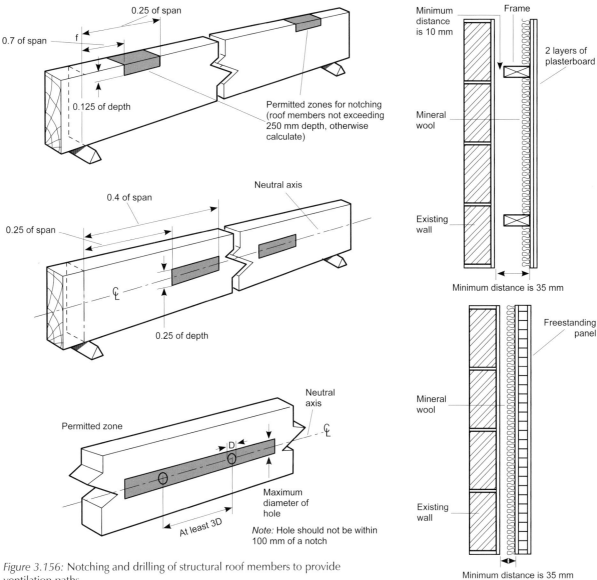

Figure 3.156: Notching and drilling of structural roof members to provide ventilation paths

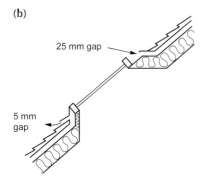

Figure 3.157: Ventilation gaps around (a) dormers, (b) roof slope windows

Figure 3.158: Methods of providing sound insulation to a separating wall © Crown copyright. Reproduced from Approved Document E of the Building Regulations (England & Wales) under the terms of the Click-Use Licence

or wood-based board with a minimum mass/unit area of 15 kg/m².

FIRE PROTECTION AND MEANS OF ESCAPE

An overriding consideration in a loft conversion is ensuring adequate means of escape from fire.

The aim of the requirements on fire is:

• to warn the occupants that there is a fire,
• to protect them from its effects,
• to allow them to escape from the building.

Early warning

In the early stages of a fire, the main hazard to the occupants is smoke and combustion gases. These very quickly spread from the source area into corridors and up stairwells.

Fire safety (particularly relevant to rooms in lofts), compartmentation and the need for cavity barriers in concealed spaces and control of internal and surface fire spread is covered in the supporting documents of the national building regulations:

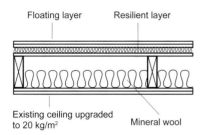

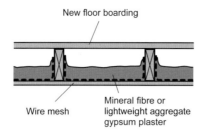

Figure 3.159: Floating floor construction to improve sound insulation
© Crown copyright. Reproduced from Approved Document E of the Building Regulations (England & Wales) under the terms of the Click-Use Licence

- England & Wales, AD B1[44],
- Scotland, Section 2 of the Technical Handbooks, Domestic and Non-domestic[45],
- Northern Ireland: Technical Booklet E[46].

The ceiling below the loft, in combination with the new floor above it and the structure supporting it, must provide 30-minute fire resistance. Provided there is 9.5 mm plasterboard on the ceiling with gypsum plaster finish, the half–hour resistance can be achieved by laying mineral fibre between the joists before adding the floor. The layer must be not less than 60 mm thick and fixed to the joist sides. It is usual to support this layer on wire mesh laid between the joists (Figure 3.160).

The new floor must be tongued-and-grooved, minimum 25 mm nominal boarding or 15 mm plywood or chipboard and must extend into the eaves to the inside face of the external walls. Fixings should penetrate into the joist sides to a minimum depth of 20 mm.

If the ceiling below is made of materials other than plasterboard, such as wet plaster on battens, or the joists are exposed, other methods of providing additional protection will be necessary (see BRE Digest 208[124]).

The walls of the new loft rooms adjoining the stairs must be of fire-resisting construction to comply with regulations. Separating walls between dwellings need 60-minute fire resistance; those in the loft may have to be upgraded and any

Figure 3.160: Method of providing 30-minute fire resistance to a new loft floor

holes or gaps in existing walls filled in, up to the underside of the roof covering. Readers should also see the guidance given in Chapter 3.1.

All existing doors at ground and first floors opening onto the hall or landing must be fitted with a self-closing device, either a closer or rising butts. Any new doors giving access to the hall and landing must be fire-resisting. Existing glazed doors, apart from bathroom and WC doors, may need to have fire-resisting glass. See also Chapter 8.2 in the BRE companion book, *Walls, windows and doors*[123].

Means of escape
There are two acceptable approaches for dealing with means of escape.
- The entire house may be considered as a new three-storey house. This involves upgrading or replacing the doors and walls enclosing the stairway such that the escape route from the loft room is protected.

- The loft room is separated from the existing house by fire-resisting construction and a window, suitable for rescue, is provided. Even following this approach, some upgrading of the existing escape route is still often necessary.

Figure 3.161 shows the position of the window on the roof slope. The window should have an unobstructed openable area of at least 0.33 m² and be at least 450 mm high and 450 mm wide. The bottom of the openable area should be not more than 1100 mm above the floor and no more than 1.7 m from the edge of the roof.

Each window that is provided for rescue purposes should be accessible for rescue by ladder from the ground. This means that there must be pedestrian access to the ground below each window, and that use of the ladder is not obstructed by a ground-floor extension. A door to a suitable roof terrace is also acceptable, as is escape across the flat roof of a ground-floor extension, provided the roof is fire resistant and has guardrails.

The existing stairway at ground and first-floor levels must be enclosed. The stairway must either open directly to a final exit or give access to at least two escape routes, each leading to a final exit and separated from each other.

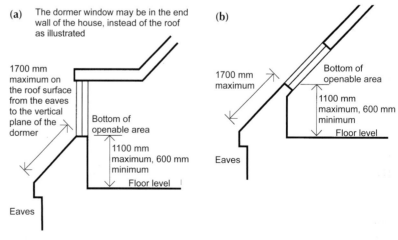

Figure 3.161: Position of (a) dormer window, (b) roof light/roof window suitable for emergency escape from a loft conversion of a 2-storey house
© Crown copyright. Reproduced from Approved Document E of the Building Regulations (England & Wales) under the terms of the Click-Use Licence

3.10 REFERENCES

[1] **BRE.** Assessing traditional housing for rehabilitation. BRE Report BR 167. Bracknell, IHS BRE Press, 1990

[2] **AMA Research.** Roofing Market Report UK 2007. Cheltenham, AMA Research, 2007. Available from www.amaresearch.co.uk

[3] **Communities and Local Government.** English House Condition Survey 2005. London, CLG, 2007 (Analyses of data were carried out especially for *Roofs and roofing*. Further information may be obtained from BRE Technical Consultancy)

[4] **Northern Ireland Housing Executive.** Northern Ireland House Condition Survey 2006. Belfast, Northern Ireland Housing Executive, 2008. Available as a pdf from www.nihe.gov.uk

[5] **Department of the Environment.** English House Condition Survey 1991. London, The Stationery Office, 1993

[6] **Welsh Office.** Welsh House Condition Survey 1991. Cardiff, Welsh Office, 1993

[7] **Scottish Homes.** Scottish House Condition Survey 1991. Survey report. Edinburgh, Scottish Homes, 1993

[8] **Northern Ireland Housing Executive.** Northern Ireland House Condition Survey 1991. First report of survey. Belfast, Northern Ireland Housing Executive, 1993

[9] **National House Building Council.** Keeping the roof on claims. Standards Extra No 4. Amersham, NHBC, 1993

[10] **TRADA Technology.** Span tables for solid timber members in floors, ceilings and roofs for dwellings. 2nd edition. High Wycombe, TRADA, 2008

[11] **Trotman PM.** An examination of the BRE Advisory database compiled from property inspections. Proceedings of the International Symposium 'Dealing with Defects in Building', Varenna, Italy. September 1994 (copies available from the author at BRE)

[12] **British Standards Institution.** BS 8103-1: 1995 Structural design of low-rise buildings. Code of practice for stability, site investigation, foundations and ground floor slabs for housing

[13] **BRE.** Dual-pitched roofs: trussed rafters, bracing and binders. Specification. Defect Action Sheet 83. 1986. Available as complete set (141 sheets). BR 419. Bracknell, IHS BRE Press

[14] **BRE.** Dual-pitched roofs: trussed rafters, bracing and binders. installation. Defect Action Sheet 84. 1986. Available as complete set (141 sheets). BR 419. Bracknell, IHS BRE Press

[15] **British Standards Institution.** BS EN 1304: 2005 Clay roofing tiles and fittings. Product definitions and specifications. London, BSI, 2005

[16] **British Standards Institution.** BS EN 490: 2004 Concrete roofing tiles and fittings for roof covering and wall cladding. Product specifications. London, BSI, 2004

[17] **British Standards Institution.** BS 5534: 2003. Code of practice for slating and tiling (including shingles). London, BSI, 2003

[18] **Herbert MRM.** Some observations on the behaviour of weather protective features on external walls. Building Research Establishment Current Paper 81/74. Watford, BRE, 1974

[19] **BRE.** Slated and tiled pitched roofs: flashings and cavity trays for step and stagger layouts — specification. Defect Action Sheet 114. 1988. Available as complete set (141 sheets). BR 419. Bracknell, IHS BRE Press

[20] **BRE.** Roofs: eaves gutters and downpipes — specification. Defect Action Sheet 55. 1984. Available as complete set (141 sheets). BR 419. Bracknell, IHS BRE Press

[21] **British Standards Instiution.** BS EN 12056-3: 2000 Gravity drainage systems inside buildings. Part 3: Roof drainage, layout and calculation. London, BSI, 2000

[22] **BRE.** Slated or tiled pitched roofs — conversion to accommodate rooms: ventilation of voids to the outside air. Defect Action Sheet 118. 1988. Available as complete set (141 sheets). BR 419. Bracknell, IHS BRE Press

[23] **BRE.** Slated or tiled pitched roofs — conversion to accommodate rooms: installing quilted insulation at rafter level. BRE Defect Action Sheet 119. 1988. Available as complete set (141 sheets). BR 419. Bracknell, IHS BRE Press

[24] **Stirling C.** Thermal insulation: avoiding risks. BR 262. Bracknell, IHS BRE Press. 2002 edition

[25] **Harrison H, Mullin S, Reeves B & Stevens A.** Non-traditional houses: Identifying non-traditional houses in the UK 1918–75. BR 469. Bracknell, IHS BRE Press, 2004

[26] **British Standards Institution.** BS 5493: 1977 Code of practice for protective coating of iron and steel structures against corrosion. London, BSI, 1977

[27] **British Standards Institution.** BS EN ISO 12944: 2007 Paints and varnishes. Corrosion protection of steel structures by protective paint systems. 8 Parts. London, BSI, 2000

[28] **British Standards Institution.** BS EN 538: 1994 Clay roofing tiles for discontinuous laying. Flexural strength test. London, BSI, 1994

[29] **British Standards Institution.** BS 8103-1: 1995 Structural design of low-rise buildings. Code of practice for stability, site investigation, foundations and ground floor slabs for housing. London, BSI, 1995

[30] **BRE.** External and separating walls: lateral restraint at pitched roof level — specification. Defect Action Sheet 27. 1983. Available as complete set (141 sheets). BR 419. Bracknell, IHS BRE Press

[31] **BRE.** External and separating walls: lateral restraint at pitched roof level — installation. Defect Action Sheet 28. 1983. Available as complete set (141 sheets). BR 419. Bracknell, IHS BRE Press

[32] **BRE.** Erecting, fixing and strapping trussed rafter roofs. Good Building Guide 16. Bracknell, IHS BRE Press, 1993

[33] **BRE.** Quality in traditional housing. Volume 2: An aid to design. Volume 3: An aid to site inspection. BRE Report. London, The Stationery Office, 1982

[34] **BRE.** Dual-pitched roofs: trussed rafters — specification of remedial bracing. Defect Action Sheet 110. 1987. Available as complete set (141 sheets). BR 419. Bracknell, IHS BRE Press

[35] **BRE.** Dual-pitched roofs: trussed rafters — installation of remedial bracing. Defect Action Sheet 111. 1987. Available as complete set (141 sheets). BR 419. Bracknell, IHS BRE Press

[36] **BRE.** Dual-pitched roofs: trussed rafters — specification of remedial gussets. Defect Action Sheet 112. 1987. Available as complete set (141 sheets). BR 419. Bracknell, IHS BRE Press

[37] **BRE.** Trussed rafter roofs: tank supports — specification. Defect Action Sheet 43. 1984. Available as complete set (141 sheets). BR 419. Bracknell, IHS BRE Press

[38] **Blackmore P.** Slate and tile roofs: avoiding damage from aircraft wake vortices. Digest 467. Bracknell, IHS BRE Press, 2002

[39] **BRE.** Pitched roofs: sarking felt underlay — watertightness. Defect Action Sheet 10. 1982. Available as complete set (141 sheets). BR 419. Bracknell, IHS BRE Press

[40] **British Standards Institution.** BS 747: 2000 Reinforced bitumen sheets for roofing. Specification. London, BSI, 2000. *NB: This Standard has been replaced by BS EN 13707: 2004*

[41] **BRE.** Pitched roofs: sarking felt underlay — drainage from roof. Defect Action Sheet 9. 1982. Available as complete set (141 sheets). BR 419. Bracknell, IHS BRE Press

[42] **British Standards Institution.** BS 8747: 2007 Reinforced bitumen membranes (RBMs) for roofing. Guide to selection and specification. London, BSI, 2007

[43] **Building Research Station.** Principles of modern building, Volume 2: Floors and roofs. London, The Stationery Office, 1961

[44] **Communities and Local Government (CLG).** The Building Regulations 2000.
Approved Documents:
 A: Structure, 2004
 B: Fire safety, 2006
 C: Site preparation and resistance to contaminates and moisture, 2004
 E: Resistance to the passage of sound, 2003
 F: Ventilation, 2006
 K: Protection from falling collision and impact, 1998
 L: Conservation of fuel and power, 2006
 P: Electrical safety: dwellings, 2006
London, The Stationery Office. Available from www.planningportal.gov.uk and www.thenbs.com/buildingregs

[45] **Scottish Building Standards Agency (SBSA).** Technical standards for compliance with the Building (Scotland) Regulations 2009
Technical Handbooks, Domestic and Non-domestic:
 Section 1: Structure
 Section 2: Fire
 Section 3: Environment
 Section 4: Safety
 Section 6: Energy
Edinburgh, SBSA. Available from www.sbsa.gov.uk

[46] **Northern Ireland Office.** Building Regulations (Northern Ireland) 2000.
Technical Booklets:
 D: Structure, 1994
 E: Fire safety, 2005
 F: Conservation of fuel and power.
  F1: Dwellings, F2: Buildings other than dwellings. 1998
 K: Ventilation, 1998
London, The Stationery Office. Available from www.tsoshop.co.uk

[47] **British Standards Institution.** BS 5250: 2002 Code of practice for control of condensation in buildings. London, BSI, 2002

[48] **BRE.** Pitched roofs: boxed eaves — preventing fire spread between dwellings. Defect Action Sheet 7. 1982. Available as complete set (141 sheets). BR 419. Bracknell, IHS BRE Press

[49] **BRE.** Pitched roofs: separating wall/roof junction — preventing fire spread between dwellings. Defect Action Sheet 8. 1982. Available as complete set (141 sheets). BR 419. Bracknell, IHS BRE Press

[50] **National Association of Rooflight Manufacturers.** Visit www.narm.org.uk

[51] **BRE & CIRIA.** Sound control for homes. BR 238 and CIRIA Report 127. Bracknell, IHS BRE Press, 1993

[52] **British Standards Institution.** BS 1202: Specification for nails.
 Part 1: 2002 Steel nails
 Part 2: 1974 Copper nails
 Part 3: 1974 Aluminium nails

[53] **Berry RW.** Remedial treatment of wood rot and insect attack in buildings. BR 256. Bracknell, IHS BRE Press, 1994

[54] **Bravery AF, Berry RW, Carey JK & Cooper DE.** Recognising wood rot and insect damage in buildings. BR 232. Bracknell, IHS BRE Press, 1992

[55] **Harrison HW.** Steel framed and steel clad houses: inspection and assessment. BR 113. Bracknell, IHS BRE Press, 1987

[56] **Mayo AP, Rodwell DFG & Morgan JWW.** Trussed rafter roofs. Building Research Establishment Current Paper 5/83. Watford, BRE, 1983

[57] **HMSO.** The Construction (Design and Management) Regulations 2007. Statutory Instrument 2007 No. 320 Health and Safety. London, The Stationery Office, 2007

[58] **Saunders G.** Safety considerations in designing roofs. Digest 493. Bracknell, IHS BRE Press, 2005

[59] **Advisory Committee for Roof Work.** Visit www.roofworkadvice.info

[60] **HMSO.** Control of Substances Hazardous to Health (COSHH) Regulations 1988. Statutory Instrument 1988 No. 1657. London, The Stationery Office

[61] **BRE.** Pitched roofs: trussed rafters — site storage. Defect Action Sheet 5. 1982. Available as complete set (141 sheets). BR 419. Bracknell, IHS BRE Press

[62] **National Federation of Roofing Contractors.** Visit www.nfrc.co.uk

[63] **British Standards Instiution.** BS 5268: Structural use of timber
 Part 2: 2002 Code of practice for permissible stress design, materials and workmanship
 Part 3: 2006 Code of practice for trussed rafter roofs

[64] **TRADA Technology.** Bracing for non-domestic timber trussed rafter roofs. GD 8. High Wycombe, TRADA, 1999

[65] **British Standards Institution.** BS 8000-6:1990 Workmanship on building sites. Code of practice for slating and tiling of roofs and claddings. London, BSI, 1990

[66] **BRE.** Surveyor's checklist for rehabilitation of traditional housing. BR 168. Bracknell, IHS BRE Press, 1990

[67] **Hart D.** The building slates of the British Isles. BR 195. Bracknell, IHS BRE Press, 1991

[68] **Emerton G.** The pattern of Scottish roofing. Edinburgh, Historic Scotland, 2000

[69] **Walsh J.** Scottish roofing slate: characteristics and tests. Research Report. Edinburgh, Historic Scotland, 2002

[70] **HSE.** Working with asbestos cement. HSG 189/2. London, HSE Books, 1999

[71] **British Standards Institution.** BS EN 12326-1: 2004 Slate and stone products for discontinuous roofing and cladding. Product specification. London, BSI, 2004

[72] **British Standards Institution.** BS 5534-1: 1997. Code of practice for slating and tiling (including shingles). Design. London, BSI, 1997

[73] **BRE.** Slate clad roofs: fixing of slates and battens. Defect Action Sheet 142. 1990. Available as complete set (141 sheets). BR 419. Bracknell, IHS BRE Press

[74] **BRE.** Pitched roofs: renovation of older type timber roofs — re-tiling or re-slating. Defect Action Sheet 124. 1988. Available as complete set (141 sheets). BR 419. Bracknell, IHS BRE Press

[75] **BRE.** Pitched roofs: re-tiling or re-slating of older type timber roofs. Defect Action Sheet 125. 1988. Available as complete set (141 sheets). BR 419. Bracknell, IHS BRE Press

[76] **British Standards Institution.** BS 473; 550:1990 Specification for concrete roofing tiles and fittings. London, BSI, 1990

[77] **British Standards Institution.** BS EN 491: 2004 Concrete roofing tiles and fittings for roof covering and wall cladding. Test methods. London, BSI, 2004

[78] **British Standards Institution.** BS 690-4: 1974 Asbestos-cement slates and sheets. Slates. London, BSI, 1974

[79] **British Standards Institution.** BS 4624: 1981 Methods of test for asbestos-cement building products. London, BSI, 1981

[80] **British Standards Institution.** BS 6432: 1984 Methods for determining properties of glass fibre reinforced cement material. London, BSI, 1984

[81] **British Standards Institution.** BS EN 1170-3: 1998 Precast concrete products. Test method for glass-fibre reinforced cement. Measuring the fibre content of sprayed GRC. London, BSI, 1998

[82] **British Standards Institution.** BS 476-3: 2004 Fire tests on building materials and structures. Classification and method of test for external fire exposure to roofs. London, BSI, 2004

[83] **British Standards Institution.** BS 680-2: 1971 Specification for roofing slates. Metric units. London, BSI, 1971

[84] **Blanchard IG & Sims I.** European testing of roofing slate. Proceedings of ICE: Construction Materials 2007: 150(1): 1–6

[85] **Schaffer JG.** The weathering of natural building stones. Shaftesbury, Donhead, 2004

[86] **British Standards Institution.** BS EN 12588: 2006 Lead and lead alloys. Rolled lead sheet for building purposes. London, BSI, 2006

[87] **British Standards Institution.** BS 6915: 2001 Design and construction of fully supported lead sheet roof and wall coverings. Code of practice. London, BSI, 2001

[88] **British Standards Institution.** BS EN 1653:1998 Copper and copper alloys. Plate, sheet and circles for boilers, pressure vessels and hot water storage units. London, BSI, 1998

[89] **British Standards Institution.** BS EN 1654: 1998 Copper and copper alloys. Strip for springs and connectors. London, BSI, 1998

[90] **British Standards Institution.** BS EN 1172: 1997 Copper and copper alloys. Sheet and strip for building purposes. London, BSI, 1997

[91] **British Standards Institution.** BS EN 1652: 1998 Copper and copper alloys. Plate, sheet, strip and circles for general purposes. London, BSI, 1998

[92] **British Standards Institution.** CP 143-5: 1964 Code of practice for sheet roof and wall coverings. Code of practice for sheet roof and wall coverings. Zinc. London, BSI, 1964

[93] **British Standards Institution.** BS EN 988:1997 Zinc and zinc alloys. Specification for rolled flat products for building. London, BSI, 1997

[94] **British Standards Institution.** BS EN 485: Aluminium and aluminium alloys. Sheet, strip and plate. London, BSI, 1994–2008

　　Part 1: 2008 Technical conditions for inspection and delivery

　　Part 2: 2008 Mechanical properties

　　Part 3: 2003 Tolerances on dimensions and form for hot-rolled products

　　Part 4: 1994 Tolerances on shape and dimensions for cold-rolled products

[95] **British Standards Institution.** BS EN 13501-1: 2007 Fire classification of construction products and building elements. Classification using data from reaction to fire tests. London, BSI, 2007

[96] **British Standards Institution.** BS EN 13501-5: 2005 Fire classification of construction products and building elements. Classification using data from external fire exposure to roofs tests. London, BSI, 2005

[97] **British Standards Institution.** DD ENV 1187: 2002 Test methods for external fire exposure to roofs. London, BSI, 2002

[98] **British Standards Institution.** BS EN 13707: 2004 Flexible sheets for waterproofing. Reinforced bitumen sheets for roof waterproofing. Definition and characteristics. London, BSI, 2004

[99] **Ashurst J & Ashurst N.** Practical building conservation. Volume 4: Metals. English Heritage Technical Handbook. Aldershot, Gower Technical Press, 1988

[100] **British Standards Institution.** BS EN 544: 2005 Bitumen singles with mineral and/or synthetic reinforcements. Product secification and test methods. London, BSI, 2005

[101] **British Standards Institution.** BS 5707: 1997 Specification for preparations of wood preservatives in organic solvents. London, BSI, 1997

[102] **British Standards Institution.** BS 5589: 1989 Code of practice for preservation of timber. London, BSI, 1989

[103] **European Union of Agrément (UEAtc).** General directive for the assessment of roof waterproofing systems. Method of Assessment and Test (MOAT) No 27. Paris, UEAtc, 1983

[104] **Walker B.** Corrugated iron and other ferrous cladding. Technical Note 29. Edinburgh, Historic Scotland, 2004

[105] **British Standards Institution.** BS 5427-1: 1996 Code of practice for the use of profiled sheet for roof and wall cladding on buildings. Design. London, BSI, 1996

[106] **BRE.** Overroofing: especially for large panel system dwellings. BR 185. Bracknell, IHS BRE Press, 1991

[107] **Davey N.** A history of building materials. London, Phoenix House, 1961

[108] **Hastings B.** Specifications on thatch. Andover, Thatching Advisory Service, 1985

[109] **Communities and Local Government.** English House Condition Survey 1995. London, CLG, 1996

[110] **Brockett P & Wright A.** The care and repair of thatched roofs. Technical Pamphlet No. 10. London, Society for the Protection of Ancient Buildings (SPAB) and Rural Development Commission, 1986

[111] **Historic Scotland.** Thatch and thatching techniques: A guide to conserving Scottish thatching techniques. Technical Advice Note 04. Edinburgh, Historic Scotland, 1996

[112] **Historic Scotland.** Scottish turf construction. Technical Advice Note 30. Edinburgh, Historic Scotland, 2006

[113] **BRE.** Ventilating thatched roofs. Good Building Guide 32. Bracknell, IHS BRE Press, 1999

[114] **Dorset Building Control Technical Committee.** Thatched buildings: 'The Dorset Model'. New properties and extensions. 1999. Available from www. dorset-technical-committee.org.uk

[115] **Wright A.** Craft techniques for traditional building. London, Batsford, 1991

[116] **BRE.** Control of lichens, moulds and similar growths. Digest 370. Bracknell, IHS BRE Press, 1992

[117] **National Council of Master Thatchers Associations.** Visit www.ncmta. co.uk

[118] **Scottish Homes.** Scottish House Condition Survey 2004. Survey report. Edinburgh, Scottish Homes, 2005

[119] **The Scottish Government.** The Scottish Housing Quality Standard 2004. www.scotland.gov.uk

[120] **BRE.** Loft conversion. Part 1: Structural considerations. Part 2: Safety, insulation and services. Good Building Guide 69. Bracknell, IHS BRE Press, 2006

[121] **HMSO.** Party Wall, etc. Act 1996. 1996 Chapter 40. London, The Stationery Office, 1996

[122] **BRE.** Remedying condensation in domestic pitched tiled roofs. Good Repair Guide 30. Bracknell, IHS BRE Press, 2001

[123] **Harrison HW & de Vekey RC.** Walls, windows and doors: Performance, diagnosis, maintenance, repair and the avoidance of defects. BRE Building Elements. BR 352. Bracknell, IHS BRE Press, 1998

[124] **BRE.** Increasing the fire resistance of existing floors. Digest 208. Bracknell, IHS BRE Press, 1988

FURTHER READING

BRE. Improving the sound insulation of separating walls and floors. Digest 293. Bracknell, IHS BRE Press, 1985

BRE. Improving sound insulation. Good Repair Guide 22 (2 Parts). Bracknell, IHS BRE Press, 1999

British Standards Institution. BS 5268: Structural use of timber
Part 4: Fire resistance of timber structures
4.1: 1978 Recommendations for calculating fire resistance of timber members
4.2: 1990 Recommendations for calculating fire resistance of timber stud walls and joisted floor constructions

British Standards Institution. BS 5588-1: 1990 Fire precautions in the design, construction and use of buildings. Code of practice for residential buildings. London, BSI, 1990

British Standards Institution. BS 7671: 2008 Requirements for electrical installations. IEE Wiring Regulations. Seventeenth edition. London, BSI, 2008

Cox J & Letts JB. Thatch. Part 1: Thatching in England 1940–1994. English Heritage Research Transactions, Volume 6. Swindon, English Heritage, 2000

Devon County Council. Thatch in Devon. Exeter, Devon County Council, 2003. Available from www.devon.gov. uk/thatching.pdf

Doran SM. Timber frame dwellings: Conservation of fuel and power: AD L1A guidelines. Special Digest 2. Bracknell, IHS BRE Press, 2006

Doran S. Timber frame dwellings: Section 6 of the Domestic Technical Handbook (Scotland): Energy. Special Digest 6. Bracknell, IHS BRE Press, 2008

English Heritage. Thatch and thatching: a guidance note, English Heritage, 2000. Available as a pdf from www.newforestnpa.gov.uk

Mindham CN. Roof construction and loft conversion. 4th edition, Oxford, Blackwell Publishing, 2006

Moir J & Letts J. Thatch. Part 1: Thatching in England 1790–1940. English Heritage Research Transactions, Volume 5. Swindon, English Heritage, 1999

Sanders C. Modelling and controlling interstitial condensation in buildings. Information Paper IP 2/05. Bracknell, IHS BRE Press, 2005

Sanders C. Modelling condensation and air flow in pitched roofs. Information Paper IP 5/06. Bracknell, IHS BRE Press, 2006

TRADA Technology & Institute of Building Control. Loft conversion guide. High Wycombe, TRADA Technology, 1996

Williamson L. Loft conversions: planning, managing and completing your conversion. Marlborough, Crowood Press, 2000

4   SHORT-SPAN DOMESTIC FLAT ROOFS

As has already been described in Chapter 2.11, flat roofs may be either warm deck or cold deck, with warm deck having two options, sandwich or inverted. The sandwich is so named because thermal insulation is sandwiched between the vapour control layer laid on the deck and the weatherproof membrane laid above it. The inverted roof is so-called because the weatherproof layer, which also acts as the vapour control layer, is not in its usual position, being buried beneath the thermal insulation and ballast.

The choice of waterproof membrane for covering flat roofs from medieval times was confined,

in practice, to lead (Figure 4.1). Later came the alternatives of mastic asphalt and built-up felts, with the former normally used on concrete decks, and the latter on timber decks. However, new products have been developed, including improved formulations of bitumen felt and synthetic polymers. Recent years have therefore seen a growth in the use of plastics and rubbers at the expense of traditional felts (Figure 4.2).

These have been followed by liquid-applied coatings which are principally used in the repair of failed membranes or in upgrading thermally deficient flat roofs. Liquid roofing systems are based on a

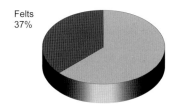

Figure 4.2: Mix of roofing membranes and felts, 2006. Data from *Roofing Market Report*[1]

wide range of materials including bitumen, thermosetting resins and acrylic-based coatings. They can be applied with a spray, roller or brush, producing a seamless roof

Figure 4.1: A well laid lead roof

which, provided workmanship can be guaranteed, particularly the avoidance of pin-holes, should reduce the chance of failure when compared with joints in sheeted membranes. They are also useful for refurbishing three-dimensional curved roofs.

Liquid-applied coatings include foamed in-situ polyurethane applied by special spray equipment to a depth of around 50 mm. A final surface protective layer is then sprayed on. Third-party certificates exist for spray-applied products and BRE receives a steady flow of enquiries about their use.

Suitable liquid roofing materials can be applied to existing structures, which makes them appropriate to remedial work since they can be applied to a range of common substrates in either pitched or flat configurations. When properly applied, liquid roofing is said to have good longevity, for example in excess of 25 years. These higher performing materials will normally be covered by third-party certification.

The European Liquid Waterproofing Association[2] lists seven generic types:

- polymer-modified bitumen emulsions and solutions,
- glass-reinforced resilient unsaturated polyester resins,
- flexible unsaturated polyesters,
- hot-applied polymer-modified bitumens,
- polyurethanes,
- bitumen emulsions and solutions,
- water-dispersible polymers.

Liquid roofing systems are said to offer a number of benefits including:

- seamless membrane, so reduced risk of joint failure,
- minimum load increase on structure,
- minimum disruption during installation.

High quality liquid-applied coatings are also available based on glass-fibre and resin, or elastomeric polymers; both types have been used on domestic new-build work.

HOUSE CONDITION SURVEYS

The English House Condition Survey 1991[3] revealed some data about the condition of structures for dwellings in England that had flat roofs. However, these data did not distinguish between roofs having different materials in their coverings, so, for convenience, the figures that follow are for all roofs in this category (ie short-span domestic flat roofs). Nor do the data distinguish between warm-deck and cold-deck structures though a very great proportion of flat roofs at this time would have been cold deck. Many coverings in these categories would have reached, or be reaching, the end of their expected lives, though in the case of asphalt coverings there was a curious peak for roofs built between 1965 and 1980 where nearly one-quarter of roofs were considered to need attention.

Table 4.1 shows data for dwellings in the UK in 1991[3–6] having flat roofs over the whole or part of their area, and the proportion that showed faults in the underlying structure or deck.

So far as the finishes are concerned the data in England in 1991[3] did not distinguish between pitched and flat but, overall, about 1 in 6 or 7 built-up roofs were in need of repair. About 1 in 8 asphalt flat roofs also needed attention.

In the data for England in 2005/6[7], the information has been analysed separately for houses and for flats and has been divided into roofs that were entirely flat (100%), predominantly flat (> 50%), those with some flat roof (< 50%) and those with no flat roof.

In England in 2005/6 there were 760,859 dwellings with entirely flat roofs:

- 178,994 houses (1.0%) and 734,034 (19.2%) flats with mainly flat roofs,
- 3,923,446 houses (21.6%) and 511,571 flats (13.4%) with some flat roofs.

Overall, some 6.7% of flat roof coverings required urgent attention in 2005 in England.

Table 4.1: Reported problems with domestic flat roofs in the UK. Data from national House Condition Surveys circa 1990[3–6]

	Number of flat roofs (whole or part)	Proportion with a problem (approx.)
England:		
Houses	3.15 million (20% of total stock)	1:34
Flats	1.16 million (33% of total stock)	1:50
Wales:	No data	—
Scotland:		
Houses and flats	120,000 (6–7% of total stock)	
Built-up felt	60,000 3–3.5% of total stock	1:5
Mastic apshalt	60,000 3–3.5% of total stock	1:6 or 1:7
Northern Ireland:		
Houses and flats	4,500	No data

Comparable data for Wales and data for 2005 in Scotland were not available for this book.

In 1991 in Northern Ireland the majority of flat-roofed houses were covered with felt.

In Northern Ireland in 2005/6[8] there were:

- 84,309 houses (13.0%) and 3,780 flats (6.7%) with some flat roofs,
- 6,898 houses (1.1%) and 5,360 (9.6%) flats with mainly flat roofs,
- 9,781 dwellings with entirely flat roofs.

Overall, some 3.6% of flat roof coverings required urgent attention in 2005 in Northern Ireland.

Some general information about defects in flat roofs can be found in the BRE Report *Assessing traditional housing for rehabilitation*[9].

4.1 BUILT-UP FELT

The basic materials used in built-up felt roofs have been described in Chapter 3.4 for pitched roofs. When, though, the pitch of the roof is nominally flat, the service conditions the materials must face are more onerous than on pitched roofs.

There is a wide range of quality and performance in the different products available on the market. However, in practice only a few combinations of materials are recommended; for example:

- glass-fibre and polyester-based felts to BS 747[10] are sometimes used in combination,
- BS 8217[11] polymer-modified materials now take a significant proportion of the market.

Oxidised bitumen gradually embrittles on exposure to solar radiation, and this effect can be controlled by the addition of:

- APP (atactic-polypropylene) or
- SBS (styrene-butadiene-styrene).

APP felts are normally torched or hot-air welded; SBS felts are stuck with hot bitumen though there are some SBS felts which can be torched. It is important to recognise the difference in performance at different temperature ranges.

BS 747 was amended in 1986 to include class 5 felts with a polyester base. The three types of class 5 felts in common use are given in Box 4.1. BS 747 was further amended in 2000 to include different weights of polyester bases (eg 5B/180).

A new European Standard, BS EN 13707[12], has been introduced and, under the rules of Europe, any conflicting National Standard has to be withdrawn. This means that BS 747: 2000[10] is

---

Box 4.1: The 3 types of Class 5 felts in common use

Class 5B

A felt that is impregnated and coated with oxidised bitumen, and used as the top layer of a membrane with a surface protection of bitumen and mineral aggregate

Class 5E

A cap sheet with a top layer of bonded mineral granules

Class 5U

An underlayer

---

superseded although reference to it is expected to continue in the short term.

BS EN 13707[12] gives only results to test methods and does not contain any system for determining the performance of the bitumen sheets to which it refers. It does, however, cover the use of all products so this includes bitumen sheets made using polymer-modified bitumen.

BS 8747[13] has been produced to enable products tested according to BS EN 13707 [10] to be used in the UK.

To this end, BS 8747[13] gives an SnPn rating where S_1P_1 has the lowest performance and S_5P_5 has the highest. S is based on strength and P on static and dynamic impact loading. For the sake of continuity, guidance continues to be given with reference to BS 747. BS 8747 does give guidance for an equivalent SP rating.

Polyester felt membranes normally comprise two layers, though an additional first layer is

---

needed where partial bonding or nailing is used.

Reinforced bituminous membranes that can be applied in a single sheet are now coming into wider use. The details associated with these membranes are similar to those for built-up membranes except for the fact that most are fixed mechanically at the edges of sheets, with adjoining sheets overlapped by 100 mm and hot-air welded into place.

The European Union of Agrément has issued guidance for its members in assessing new developments in APP felts[14].

Polymeric single-layer membranes until now have been used rather more frequently in larger commercial and industrial buildings and are therefore covered in Chapter 5.2.

CHARACTERISTIC DETAILS

Basic structure

Timber decks are usually found on houses and concrete decks on blocks of flats, though occasionally examples can be found the other way round. In timber decks, the simplest way of ensuring that basic strength and stability requirements are met is to ensure that members are sized in accordance with the appropriate national building regulations[15–17]. This also reduces to acceptable limits deflections which frequently are a cause of defects that develop later. BRE investigators have found that reinforced concrete rarely deflects to the extent that problems ensue for the coverings (Figure 4.3).

Tongued-and-grooved timber boarding is better than plain-edged boarding since it produces a level surface.

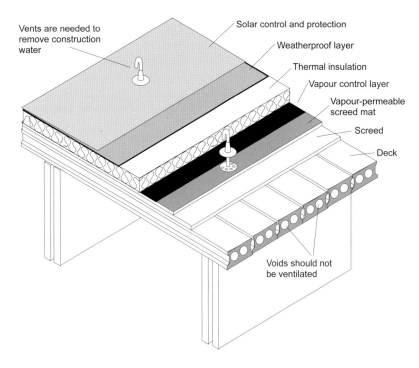

Vents are needed to remove construction water

Solar control and protection

Weatherproof layer

Thermal insulation

Vapour control layer

Vapour-permeable screed mat

Screed

Deck

Voids should not be ventilated

Figure 4.3: A typical concrete warm deck roof. If construction water is not allowed to dry out before covering the roof, vents will be needed above the screed to prevent high pressures both blistering the covering and driving moisture downwards through the deck and ceiling

Until recently, metal decks have rarely been used in housing. They are dealt with in Chapter 5.2. Plywood decks suitable for use with this category of roofing are described in Chapter 3.4.

Eaves and verges
Welted drips of at least 50 mm depth and projection should be provided for eaves and verges (Figure 4.4). These are normally fixed over a preservative-treated timber batten.

Abutments
A generous fillet should be provided at abutments or changes of level for all kinds of flat roof coverings (Figure 4.5). Built-up membranes, for example, which are turned through a right angle without employing a fillet, are very likely to crack at the angle even if subjected to only a very small strain.

Roof lights
Roof lights suitable for installation into pitched roofs covered with felts are not always suitable for use on flat roofs. Lights specifically designed for flat roofs, ie with deep all-round kerbs, should be used.

Thermal bridging can occur at kerbs, even when insulated (Figure 4.6). Condensation may form on single-glazed roof lights.

Service perforations
The aim in both new design and in refurbishment of older flat-roofed dwellings should be to keep the

Vapour control layer

Thermal insulation

Timber deck on timber joists

No ventilation at eaves

External wall support

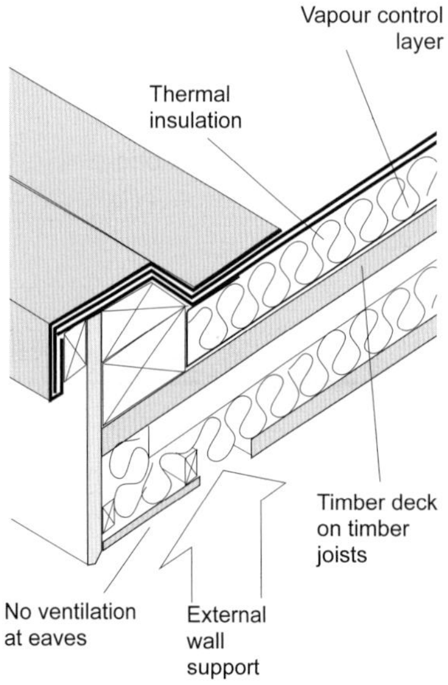

Figure 4.4: Detail of eaves in the timber deck of a warm roof. The external wall is not shown

flat roof as free as possible from services; it is these interruptions in the weathertight layer that make it vulnerable to rain penetration. Heating system flues should be routed through external walls in preference to roofs wherever possible. Alternatively, outlets should be adequately detailed to allow for the effects of both movement and heat (Figures 4.7 and 4.8).

Rainwater disposal
At one time, BRE recommended minimum falls of 1 in 60 to 1 in 80 for flat roofs to assist drainage. However, BRE Digest 144[18] reported that there was no evidence from a survey of the performance of membranes in Crown Estate buildings that falls had any effect on the life of the coverings. In any case, a truly flat surface is not likely to be achieved in practice and ponding is unlikely to occur when a fall sufficient to drain depressions is provided. Ponding is less harmful to coverings than was at one time supposed; but if a leak occurs in a ponded area, water will enter the building in greater quantities than through an efficiently drained surface. Ponding of up to 75 mm depth has been reported to BRE.

BS 6229[19], the code of practice for flat roofs, recommends a design fall of 1 in 40 so that once deviations

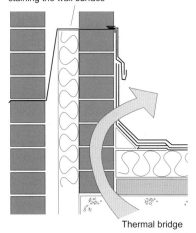

A tray turned outwards (so that water flows to the outer leaf) risks the water staining the wall surface

Thermal bridge

Figure 4.5: A typical abutment detail resulting in a thermal bridge. The inside face should be insulated as shown in Figure 4.6

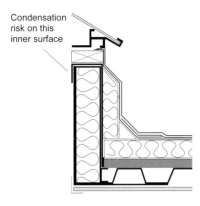

Figure 4.6: An insulated kerb to a roof light in a warm roof. If the kerb is insufficiently insulated, condensation can occur on the inside face; this may be confused with condensation from the lower edge of a single-glazed roof light above the kerb

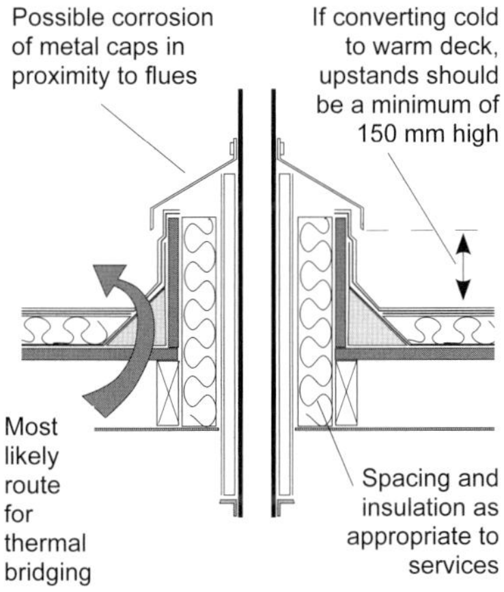

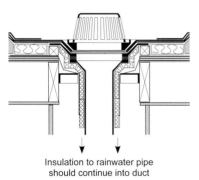

Figure 4.7: Service outlets need to be adequately capped over upstands

Figure 4.8: Rainwater outlets penetrating warm roofs should be thermally insulated

have been taken into account, a 1 in 80 fall will be achieved.

These falls will allow a roof to drain readily to the outlets and into the rainwater dispersal system. There is a view that rapid drainage from roofs is a contributory factor to localised flooding and that retaining water on the roof, or holding it in a buffer tank would have a beneficial effect. These issues are also being examined for the usage of 'grey' water in buildings for toilet flushing.

So, although there may not be the same strict requirement for adequate falls as there used to be in the days when oxidised bitumen felts were the only solutions available, a slight fall is a useful insurance.

MAIN PERFORMANCE REQUIREMENTS AND DEFECTS

Strength and stability

The fault that was seen to occur most frequently in the BRE studies of rehabilitation work on housing with flat roofs was sagging of roof beams in timber roofs causing ponding and subsequent penetration of the upper decks[9] (Figure 4.9). Usually the cause was undersized joists.

Correct bonding of the weatherproof surface to the deck, in the case of both cold and warm deck roofs, is of crucial importance to the performance of the covering under wind suctions and movements of the deck. On timber decks the best practice is to nail or spot-bond the first layer, with subsequent layers in most cases fully bonded. This will allow movement of the deck to take place without disrupting the covering. On concrete decks the

first layer should also be partially bonded.

In the case of plywood decks it is important that the spans selected, or the thickness of the plywood, do not allow deflection sufficient to exceed the slope of the roof and therefore to cause ponding of rainwater on the surface. Deflection can follow from condensation collecting on the underside of the waterproof layer leading in turn to moisture movement in the deck. Deflections surrounding rainwater outlets need to be minimised and a flexible connector to the rainwater outlet specified.

Wind loads creating suction on flat roofs can be substantial (Figure 4.10). Further information on the design of flat roofs to resist wind loads is available in the various parts of BRE Digest 346 (8 Parts)[20].

Dimensional stability

So far as the structure of a small-span domestic flat roof is concerned, there can be considerable exposure to solar radiation; this could lead to thermal movement in certain kinds of design, such as a concrete slab, unless the roof is very small. The temperature reached on roof surfaces may be increased by dark-coloured coverings. Movements are unlikely to take place in warm deck or inverted roofs constructed since about the 1980s because the deck is protected from thermal gains. Flat roofs of conventional cold deck construction, whether of light or heavy construction, will, however, experience wide ranges of temperature, and this may result in movement of the deck and disruption of the coverings.

In the survey of flat roofs described in BRE Digest 144[18], which did not of course include any examples of BS 747 class 5 felts, about half of the failures in built-up felt coverings were attributed to splits produced by locally concentrated movements of the substrate. Very few were attributed to ageing of the felt. Blisters and ridges in the felt are rarely directly responsible for leakage, but they may be vulnerable to damage and may impede drainage. If a bituminous felt membrane is not

Figure 4.9: Ponding of rainwater due to a sagging deck

Figure 4.10: Gale damage to a built-up felt roof. Each layer has failed in turn. From right to left: cap sheet, lower sheets, underlay, timber boarding, firring pieces, and even joists. Some of the boards have pivoted on the remaining nails

(a)

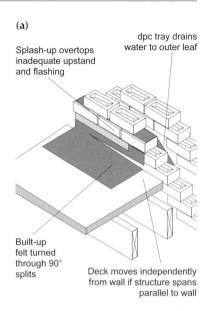

Splash-up overtops inadequate upstand and flashing

dpc tray drains water to outer leaf

Built-up felt turned through 90° splits

Deck moves independently from wall if structure spans parallel to wall

(b)

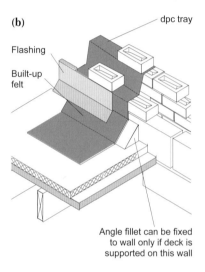

dpc tray

Flashing

Built-up felt

Angle fillet can be fixed to wall only if deck is supported on this wall

Figure 4.12: Defects found at built-up felt abutments (a) and how they may be avoided (b)

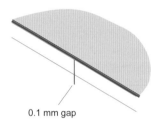

0.1 mm gap

Fully bonded membrane before movement

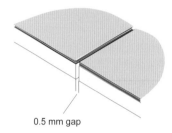

0.5 mm gap

Fully bonded membrane after movement and failure (membrane stretched by 400%, so that it splits)

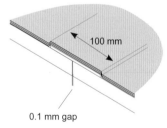

100 mm

0.1 mm gap

Partly bonded membrane before movement

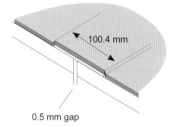

100.4 mm

0.5 mm gap

Partly bonded membrane after movement. No failure (membrane stretched by 0.4% so that it does not split)

Figure 4.11: Provisions necessary to isolate built-up felts from movements in the substrate

isolated from localised movements in the substrate, it is likely to split and leak (Figure 4.11).

Fatigue failure of the membrane can occur following rapid changes in temperature. The higher the levels of thermal insulation in a sandwich warm deck flat roof, the higher the thermal stresses and the greater the risk of disruption. In these circumstances, using BS 747 class 5 felts will provide higher tensile strengths, though the surface

should have a reflective finish which is normally achieved by embedding mineral granules in the surface during manufacture.

Exclusion and disposal of rain and snow

As might be expected, many of the older roofs covered in three-ply built-up felt that were seen by BRE investigators during the surveys of rehabilitated housing had problems of rainwater penetration. Patching

occurred frequently. Roofs of this kind, using organic fibre felts over boards of high moisture or thermal expansion characteristics, have not generally been specified since the mid-1970s. Although cases of leaks in asphalt roofs were fewer, they were invariably found in the very oldest of such roofs.

Abutments have been a common source of problems (BRE Defect Action Sheet 34[21] and Figures 4.12 and 4.13).

Extruded aluminium or plastics edge trims in short lengths, used at the eaves of built-up roofing, should be securely fixed as closely as practicable on each side of joints in the lengths of trim. Otherwise, thermally induced size changes are very likely to initiate a crack in the

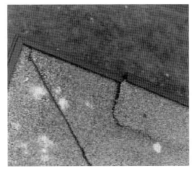

Figure 4.14: This eaves trim was not securely fixed down to the deck. Consequently, outward movement has torn the built-up felt. The shorter the length of metal or plastics, the less will be the thermal movements. Fixing should be made at both ends of lengths, close to joints in the trim

Figure 4.15: Laying polyisocyanurate boards to create a warm deck roof

Figure 4.13: A defective abutment detail. Looking down on the canopy to a doorway it can be seen that a fillet has been provided at the junction between the canopy and the wall but there is no flashing

membrane under each joint (Figure 4.14).

Energy conservation and ventilation

In a warm deck roof (Figure 4.15) the thermal insulation is placed above a vapour control layer. The integrity of this vapour control layer is crucial to subsequent performance in reducing the risk of condensation. If the layer is of polyethylene, non-recycled material could prove more durable than recycled though current advice is that there is no detectable difference between the two in permeability to water vapour.

In a cold deck roof the thermal insulation is placed at ceiling level with the void between the ceiling and the deck ventilated to the outside to remove any condensation. Ventilation requirements for a cold deck flat roof are a continuous 25 mm gap at eaves for spans up to 5 m, and 30 mm for 5–10 m. There should be free air space above the insulation of 50 mm for spans up to 5 m and 60 mm for 5–10 m.

The following faults have frequently been seen by BRE investigators on site.
- Gaps in thermal insulation (in some cases up to 20 mm)

leading to thermal bridging and condensation
- No vapour control layer beneath a cold deck roof
- No ventilation provided to a cold deck roof. Cases have been seen where insulation fully fills the space between deck and ceiling, thus negating the provision of ventilation at eaves. Flat roofs over dormers set into pitched roofs are often poorly ventilated

When improving the performance of flat roofs which involves upgrading thermal insulation by placing it above the waterproof membrane, either converting the roof to a conventional warm deck or to an inverted warm deck, it is important to ensure that there is no risk of condensation below the waterproof membrane (Box 4.1). Calculations should be carried out to check that, by adding insulation, the air below the waterproof membrane does not reach dewpoint. In general, at least 75% of the total thermal resistance of the upgraded roof should be on the outside of the membrane.

Where a vapour control layer is required the best performance has been given by two layers of a felt to BS 747 (or type 5) bonded in hot bitumen and fully supported.

In the case of a cold roof that cannot be converted to a warm deck roof, the vapour control layer will often be a polyethylene sheet with sealed joints. The performance, though, of a vapour control layer will

be compromised by penetration by services and fixings, and therefore can be only of limited value.

Control of solar heat

In both warm deck sandwich and cold deck construction, the weatherproof membrane is exposed to solar radiation. Temperatures can range from $-20\ ^{\circ}\text{C}$ to $80\ ^{\circ}\text{C}$ where directly exposed to the sun. In the inverted roof the membrane is shielded from solar radiation, and temperature effects are greatly reduced or even eliminated altogether (but care is required at upstands and perimeters).

Where built-up roofing is used, consideration may be given to using white or light-coloured mineral surfacing of the cap sheet to reduce solar heat gains. Such protection will enhance durability. The use of solar-reflective dressings or paints was discussed in the section on prediction of temperatures in Chapter 2.5.

Fire precautions and lightning protection

The Building Regulations 2000 (England & Wales) Approved Documents B1 & B2[15] state that to achieve an AA rating, built-up bitumen felt flat roofs (irrespective of the felt specification; Figure 4.17) need to be laid on decks of:
- 6 mm plywood,
- 12.5 mm wood chipboard,

Box 4.1: Checks and actions for converting cold deck roofs to warm deck

(a)

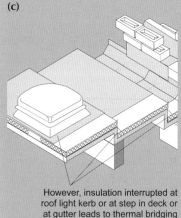

Defective cold deck with poor ventilation, vapour control layer absent or ineffective, with water vapour condensing below covering

(b)

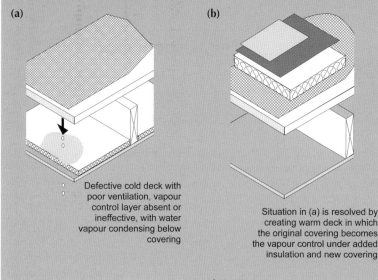

Situation in (a) is resolved by creating warm deck in which the original covering becomes the vapour control under added insulation and new covering

(c)

However, insulation interrupted at roof light kerb or at step in deck or at gutter leads to thermal bridging

(d)

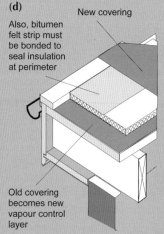

Also, bitumen felt strip must be bonded to seal insulation at perimeter

New covering

Old covering becomes new vapour control layer

Figure 4.16: Points to watch in converting cold deck flat roofs to warm deck

Before deciding to convert existing cold deck flat roofs to warm deck it is important to check that:
- the roof structure and deck are sound
- the upstands are sufficiently high to accommodate the extra layers
- there is no existing thermal insulation below the deck. If insulation is present, ideally it should be removed so that all insulation in the new roof is above the vapour control layer (Figure 4.16b). If removal is not practical, the insulation installed above the deck may not entirely remove the condensation risk
- the existing covering is free from irregularities which might cause new insulation to lie unevenly.

If, having made the checks, the decision is made to convert, the following actions should be taken:
- ascertain the manufacturers' instructions for the chosen products as instructions do vary
- specify that boards are to be kept as dry as possible
- check that the insulation covers every surface in the roof which provides a pathway for heat loss; in other words, there are no thermal bridges in the construction (Figure 4.16c)
- specify that the vapour control layer (ie the former roof covering) is to be sealed to the new covering at the perimeter of the insulation (Figure 4.16d)
- check that existing ventilation paths are sealed after ensuring that the structure is dry

- 16 mm (finished) tongued-and-grooved softwood,
- 19 mm (finished) plain-edged softwood,
- screeded woodwool slab,
- profiled fibre-reinforced cement deck,
- profiled steel with or without insulating board,
- profiled aluminium deck with or without insulating board,
- in-situ concrete or clay pot or precast concrete,

and have a finish over the whole surface of:
- 12.5 mm depth of bitumen-bedded stone chippings,
- bitumen-bedded non-combustible tiles,
- sand and cement screed, or
- macadam (Figure 4.17).

Lightning protection should comply with BS 6651[22]. This usually means in the case of flat roofs, a grid of conductors is laid near the eaves with intermediate conductors for larger spans.

Sound insulation
Readers should refer to this section in Chapter 3.3.

Durability
Built-up bitumen felt may suffer from:
- splitting due to movement in the deck or insulation below (particularly differential movement at upstands),
- cracking due to exposure to sunlight (Figure 4.18),
- blistering due to moisture entrapment,
- incomplete bonding of layers,
- rucking due to poor laying.

The durability of coverings is dealt with in BRE Digest 144[18], with some additional information in BRE Digest 372[23]. The natural ageing of oxidised bitumen felts limits their useful life to around 20 years, though built-up polyester membranes have performed better than older types of felt to BS 747. In particular, there is no evidence of failures caused by fatigue cracking of the membranes when fully bonded to foamed plastics insulation

AA Finish of 12.5 mm bitumen-bedded
stone chippings

Bitumen-bedded non-
combustible tiles

Sand:cement screed or
macadam

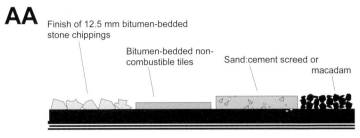

Decks of most materials

Figure 4.17: Expected fire performance of built-up felt coverings

Figure 4.19: Evidence of more than one unsatisfactory and totally unskilled attempt at repair with poor adhesion of the patch on the upstand and bitumen partially blocking the grating

boards in warm deck designs. BRE is unaware of failures of the membranes for the oldest roofs using class 5 polyester-based felts which date from the mid-1970s*.

In 1972, BRE analysed the records of a sample of flat roofs up to 40 years of age which confirmed that the probability of failure increased with age; this might be expected for roofs constructed with the materials available at the time. There was a sharp rise in maintenance costs after 12 years. Although materials offering better performance have now become available, built-up roofing remains heavily dependent for durability on the quality of workmanship and it is therefore not easy to give any guarantee of lives for these coverings.

Although manufacture in the UK has ceased, built-up felts of the older BS 747 classes can still be found, particularly on small areas of flat roofs, and it behoves supervisors to check that detailing in

---

* At least until the mid 1990s.

Figure 4.18: Cracking of bitumen felt roof covering and upstand due to exposure to sunlight

rehabilitation work is being correctly observed. Cases have been seen by BRE site investigators where felts were being installed with right-angled bends (ie without triangular fillets) giving rise to concern about durability. Patch repairs using new felts can cause older weak felts to be overstressed and induce failure around the patch.

One cause of early failure of built-up flat roofs is lack of provision for differential movement between the deck and the covering. This is normally provided by a partially bonded or mechanically fixed layer to which the upper sheets are bonded. A further cause of premature failure of repairs to the built-up covering has been put down to lack of cleaning of the original surface and lack of proper preparation before application of new material.

Ease of maintenance

It is good practice to examine flat roofs annually, noting any blisters in exposed weatherproof membranes, and any signs of disturbance at parapets and changes of level. Rainwater outlets must be cleared (Figure 4.19).

Liquid-applied membranes, some of them reinforced with a glass-fibre or polyester base, may be used in remedial work to provide a temporary cure for small leaks, though compatibility of the new material with the old roof should be ascertained from the manufacturer. Choosing a product that has third-party certification is advisable.

Splits in felt can be repaired satisfactorily provided the methods or workmanship that led to the original error are not repeated. Faults in the structure itself that lead

to splitting should be corrected. The first strip is either not bonded, or bonded only at one edge. Subsequent layers may be fully bonded provided the minimum dimensions given in Figure 4.20 are observed.

Moss, and even plants, will grow in the detritus that builds up over time amongst the stone chippings with which so many built-up felt roofs are finished; this unwanted material will need clearing away from time to time if drainage is not to be impeded (Figure 4.21).

WORK ON SITE

Access, safety, etc.
Safety advice in connection with work on flat roofs that are more than 2 m above ground is available from HSE[24]. Contractors are also responsible for the safety of operatives and of the public in connection with their work on the site, including the security of access equipment. The Flat Roofing Alliance provides technical guidance on installation[25].

Responsibility for safety precautions applies to managerial, supervisory and inspecting staff as well as to site operatives.

Storage and handling of materials
Some thermal insulation materials (eg expanded polystyrene) can be damaged by the heat involved in laying hot bitumen. Such insulation needs to be protected with a layer of, for instance, fibre or cork board.

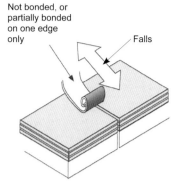

Not bonded, or partially bonded on one edge only

Falls

Fully bonded

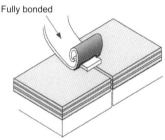

Figure 4.21: Moss growing on a built-up felt roof covered with small stone chippings. Although the moss and detritus interfere with the free flow of water, on such flat surfaces the detritus does not tend to migrate to the rainwater outlets. The felt flashing continues up behind the lowest course of tiles though the soft mortar at the tuck-in is eroding

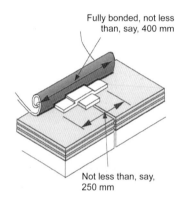

Fully bonded, not less than, say, 400 mm

Not less than, say, 250 mm

Figure 4.20: Stages for the repair of a split in a felt roof

Restrictions due to weather conditions

A flexible roof membrane of any type will not perform well if laid on a wet or frosted deck. Adhesion will be poor and trapped moisture may later cause blisters. Also, the risk of cracking membranes as they are unrolled increases with falling temperature. Low temperatures may also cause lack of adhesion even when the adhesives are hot-applied.

The following precautions should be observed:
* sheet membranes are not handled or laid in winds of more than 16 m/s,
* sheet membranes are not handled or laid when the air temperature is less than 5 °C for heavy sheets and 0 °C for lighter types,

* sheet membranes are not handled, laid or jointed when any rain, sleet or snow is falling.

Further precautions are listed in BRE Defect Action Sheet 63[26].

Workmanship
Although all kinds of built-up roofing are dependent on the quality of workmanship for adequate performance, this is especially true of torched sheeting. Also, all polymeric materials should be installed only by fully trained operatives. The Flat Roofing Alliance[25] and National Federation of Roofing Contractors[27] have lists of contractor members who have been vetted by the industry.

Supervision of critical features
Adequate drying time for concrete decks needs to be allowed or there will be the risk of trapped water. In general, if protected from further wetting, construction water dries out from concrete at a rate of about 25 mm per month, depending on circumstances. If in doubt, a test can be carried out to check for the dryness of slabs.

In the experience of BRE, it is sometimes quite difficult to ensure that partial bonding of the built-up roofing to the deck is achieved. In these circumstances, using a perforated first layer, laid unbonded,

is useful; the bond of subsequent layers can then be made only through the holes in the membrane (Figure 4.22).

In general, it is recommended that all layers should be laid to break joint (ie the joints should be staggered).

On metal decks placing the thermal insulation diagonally has been used as a means of reducing the build-up of movements and has been reported to be beneficial.

Lists of important checks to be made when constructing warm deck sandwich, inverted and cold deck flat roofs are given in Box 4.4.

Figure 4.22: Where deck movements may be significant, the first layer is of perforated material loose laid to ensure only partial bonding of subsequent sheets to the deck with adhesive. Where high-performance felts are used, there is less need for partial bonding of the first layer provided deck movements are likely to be small

Box 4.2: New Standards

New European product standards have superseded old British Standards. These are:

- BS EN 13707: 2004 Flexible sheets for waterproofing. Reinforced bitumen sheets for roof waterproofing. Definition and characteristics[12]
- BS EN 13956: 2005 Flexible sheet for waterproofing. Plastic and rubber sheets for roof waterproofing. Definitions and characteristics[28]

A new guidance document, BS 8747[13], has been produced to enable the use of BS EN 13707 to be matched with the old Standard BS 747[10].

Box 4.4: Checklists for construction of flat roofs

Warm deck sandwich

- Vapour control layers are continuous and bonded to decks
- Vapour control layers are turned up at perimeters and are bonded to weatherproof layers
- Appropriate insulation materials are used (eg not extruded polystyrene)
- Insulation is kept dry until roofs are sealed
- Roof insulation links with wall insulation
- Insulation is continuous without gaps
- Expanded polystyrene is protected from hot bitumen with fibreboard (There are alternative insulants available that are not in need of such protection but they will need protection from the weather)

Inverted

- Extra insulation is placed beneath main layers where specified
- Drainage layers are fitted if specified
- Drainage is free from obstruction
- Half-lapped insulation boards are fitted
- Ballast is applied immediately after laying insulation
- Design loads can effectively be carried through all the layers (eg without exceeding the crushing strength of the insulation)
- Upstands are protected from solar gain

Cold deck

- Continuity of vapour control layers is as good as can be obtained
- Ventilation openings are as specified (top vents alone will not be sufficient)
- Roof insulation laps with wall insulation

Box 4.3: Inspection of built-up felt roofs

In any investigation of faults in flat roofing, of whatever design, it will be important to determine the method of support of the deck and the effects this will have on restraint of the deck against thermal movement. The direction in which size changes are occurring will often be indicated by the angle of rucks in the waterproof covering at perimeters. A straight edge applied across cracks in the membrane will reveal whether distortion or deflection of the deck panel is a more likely cause than linear thermal size changes.

Where lap joints have opened or lifted in an older membrane, filling the gap with a sealant is rarely effective, and usually means replacement with a newer type of membrane.

In external inspections of existing built-up roofs, the problems to look for are listed below.

- Signs of poorly executed previous repairs

- Sagging (usually suggesting failure of decking between joists)
- Excessive deflections in the deck
- Blistering, splitting or rucking of coverings (Figure 4.23)
- Gaps between roof and adjacent walls
- Inadequate falls or drainage
- Ponding
- Poor adhesion of covering to decks
- Failed adhesion of lap joints
- Mechanical damage
- Inadequate solar protection

Figure 4.23: Tearing of built-up felt over the arris of a verge kerb

- Flashings absent or damaged
- Poor detailing of flashings
- Inadequate flashings to upstands
- Insufficient height of upstands

If faults are suspected during external inspection of built-up roofs, and of the rooms below flat roofs, some opening up of the roof spaces should be considered. In these cases, the problems to look for are listed below.

- Inadequate thermal insulation
- Inadequate cross-ventilation in cold roofs
- Presence of dampness, staining, mould growth, wood rot and insect attack
- Condensation
- Decks of inadequate construction (eg rotation of joists or corroded hangers)
- Absence of fire stops at separating walls

4.2 MASTIC ASPHALT

Table 4.2 shows data for dwellings in England in 1990[3] having mastic asphalt flat roofs over the whole or part of their area, and the proportion that showed faults in the underlying structure or deck. This information is no longer available but it seems unlikely that the fault rate has improved much in the intervening years.

There are observable differences in the performance of the older natural and synthetic materials. Much has been learned about the reasons for these differences since the 1939–45 war and failures should not now occur.

The more straightforward details on timber decks are dealt with in this chapter, but the more unusual (eg on metal decking typical of schools construction) are dealt with in detail in Chapter 5.2. Mastic asphalt is covered in BS 8218[29]. BS 6229[19] is also relevant. Materials are dealt with in BS 6925[30] and BS 5284[31]. Further guidance is available from the Mastic Asphalt Council[32].

CHARACTERISTIC DETAILS

Basic structure
Readers should refer to this section in Chapter 4.1.

Eaves and verges
Eaves and verges have been seen occasionally with the mastic asphalt dressed down over the fascia to form a drip or, more commonly, with the verge covered with a metal flashing or strip of high-performance elastomeric felt dressed down to the drip or gutter (Figure 4.24). The former detail is rarely applied because of the skill needed to strike off the mastic asphalt to a straight line to form the drip.

	Number of flat roofs (whole or part)	Proportion with a problem (approx.)
England:		
Houses	309,000	1:8
Flats	371,000	1:7

Table 4.2: **Reported problems with domestic mastic asphalt flat roofs in England. Data from English House Condition Survey 1991[3]**

Where access is provided to the roof (eg where it is used as a means of escape) handrails must be installed. These are commonly a source of problems and solutions that avoid perforating the asphalt are preferred (Figure 4.25).

Abutments
Although asphalt is not normally found on pitched roof surfaces, it is sometimes used on relatively small changes of level that involve covering vertical or steeply sloping surfaces more than about 150 mm in height. On these surfaces the asphalt must be chased into a substantial rebate (25 × 25 mm) and provided with a suitable key (eg expanded metal) firmly fixed to the substrate. Otherwise, sagging of the asphalted covering and subsequent disruption will probably occur. It is also good practice to keep the cavity tray at least 75 mm above the rebate position so that damage to the tray when cutting the rebate is avoided. Thermal bridging can be a problem too (Figures 4.26 and 4.27).

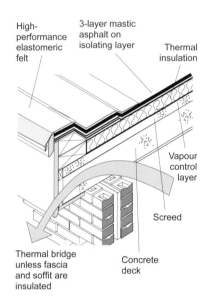

High-performance elastomeric felt

3-layer mastic asphalt on isolating layer

Thermal insulation

Vapour control layer

Screed

Thermal bridge unless fascia and soffit are insulated

Concrete deck

Figure 4.24: Eaves detail for a warm roof over a concrete deck

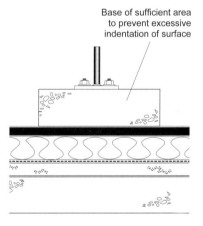

Base of sufficient area to prevent excessive indentation of surface

Figure 4.25: It may be possible to avoid pedestrian barriers (where they are to be situated away from eaves) from perforating the thermal insulation of a warm deck roof by fixing the standards to separate bases which spread the load

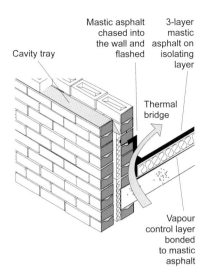

Cavity tray

Mastic asphalt chased into the wall and flashed

3-layer mastic asphalt on isolating layer

Thermal bridge

Vapour control layer bonded to mastic asphalt

Figure 4.26: The abutment detail in the design of this warm roof has failed to eliminate thermal bridging

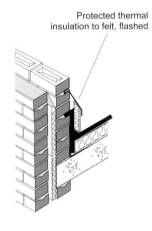

Protected thermal insulation to felt, flashed

Figure 4.27: Detailing for an insulated upstand which maintains the integrity of the warm roof design

Roof lighting and service perforations

Kerbs to roof lights should be treated as if they were kerbs to abutments. Upstands of at least 150 mm are required, and more if drifting snow is to be kept at bay (Figure 4.28; see also Figure 4.29).

Rainwater disposal

Where sumps are formed there needs to be continuity in thermal insulation (Figure 4.30). In asphalt roofs, ponding can be the cause of slight crazing of the surface. In the main this does not warrant repair or replacement on technical grounds, though it may do so on grounds of appearance where the affected

roofs are open to view from higher buildings (Case study 4).

See also the section on Rainwater disposal in Chapter 4.1 which, for the most part, is relevant to this section.

Movement joints

Although asphalt itself does not require the provision of movement joints (any minor movements being taken up by the material), movement joints in the structure will nevertheless need to be provided through the asphalt. Since it is comparatively rare that movement joints will be needed in domestic construction, discussion of this topic will be found in Chapter 5.2 in the context of medium-span roofs on commercial and public buildings.

MAIN PERFORMANCE REQUIREMENTS AND DEFECTS

Choice of materials for structure

Timber decks experience moisture movements at joints, though normal laying procedures for mastic asphalt with suitable underlays and partial bonding should counteract any effects on the covering. Movements at joints between timber sheet materials may be larger than those between boards.

Resistance to indentation

Where personal access to the roof is to be provided, the top layer of asphalt should be of the correct specification for foot traffic. This is usually achieved with the addition of coarse aggregate to the mix. Unless the load can be spread over a substantial area, it is better that standing loads on the asphalt are avoided. Fixtures unavoidably taken through the surface should be provided with a weathered upstand.

Exclusion and disposal of rain and snow

Rainwater disposal from mastic asphalt roofs is normally via hoppers through a parapet or, occasionally, via gutters attached to the eaves also through the centre of the roof, particularly in blocks of flats. Normally, the top of the slope at an abutment will have an upstand; the

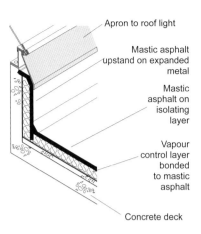

Apron to roof light

Mastic asphalt upstand on expanded metal

Mastic asphalt on isolating layer

Vapour control layer bonded to mastic asphalt

Concrete deck

Figure 4.28: Kerbs to roof lights need to be insulated

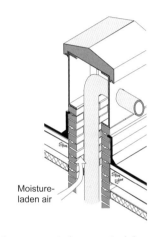

Moisture-laden air

Figure 4.29: A duct terminal through an existing cold deck roof can lead to massive heat losses; and insulation without a vapour control layer can lead to condensation

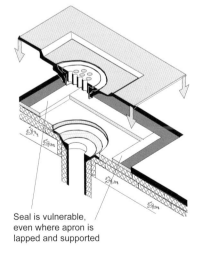

Seal is vulnerable, even where apron is lapped and supported

Figure 4.30: A rainwater outlet with a sump and grating in a warm roof

flashing over the upstand must not be in tight contact with the asphalt to avoid capillary attraction of rainwater under the apron.

The parting of asphalt upstands from their adjoining abutments and parapets is one of the most common problems seen with asphalt coverings.

Energy conservation and ventilation

Most roofs covered in mastic asphalt are existing roofs built to a cold deck design. Warm deck and inverted roofs, introduced in the early 1980s, are still relatively uncommon. If the asphalt is in good condition and the structure strong enough and the parapets high enough to take an increased upstand, an existing cold deck roof can be converted to

inverted warm deck with the existing covering continuing to function as both weatherproof layer and vapour control layer. The procedure is described in Chapter 4.1.

Before laying asphalt on a screed, water trapped within the construction must be allowed to escape or evaporate. If there is any possibility that water could be trapped within the screed, hooded ventilators can be inserted through the asphalt to get rid of excess water vapour, otherwise blistering of the surface will result.

Control of solar heat gain

An inverted roof overcomes most solar heat gain problems

On asphalt membranes, thermal size changes can be discounted as the cause of star-shaped cracks or of cracks forming a roughly circular pattern. This is much more likely to be due to moisture entrapment or blistering.

BRE Digest 144[18] suggests that the most credible cause of blistering of abutment edges during hot weather is vapour pressure from previous water leaks into the structure. Once the blister has formed, the heat of the sun will soften the asphalt and sagging ensues (Figure 4.31).

BS 8218[29], the code of practice for mastic asphalt, recommends a surface treatment using stone chippings, stone aggregate, light-coloured pedestrian tiles, concrete paving slabs or solar-reflective paint be used to reduce solar gain but see Case study 5. Vertical surfaces can slump so they should be protected using a solar-reflective paint (or in the case of an inverted roof, insulated metal overflashing).

BRE have known mastic asphalt upstands to sag when they are installed immediately below metal-framed doors/windows giving access to balconies (usually south facing).

Fire precautions

The Building Regulations 2000 (England & Wales) Approved Documents B1 & B2[15] state that to achieve an AA rating, mastic asphalt roof coverings need to be fully supported on timber rafters, with or without underfelt, and laid on:

Figure 4.31: Blistering and sagging in an asphalt skirting

- decks of timber joists and tongued-and-grooved or plain-edged boarding,
- steel or timber joists with deck of woodwool slabs, wood chipboard, fibre-insulating board or 9.5 mm plywood, or
- concrete or clay pot slab (in-situ or precast) or
- non-combustible deck of steel, aluminium, or fibre-reinforced cement (with or without insulation).

Sound insulation

A flat timber joist roof covered in asphalt on boarding, with a 12 mm plasterboard ceiling and thermal insulation, will give around 30 dB(A). If on a 100 mm concrete deck it will give around 45 dB(A).

Readers should also see the section on Sound insulation in Chapter 3.3.

Durability and ease of maintenance

The 1972 BRE study of the performance of flat roofing materials referred to in Chapter 4.1 found that the failure rate of asphalt roofs during the first 40 years of life did not increase with age[18]. Asphalt roofs, if designed and laid correctly, should last 50 to 60 years given regular maintenance at intervals not exceeding six years (Figure 4.32).

Older roofs covered with mastic asphalt may have cracked as the result of movement of decks, blistered because of entrapped moisture, sagged due to lack of key at vertical upstands and abutments, or have been indented due to point

Case study 5: Domestic flat roof covered in mastic asphalt

A warm deck sandwich roof had been built with lightweight concrete deck planks covered with a hot bitumen-bonded vapour control layer, pre-felted thermal insulation laid to falls, fibreboard bonded in hot bitumen, sheathing laid loose, asphalt in two layers, and a final covering of solar-reflective paint with some areas laid with promenade tiles.

Inspection, seven years after the roof had been laid, revealed that the asphalt had rippled with the outline of the insulation panels showing through. At the upstands the asphalt had sagged and some cracking of the angle fillet had occurred. Some promenade tiles had ridden up over adjoining tiles.

Metallic solar-reflective paint is known to cause stress problems in mastic asphalt because of their differing responses to the range of temperatures experienced by roofs. This range can be of the order of −10 °C to 80 °C, and even higher with well-insulated roofs. For this warm deck roof the paint provided no heat sink which would have been afforded by, say, 10–14 mm stone chippings.

Promenade tiles normally provide solar protection. The reason for their riding up came from being tightly butted on laying, whereas an allowance of a few millimetres between the tiles would have been adequate.

The client was recommended to have the roof taken up since leaks had saturated the fibreboard. However, the roof could be rebuilt to the original specification with some alteration to edge details and the substitution of a stone chipping solar-protection layer.

Figure 4.32: Signs of slight ponding on an asphalt roof: usually no action is necessary if the situation is visually acceptable

loading as a result of lack of care during maintenance.

Problems were experienced at one time in Scotland where mastic asphalt laid over well insulated roofs cracked from the low ambient temperatures experienced. This particular problem was solved by a change of specification of the asphalt.

One asphalt roof seen by BRE investigators on a block of flats was estimated to be about 80 years old. This roof covering had some cracks up to 12 mm wide and

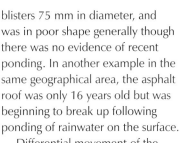

blisters 75 mm in diameter, and was in poor shape generally though there was no evidence of recent ponding. In another example in the same geographical area, the asphalt roof was only 16 years old but was beginning to break up following ponding of rainwater on the surface.

Differential movement of the deck and the asphalt layer can give rise to failure if not properly allowed for in the design (eg by a loose-laid isolating membrane).

Fortunately, mastic asphalt is easily repaired since the areas affected can be poulticed with hot asphalt, though never heated directly with a blow torch.

Newer roofs of mastic asphalt are likely to suffer much less than older roofs as a result of the introduction of polymer-modified asphalts. BRE tests showed that the polymer-modified material performed much better than existing types in both thermal shock and low temperature bending tests[33]. It was concluded that the material should perform well in temperatures down to −20 °C.

Where the roof is exposed to acids, an asphalt with improved acid-resisting qualities may be used, though it is more likely to be needed on floors than on roofs or where a roof is used, for instance, for car parking.

Modifications for climate change

Mastic asphalt has a reputation for providing a waterproof covering for many years. It is always used with a sheathing felt below it, regardless of the substrate. It is particularly well suited for use as the membrane in inverted warm deck roofs. It is also common to use polymer-modified asphalt in preference to those meeting the requirements of BS 6925[30]. There are no standards relating to the performance of polymer-modified mastic asphalt so products supported by third-party certification are recommended. No specific changes in specifications to accommodate climate change can be made.

WORK ON SITE

Access, safety, etc.

See the same section in Chapter 3.1.

Workmanship

To ensure watertightness of the structure at the earliest possible stage and to enable drying out to take place, asphalting of the roof slab should be carried out at the earliest opportunity, noting that trapped and construction water are first required to dry out. Installation or replacement of a typical mastic asphalt roof is described in Box 4.5.

Great care will be required to prevent asphalt-heating apparatus from igniting flammable parts of the structure; also to prevent hot asphalt coming into contact with heat-sensitive thermal insulation such as expanded polystyrene.

In any repair work, blow torches should not be used on asphalt since the material will be carbonised. If asphalt has been overheated during the laying process, a more brittle material results and any cracking due to low temperatures is exacerbated. It is possible to modify the asphalt by the addition of a suitable polymer to make it less vulnerable to overheating.

Box 4.5: Installation of a typical mastic asphalt roof

The roof must be thoroughly cleaned and levels checked. Levelling gauges are normally fixed to the perimeter upstand walls and to the core to indicate finished levels; profiles are cut and fixed to mark falls. A separating membrane is laid on the deck, formed either of felt or of glass-fibre tissue. The quantities of molten asphalt handled must be small enough to prevent it setting before the next batch has been laid. Asphalt is laid in bays in parallel strips starting from the lowest point of the roof. Each bucket run overlaps the one before and a hot float is used to smooth and level the asphalt. Once the first coat is laid and set, a second coat is laid. The work is carried out exactly as the first coat except that strips are formed to lap the first coat.

If a third coat is necessary, it is laid in the same way as the first and second coats with strips lapping the second coat.

Once the main roof area is completed, rainwater outlets, upstands and fillets can be formed. Completion of rainwater outlets, upstands and fillets depends on completion of any masonry in the parapet, masonry cladding to the roof core, and the fitting of rainwater hoppers and pipes. Upstands and fillets are formed by cutting chases for the top edge of the asphalt, priming the walls with a high-bond primer, fixing expanded metal where necessary to reinforce the angles and building up the asphalt fillets layer by layer. A sheathing underlay should not be used on upstands.

Finally, solar protection is provided by applying a sand or spar finish to the roof surface and, especially, to any upstands. This is done by spreading the material evenly over the surface and brushing away the excess or by rubbing it in.

Box 4.6: Inspection of mastic asphalt flat roofs

In addition to most of the external items listed in Box 3.11 at the end of Chapter 3.1, those listed in Box 4.3 in Chapter 4.1 are also relevant and are repeated below for convenience. The problems to look for are listed below.

- Sagging (usually suggesting failure of decking between joists)
- Sagging and detachment of upstands
- Blistering, splitting or rucking of coverings
- Inadequate falls or drainage
- Ponding
- Poor adhesion of covering to decks
- Mechanical damage
- Inadequate solar protection
- Flashings absent or damaged
- Inadequate flashings to upstands

If faults are suspected during external inspections of asphalt roofs, and of the rooms below asphalt roofs, some opening up of the roof spaces should be considered in which case the problems to look for are listed below.

- Inadequate thermal insulation
- Inadequate cross-ventilation in cold roofs
- Presence of dampness, staining, mould growth, wood rot and insect attack
- Condensation
- Decks of inadequate construction
- Absence of fire stops at separating walls

4.3 INVERTED FLAT ROOFS

In a warm deck inverted roof design the waterproof membrane is fully protected from temperature excesses and from mechanical damage from access. Since the waterproof membrane is on the warm side of the thermal insulation, it acts as a vapour control layer and there is no need for a separate layer to perform this function. There is much to be said in favour of this form of design.

The disadvantages of an inverted design include the fact that the waterproof membrane is not accessible for inspection. Other considerations for an inverted roof are that:

- the insulation must be of suitable compressive strength and be ballasted or otherwise fixed down to retain it in position under wind suction,
- it must not be water-retaining in itself if its thermal properties are not to be impaired, and
- it must be durable against frost attack.

In spite of these requirements, however, the inverted roof has become popular because its advantages outweigh its limitations. There is a fuller discussion of the advantages and disadvantages in BRE Digest 312[34].

CHARACTERISTIC DETAILS

Abutments
Many of the details appropriate to the inverted roof will follow normal practice adopted for other kinds of coverings except that it is important to allow for the greater depth of the weathering layer, insulation and ballast, especially at abutments and when converting existing roofs to inverted roofs. This greater depth

should also allow for the thicker layers needed to compensate for cooling due to rain percolation (Figure 4.33).

Skirtings at abutments are often the points where deficiencies are first revealed. The greatest risk is of a thermal bridge at the junction (Figure 4.34).

Rainwater disposal
There is no possibility, with the inverted roof, of providing for all drainage to take place on the upper surface. Inevitably, some rainwater will percolate underneath the insulation to the waterproof layer below. It is this level that drainage outlets must serve. Since the level is usually well below the top surface, some means of shielding the outlet from filling with detritus that impedes drainage will be needed. Wherever possible, however, drainage from the top surface should also be provided for as it improves the thermal efficiency of the covering (Figure 4.35).

MAIN PERFORMANCE REQUIREMENTS AND DEFECTS

Choice of materials for structure
The additional load of the ballast needs to be allowed for, especially when converting existing roofs to inverted.

Strength and stability
Thermal insulation boards should not be bonded to the waterproof membrane since thermal movements of the former could damage it. As the thermal insulation is protected only by ballast in many cases, some protection against maintenance traffic and ladders used

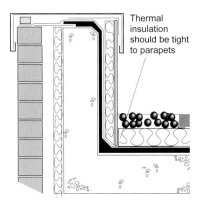

Figure 4.33: A parapet detail with an externally insulated upstand on a mastic asphalt-covered inverted roof

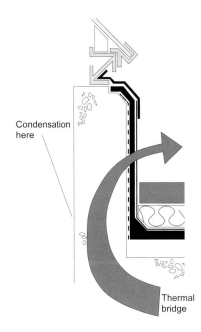

Figure 4.34: An abutment detail with an uninsulated upstand. This detail risks condensation

Figure 4.35: A rainwater outlet on an inverted roof. Saw cuts and cracks are evident in the cement surface to the insulation boards. Cement surfaces often can be felt to deform when walked on, and access should be limited to essential maintenance. The stones around the grating should be sufficiently large not to enter the drain

by window cleaners on abutment glazing will need to be provided for. The simplest means is slabbing a walkway on which the feet of ladders can be placed. However, point loading of slab supports should be checked.

Dimensional stability

The insulation will provide shielding of the roof from excessive swings in temperature which should reduce thermal movements of the deck but care needs to be taken with detail design of upstands.

Exclusion and disposal of rain and snow

As already noted, one main disadvantage of the inverted type of roof is that the weatherproof layer is not available for easy inspection. This may make tracing the source of leaks even more difficult than it usually is. There is a possibility that detritus may be washed down to become trapped between the insulation and the weatherproof layer. In time, constant movement could cause perforation of the membrane, especially if it is thin. It is a good idea, therefore, to install a suitable porous textile membrane immediately under the ballast layer to filter out the detritus and

Figure 4.36: Thermal insulation slabs covered with a thin layer of mortar for ballast. The slabs and mortar covering are vulnerable to breakage during handling and installation

help to protect the weatherproof membrane.

Energy conservation and ventilation

Boards or slabs of extruded polystyrene or compressed mineral fibres, sometimes pre-surfaced with an outer wearing layer, are suitable insulants (Figure 4.36). However, the combustibility of foamed polystyrene may limit its use in particular circumstances. Also to be borne in mind is that, until recently, extruded polystyrene foams were manufactured with CFCs (chlorofluorocarbons) or HCFCs (hydrochlorofluorocarbons) as blowing agents, and these materials

are subject to restriction. Cellular glass is not recommended for this application.

Another possible problem ensuing from the use of an inverted roof design is the creation of thermal bridges under certain conditions. Rainwater will percolate between the joints of the thermal insulation boards leading to thermal bridging and, theoretically, the risk of condensation under the waterproof layers (Figure 4.37). This effect is localised, intermittent and independent of insulation thickness. The phenomenon occurs in lightweight decks rather than heavyweight decks where temperature swings are more rapid. The reduction in thermal insulation value of the whole roof can be compensated for, in the longer term in new designs, by making an additional allowance of about 20% to the thermal insulation underneath the membrane. Laying the thermal insulation in two layers to break joint is unlikely to be beneficial since the risk is from cold rainwater filling the joint. For new construction, a thin layer of insulation (a material of, say, 0.15 m^2K/W) under the waterproof layer is beneficial. Remedial work to reduce the risk of condensation, however, is hardly practicable without taking up the ballast and existing insulation.

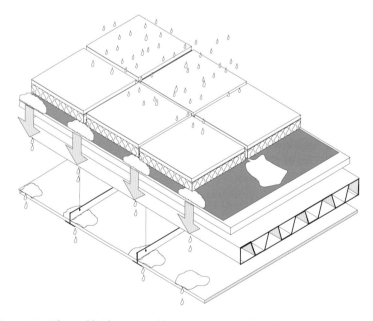

Figure 4.37: Thermal bridges caused by rainwater percolating between slabs of insulation on an inverted roof over a lightweight deck can lead to condensation on the ceiling beneath

Control of solar heat, air temperature, etc.

One of the main advantages of the inverted roof is that the weatherproof membrane is protected from sunlight and extremes of temperature.

Fire precautions

Where foamed polystyrene boards are used as the insulant, the external fire behaviour of the roof is usually taken care of by the ballast and an AA rating should be achievable. In very small roofs, where edge fixing replaces ballast, further fire protection to the surface of the insulation may be needed if the roof is less than a certain distance from the site boundary.

Sound insulation

The heavier the ballast, the better the sound insulation.

Readers should also refer to this section in Chapter 3.3.

Durability and ease of maintenance

Insulation materials used in inverted roofs need to have good frost resistance and not to take up water (ie they should be closed cell if foamed). It will be an advantage also if they are not subject to degradation by ultraviolet light since the exposed edges at joints may be vulnerable.

BRE site inspections have revealed a tendency for large-stone ballast to cause indentations when laid directly on asphalt at abutments though there has been no evidence of resulting rain penetration. There may also be a problem with the growth of plant root systems underneath, or even within, the thermal insulation laid over inverted

Figure 4.38: Tree growth on an inverted flat roof. These growths can have extensive root systems beneath the insulation and should be removed. Roof gardens are dealt with separately in Chapter 5.4

roofs. Care should be taken to ensure that these growths are not excessive (Figure 4.38).

WORK ON SITE

Access, safety, etc.

Readers should refer to this section in Chapter 4.1.

Storage and handling of materials

Thermal insulation sheets, factory coated with a thin cement render, are liable to break and must be handled carefully. Once in place, however, cracks in the render are not of great significance.

Restrictions due to weather conditions

Converting an existing conventional flat roof to an inverted roof is probably one of the building operations on flat roofs least sensitive to weather conditions. Membranes that are wet will not

Box 4.7: Inspection of inverted flat roofs

In addition to the external items listed in Box 4.3 at the end of Chapter 4.1, the problems to look for are listed below.

- Inadequacy of ballast, especially where affected by wind scour
- Thermal insulation retaining water
- Large gaps in insulation
- Shielding of rainwater outlets
- Condensation
- Thermal bridges, especially at abutments
- Absence of textile detritus filters
- Crushed insulation caused by overloading
- Poorly placed slab supports

prevent remedial work as the thermal insulation is not bonded to it and water drainage in any case has to take place at this level.

Workmanship

Large open joints are to be avoided as far as possible since they present potential thermal bridges. Most manufacturers recommend lapped or tongued-and-grooved joints. Special care is required near to abutments, taking the insulation as close to the upstand as possible. It is best to carry insulation up to the top of the membrane upstand and cover with a metal capping. Where slabs are used close to abutments, ballast should replace the slabs for the last 200 mm to allow for the possibility of movements but ensuring that thermal bridges through the thermal insulation layer are avoided.

4.4 FULLY SUPPORTED METAL ROOFS

When metal roof coverings are used in the nearly flat situation, some modification of details is required from those used on pitched roofs since the water loads will tend to be greater. This particularly affects treatment of the joint across the roof slope more than the joint in line with the roof slope; in the former circumstance, an extra drip or weir may need to be introduced to maintain integrity of the covering. It follows that the steepest possible pitch should be used for these coverings. A minimum finished fall of 1 in 60 is recommended in BS 6229[19] for roofs using aluminium, copper or zinc. For lead the recommendation is 1 in 80.

CHARACTERISTIC DETAILS

Basic structure

Flat roof decks are constructed in two main ways:
- roof joists laid to fall,
- roof joists laid flat and firring pieces used to create the slope.

There are many timber-joisted, metal-clad roofs of cold deck design that are poorly insulated. There was a brief discussion in the section on thermal insulation in Chapter 2.11 of the difficulties in achieving adequate ventilation, even where the eaves and verges are unobstructed; it is even more difficult where obstructions exist. If the roof joists span parallel to the obstructing walls, cross-ventilation can be achieved through the gaps between the joists. If, however, the joists span between the obstructing walls (Figure 4.39), cross-ventilation is difficult, perhaps impossible, unless the firrings have been laid across the joists rather than in line.

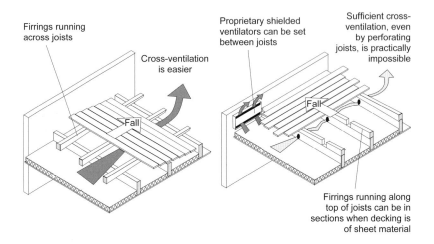

Figure 4.39: Typical firrings used to create falls on flat roofs of timber construction and measures for ventilation

If conversion to a warm deck is not an option, as will be the case with many metal roofs, cross-ventilation can be only marginally enhanced if the ceiling and existing insulation are removed and vents are cut through the joists. With this solution the joists may need strengthening before the new insulation and ceiling are replaced. Alternatively, proprietary ventilators are available which fit alongside each abutment so that some ventilation is achieved between the joists, albeit that they (the ventilators) are shielded by the abutment. In one case known to BRE, a surveyor, in desperation and left with no alternative, installed mechanical ventilation in a cold deck roof.

Joints along slopes

All the metals in common use can be laid to either standing seam or batten roll (Figure 4.40).

Joints across slopes

An ordinary standing seam joint across the slope should be avoided wherever possible in pitches of less than 10°. Where it is necessary to form cross-joints because of limitations in the lengths of sheets available, weirs should be provided with a substantial change of level. For this reason, to avoid such provision, spans of roofs that are intended to be covered in metal are often limited by the lengths of sheet available (Figure 4.41). (Thermal movement also restricts the length of sheets.)

Generally speaking, for all metals, the cross-joints must provide a weir of not less than 50 mm depth, preferably deeper. It is also good practice to provide for a degree of thermal movement in the joint where this can be done without detriment to the welt. For pitches of less than 10° in copper, cross-joints should be formed at not less than 3 m centres.

Abutments

Readers should refer to the same section in Chapter 3.3 and Figure 4.42.

Figure 4.40: Batten roll joints in a lead-covered roof

Figure 4.41: This lead roof dating from the 1870s is still giving good service

Figure 4.42: The leadwork at this abutment and change of level in the flat roof show evidence of adaptation over the years but the roof remains watertight

Rainwater disposal

Run-off from metal flat roofs will be less rapid than from pitched roofs. Older buildings with parapets are usually drained by means of large capacity gutters, lined in the same metal, leading to hopper heads on the outside of the walls, but

Figure 4.43: A drainage outlet behind a parapet on a domestic lead-covered roof that is approximately 120 years old

sometimes sumps are provided inside the parapets (Figure 4.43).

Readers should also refer to the same section in Chapter 3.3.

MAIN PERFORMANCE REQUIREMENTS AND DEFECTS

Strength and stability

Resistance to indentation is a crucial factor in the longevity of flat metal roofing and suitable protection (in the form, perhaps, of duckboards) needs to be provided for maintenance foot traffic.

Dimensional stability

Readers should refer to the same section in Chapter 3.3 and Figure 4.44.

Exclusion and disposal of rain and snow

Metal-lined gutters are prone to leaking, especially where blocked

by snow and ice (snow boards will be needed). Such gutters may also be a source of condensation on the underside of the metal, leading to deterioration in the gutter soles.

Weirs provide upstands at vulnerable joints across the sheets. They will need checking for signs of lifting after strong winds, especially if of lighter-gauge materials.

Suitable provision needs to be made for run-off from copper roofs that might stain other materials.

Energy conservation and ventilation

Readers should refer to the same section in Chapter 3.3.

Control of solar heat

Readers should refer to the same section in Chapter 3.3.

Fire precautions

Readers should refer to the same section in Chapter 3.3 and Figure 4.45.

Daylighting

Readers should refer to the part dealing with roof lights of the same section in Chapter 4.1.

Sound insulation

Readers should refer to the same section in Chapter 3.3.

Durability and ease of maintenance

The durability of metal-covered roofs with pitches under 10° slope will be less than that of roofs with metal coverings laid to steeper pitches. The water load on standing seams will be greater than with steeper pitches and the risk of water penetration will therefore be increased.

Particularly to be watched will be the potential build-up of detritus at changes of level. Annual inspection in the autumn is necessary to remove all organic matter that may cause attack on the metal.

Run-off from pitched roofs covered with lichens or mosses will present durability problems for any metal-covered flat roofs below. The run-off from these surfaces will be acidic.

Figure 4.44: Despite the wide spacing of these tacks, they have succeeded in holding a flashing at the junction of these flat and pitched roofs

AA

Lead sheet
on plain-edge
boarding gives

BA

Figure 4.45: Expected fire performance of sheet metals. Metals normally achieve AA. However, an exception is lead sheet on plain-edge board which achieves only BA. Lead on t&g boarding achieves AA

WORK ON SITE

Access, safety, etc.
Readers should refer to the same section in Chapter 4.1.

Restrictions due to weather conditions
Very low-pitched metal-covered roofs should not be laid in wet weather. To do so is to risk water remaining on the surface of, or within, the deck, evaporating and subsequently condensing on the underside of the metal.

Handling large sheets of metal becomes dangerous in wind speeds that are greater than 12 m/s.

Box 4.8: Inspection of fully supported metal flat roofs

In addition to most of the external items listed in Box 4.3 at the end of Chapter 4.1, the problems to look for are listed below.

- Inadequacy of thermal insulation
- Inadequacy of cross-ventilation in cold roofs
- Presence of dampness, staining, mould growth, wood rot and insect attack
- Condensation
- Decks of inadequate construction
- Absence of vents to dry out construction in wet decks or screeds
- Deformation of welts by maintenance traffic
- Absence of duckboards
- Absence of snow boards
- Detritus build-up at changes of level
- Corrosive run-off (eg from preservative-treated timber)
- Electrolytic corrosion where dissimilar metals are used.

4.5 SINGLE-LAYER MEMBRANES

Single-layer membranes are, as the name suggests, single layers of plastics or rubber sheets bonded together at seams for use as a waterproofing layer. The method of bonding may be by heat-welding, solvent-welding or adhesive for rubber membranes that can be in the form of specially developed 'tapes'.

Although these products can be used for many applications they have been specially developed for use on flat and shallow-pitched roofs. These roofing sheets are likely to include additives to enhance fire performance and resistance to ultraviolet (UV) degradation.

Single-layer membranes were developed and used in mainland Europe. They have subsequently become popular in the UK to the extent that more than half of new roofs constructed now use a single-layer membrane.

There is a wide range of materials available and these are shown in Table 4.3. Since their use in mainland Europe preceded that in the UK, nearly all products to date have been imported.

The most widely used material in the UK is PVC. There are British Standards to cover testing of individual materials but to date no British Standards cover plastics and rubber sheets for roof waterproofing. However, BS EN 13956[28] which refers to plastics and rubber sheets has recently become available and allows CE marking of these products.

Table 4.3: Materials used to manufacture single-layer membranes

Reference	Description
Materials classed as plastics	
CSM or PE-CS	Chlorosulfonyl polyethylene
EEA	Ethylene/ethyl acetate or Ethylene/ethyl acetate terpolymer
EBA	Ethylene/butyl acetate
ECB or EBT	Ethylene, copolymer, bitumen
EVAC	Ethylene/vinyl acetate
FPO or PO-F	Flexible polyolefin
FPP or PP-F	Flexible polypropylene
PE	Polyethylene
PE-C	Chlorinated polyethylene
PIB	Polyisobutylene
PP	Polypropylene
PVC	Polyvinylchloride
Materials classed as rubbers	
BR	Butadiene rubber
CR	Chloroprene rubber
CSM	Chlorosulfonyl polyethylene rubber
EPDM	Terpolymer of ethylene, propylene and a diene with residual unsaturated portion of diene in the side chain
IIR	Isobutene-isoprene rubber (butyl rubber)
NBR	Acrylonitrile-butadiene rubber (nitrile rubber)
Materials classed as thermoplastics rubbers	
EA	Elastomeric alloys
MPR	Melt processible rubber
SEBS	Styrene ethylene butylene styrene
TPE	Thermoplastics elastomers, not cross-linked
TPE-X	Thermoplastics elastomers, cross-linked
TPS or TPS-SEBS	SEBS-copolymers
TPV	Thermoplastics rubber vulcanisate

BS EN 13956[28] contains definitions and characteristics that include initial type tests to characterise the product and other test methods for purposes of factory production control (fpc). The CE mark itself contains a brief product description using those in Table 4.3 together with the results of test methods that generally relate to fire and tensile strength. The results for a wider range of tests will be given in the manufacturer's data sheet.

At the time of drafting this new edition there is no interpretative document that gives guidance concerning how a specifier might use the information that accompanies CE marking.

Since most products are imported we might expect to see CE marking on them. Although the CE marking is not compulsory in the UK, its use is likely to be adopted.

Currently, most good quality products have third-party certification. The higher quality PVC products tend also to have been assessed according to a Swiss Standard SIA 280[35].

Since there has never been a British Standard for the plastics and rubber materials for roof waterproofing, their use has historically been promoted by the Single Ply Roofing Association (SPRA)[36]. A number of guides are available from this organisation and their advice should be adopted.

CHARACTERISTIC DETAILS
Single-layer membranes can be applied by adhesion, ballasting or mechanical fixing. This means that they can be used in a number of different roof configurations which are illustrated in Figure 4.46.

The roof types are described in the following sections. Unless stated otherwise, they use insulation in the warm deck sandwich configuration.

Gravel ballasted roof
In this type of roof construction the sheet is laid loose over the insulation and held in place by ballast or paving slabs. The sheet may have a backing material applied to it for use as a separating layer in case of incompatibility between materials (eg PVC membranes should not

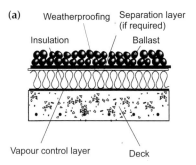

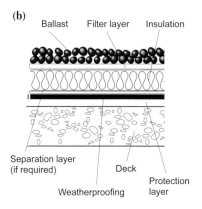

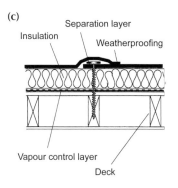

Figure 4.46: Roof configurations where single-layer membranes can be used: (a) gravel ballasted, (b) gravel ballasted inverted, (c) mechanically fastened

directly contact expanded or extruded polystyrene insulation).

The ballasting will need to be of sufficient mass to take account of wind uplift forces. The sub-structure will need to take account of the loading on this type of roof.

Gravel ballasted inverted roof
The membrane in this type of roof is loose laid over the structural deck. It is then covered by the thermal insulation (usually extruded polystyrene) and ballast or paving slabs. A filter membrane is included to prevent dirt from accessing the membrane. Dirt and debris collecting beneath the insulation has been known slowly to erode the

membrane due to repeated thermal movement of the insulation. The inclusion of the filter membrane below the ballast will reduce the chance of this occurring.

The ballast in this case has to ensure that there is sufficient loading to resist wind uplift forces and to prevent flotation of the insulation. The sub-structure will need to take account of the loading on this type of roof, including the additional load due to water retention.

Mechanically fastened roof
In this type of roof the roof sheet is held in place by mechanically fixing it to the structural deck. The fixings tend to be located in the side laps so that they are sealed. Some systems use fixing bars.

The number and spacing of fixings is determined with reference to the wind uplift requirements for the roof.

It should be noted that the fixings will penetrate the vapour control layer. It is difficult to seal these and some degree of self sealing around the fixing would therefore be advantageous. The use of such fixings over high moisture-producing areas such as swimming pools should be avoided and other methods of securing the membrane should be used.

Roof gardens (see also Chapter 5.4)
In this case, the membrane is buried in a green roof and will need to be root resistant as well as waterproof. The roof membrane will normally be loose laid and then weighted down with the other layers. It is likely that the load will be sufficient to prevent wind uplift failure.

It must be recognised that significant quantities of water can be stored on this type of roof and the sub-structure must be able to support this in addition to the other dead loads due to soil, planters, etc.

Adhered roof
The adhered roof is where the membrane system is fully adhered to the substrate, in this case an insulating material. The sheet can have a fleece backing which helps adhesion. It is usual for the other

elements of the roofing system to be adhered to each other (eg fully bonded insulation to the vapour control layer).

This type of system must develop sufficient bond strength between each of the various layers to resist wind uplift forces.

Adhered re-roof

The adhered re-roof simply covers over a roof where the previous waterproofing layer is nearing the end of its useful life or a fault has developed. The existing roof will need a careful examination to ensure the current condition is suitable for overlaying.

Providing the supporting structure can support the additional loading, a ballasted system could also be used.

MAIN PERFORMANCE REQUIREMENTS AND DEFECTS

Durability and ease of maintenance

The behaviour of each type of product is dependent on the material from which it is manufactured. PVC sheets, supplied by only a few companies, have been the market leader. The rubber material EPDM has suffered from poor seam strength when it was first introduced. However, the development of special joining adhesive tapes resulted in a marked improvement to the extent that this is no longer an issue.

Despite the claim that the PVC is manufactured using chemicals that are damaging to the environment, PVC is still the market leader. However, manufacturers are said to have developed other products known as flexible polyolefins which have more 'green' credentials.

Installers have found that it has been more difficult to achieve a satisfactory seam when using some of these materials.

The single-layer materials have either high strength or have great extensibility making them suitable for withstanding the effects of exposure on roofs for many years. There are examples of early PVC roofs that are approaching 40 years old.

Problems have been found when the relatively thin sheet has been punctured. This is usually associated with operations from following trades or later on from maintenance of plant. It is common for an extra sheet of contrasting colour and surface texture to be applied as a protective walkway.

WORK ON SITE

Each manufacturer of a single-layer membrane has developed a series of edge trims and other accessories for complete installation of the membrane. Upstands at edges of the roof or at abutments are often pre-formed and the main roof sheets joined to them by the appropriate method for the system. It is therefore imperative that the membrane is installed according to the manufacturer's recommendations.

Most of the manufacturers do not carry out the installation of their products. This process is usually undertaken by roofing contractors who have had the appropriate training from the manufacturer. These trained contractors are usually registered by the manufacturer as being qualified to install the product. One contractor can be registered with a number of different manufacturers.

4.6 REFERENCES

[1] **AMA Research.** Roofing Market Report UK 2007. Cheltenham, AMA Research, 2007. Available from www.amaresearch.co.uk

[2] **European Liquid Waterproofing Association.** www.elra.org.uk

[3] **Department of the Environment.** English House Condition Survey 1991. London, The Stationery Office, 1993

[4] **Welsh Office.** Welsh House Condition Survey 1991. Cardiff, Welsh Office, 1993

[5] **Scottish Homes.** Scottish House Condition Survey 1991. Survey report. Edinburgh, Scottish Homes, 1993

[6] **Northern Ireland Housing Executive.** Northern Ireland House Condition Survey 1991. First report of survey. Belfast, Northern Ireland Housing Executive, 1993

[7] **Communities and Local Government.** English House Condition Survey 2005. London, CLG, 2007

[8] **Northern Ireland Housing Executive.** Northern Ireland House Condition Survey 2006. Belfast, Northern Ireland Housing Executive, 2008. Available as a pdf from www.nihe.gov.uk

[9] **BRE.** Assessing traditional housing for rehabilitation. BRE Report BR 167. Bracknell, IHS BRE Press, 1990

[10] **British Standards Institution.** BS 747: 2000 Reinforced bitumen sheets for roofing. Specification. London, BSI, 2000. *NB: This Standard has been withdrawn and replaced by BS EN 13707: 2004*

[11] **British Standards Institution.** BS 8217: 2005 Reinforced bitumen membranes for roofing. Code of practice. London, BSI, 2005

[12] **British Standards Institution.** BS EN 13707: 2004 Flexible sheets for waterproofing. Reinforced bitumen sheets for roof waterproofing. Definition and characteristics. London, BSI, 2004

[13] **British Standards Institution.** BS 8747: 2007 Reinforced bitumen membranes (RBMs) for roofing. Guide to selection and specification. London, BSI, 2007

[14] **European Union of Agrément (UEAtc).** Special directive for the assessment of reinforced waterproof coverings in:
atactic polypropylene (APP) polymer bitumen (Method of Assessment and Test (MOAT) No 27);
atactic polypropylene (APP) polymer bitumen reinforced with sheets of polythene film (Method of Assessment and Test (MOAT) No 44)
Paris, UEAtc, 1983

[15] **Communities and Local Government (CLG).** The Building Regulations 2000.
Approved Documents:
A: Structure, 2004
B: Fire safety, 2006
C: Site preparation and resistance to contaminates and moisture, 2004
E: Resistance to the passage of sound, 2003
F: Ventilation, 2006
K: Protection from falling collision and impact, 1998
L: Conservation of fuel and power, 2006
P: Electrical safety: dwellings, 2006
London, The Stationery Office. Available from www.planningportal.gov.uk and www.thenbs.com/buildingregs

[16] **Scottish Building Standards Agency (SBSA).** Technical standards for compliance with the Building (Scotland) Regulations 2009.
Technical Handbooks, Domestic and Non-domestic:
Section 1: Structure
Section 2: Fire
Section 3: Environment
Section 4: Safety
Section 6: Energy
Edinburgh, SBSA. Available from www.sbsa.gov.uk

[17] **Northern Ireland Office.** Building Regulations (Northern Ireland) 2000.
Technical Booklets:
D: Structure, 1994
E: Fire safety, 2005
F: Conservation of fuel and power.
F1: Dwellings, F2: Buildings other than dwellings. 1998
K: Ventilation, 1998
London, The Stationery Office. Available from www.tsoshop.co.uk

[18] **BRE.** Asphalt and built-up felt roofings: durability. Digest 144. 1972

[19] **British Standards Institution.** BS 6229: 2003. Flat roofs with continuously supported coverings: Code of practice. London, BSI, 2003

[20] **BRE.** The assessment of wind loads. Digest 346, 8 Parts. Bracknell, IHS BRE Press, 1989–1992

[21] **BRE.** Flat roofs: built-up bitumen felt — remedying rain penetration at abutments and upstands. Defect Action Sheet 34. 1983. Available as complete set (141 sheets). BR 419. Bracknell, IHS BRE Press

[22] **British Standards Institution.** BS 6651: 1999 Code of practice for protection of structures against lightning. London, BSI, 1999

[23] **BRE.** Flat roof design: waterproof membranes. Digest 372. 1992

[24] **HSE.** Working on roofs. Construction Sheet IND(G) 284. 2008. Available as a pdf from www.hse.gov.uk

[25] **The Flat Roofing Alliance.** www.fra.org.uk

[26] **BRE.** Flat or low-pitch roofs: laying flexible membranes when weather may be bad. Defect Action Sheet 63. 1985. Available as complete set (141 sheets). BR 419. Bracknell, IHS BRE Press

[27]   **The National Federation of Roofing Contractors.** www.nfrc.co.uk

[28]   **British Standards Institution.** BS EN 13956: 2005 Flexible sheet for waterproofing. Plastic and rubber sheets for roof waterproofing. Definitions and characteristics. London, BSI, 2005

[29]   **British Standards Institution.** BS 8218: 1998 Code of practice for mastic asphalt roofing. London, BSI, 1998

[30]   **British Standards Institution.** BS 6925: 1988 Specification for mastic asphalt for building and civil engineering (limestone aggregate). London, BSI, 1988

[31]   **British Standards Institution.** BS 5284: 1993 Methods of sampling and testing mastic asphalt used in building and civil engineering. London, BSI, 1993

[32]   **Mastic Asphalt Council.** www.masticasphaltcouncil.co.uk

[33]   **Beech JC & Saunders GK.** Mastic asphalt for flat roofs: testing for quality assurance. Information Paper IP 8/91. Bracknell, IHS BRE Press, 1991

[34]   **BRE.** Flat roof design: the technical options. Digest 312. Bracknell, IHS BRE Press, 1986

[35]   **SIA.** Kunststoffdichtungsbahnen - Produkte- und Baustoffprüfung, Anwendungsgebiete. Swiss Standard SIA 280. www.sia.ch

[36]   **Single Ply Roofing Association (SPRA).** www.spra.co.uk

5   MEDIUM-SPAN COMMERCIAL AND PUBLIC ROOFS

Virtually all domestic roofs are of short span (ie less than 9 m). For the purposes of Chapters 5 and 6, then, medium-span roofs are described in the context of non-housing sectors.

No published data have been discovered on the relative proportions of pitched and flat roofs in the non-housing building stock of the UK.

However, BRE investigators have examined 39 aerial photographs of towns and cities taken during the 1960s. There was a preponderance of large towns (above 200,000 population) in this set of photographs and they were selected to include a wide range of urban areas throughout the UK, from small towns to metropolitan areas. Shops, office blocks, schools, churches, public and industrial buildings, all fairly easily recognisable, were identified on the photographs and estimates were made of the relative areas of flat and pitched roofs on those buildings. Housing was ignored.

The photographs revealed substantial variations in types of roof (eg the proportions of roofs that were flat varied in the photographs from 0 to 90%) and appeared to depend in part on the size of town or city, the extent of comprehensive redevelopment, and the relative age and dominant character of the area (especially the presence of industry).

The mean value of the areas on all the photographs was 23% flat roofs.

The examination was repeated on a further 16 similar photographs taken in the 1980s. There was a greater number of small towns (less than 50,000 population) in this set. The resulting range of roof types was, however, similarly broad, and the mean value for flat roofs was 20%.

Metropolitan areas and comprehensively redeveloped areas in medium-sized towns tended to have the highest proportions of flat roofs. Small, old towns had the lowest; planning and conservation guidelines had almost certainly been applied. Changes since 1980,

Figure 5.1: A variety of roofs is evident in this photograph of south London

such as the fashion for cottage-
type commercial roofs and the
replacement of old flat roof factories
by development areas containing
small factories with pitched profiled
sheet roofs, may have altered the
balance towards pitched roofs.

Given the obvious limitations of
the study, the figures seem to suggest
that the proportions of flat roofs
within the UK non-housing stock
differ very little from the figures for
the UK housing stock.

5.1 PITCHED ROOFS

Most medium-span pitched roof structures are in steel or timber. Aluminium structures in the past have proved unlikely to be economic for spans less than about 24 m.

So far as roof coverings are concerned the basic information for the most frequently used coverings has been given in Chapter 3 in relation to domestic construction, but it should not be assumed that domestic details invariably will be suitable for larger spans (Figure 5.2). Major differences exist, eg in relation to:

- snow and rainwater disposal,
- the need to provide larger areas of roof lighting,
- the need for increased ventilation, and
- shapes of roof that are not found frequently in housing.

Slates have been used on domes, flêche cones and spires of relatively small diameters. To conform closely with the shape of the roof shown in Figure 5.3, the slates needed to be narrow and so the side laps are smaller than would be found on roofs laid with slates of normal size. BRE has no evidence that rain penetration is a problem in these cases, though the importance of a sound underlay will be apparent. Laps in the vertical direction on convex shapes will leave the tails lifting proud of the slate below, with a risk of rain penetration, and therefore need to be very carefully considered; concave shapes present no such problems. Roofers in France have in the past contrived successfully to cover some very large and complex shapes in slate; for example, in the châteaux of the Loire.

Sheet materials, however, are being used more frequently on roofs in this category. The main materials from which profiled roofing sheets are formed include fibre-reinforced materials of various kinds, plastics and cements, steel and aluminium. Most materials can be formed into different shapes and most, too, are supplied with trims of the same material.

While steel and aluminium are likely to be the metals most commonly used for self-supporting roofing, other metals (such as austenitic stainless steels, lead and copper) may also be used where

Figure 5.3: The slating on this flêche successfully breaks most of the normal rules on side and head laps to cope with the conical shape which is convex in plan

Figure 5.2: The new unweathered slates show up on the refurbished roof of this Cornish building

the roof is fully supported. The alternatives have been described in Chapter 3. Austenitic is a term used to denote the more corrosion-resisting alloy steels with a content of not less than 18% chromium and 8% nickel. Stainless steel used in roof coverings can retain its bright finish for a few years given appropriate local conditions (Figure 5.4). Where a dull finish is required, matt finish or terne-coated (solder-coated in simple terms) stainless steels are available. Further information on stainless steels is available in BRE Digest 349[1] and BS EN ISO 9445[2].

BS 4868[3] gives information about profiles available in aluminium as well as details of materials and finishes. Zinc- and aluminium-coated steels to BS 3083[4] are dealt with in Chapter 6.5. However, profiled steel sheet is available in many different forms and specifications that are not covered currently in British Standards. Many of these products have potentially a much superior performance to the simple corrugated sheets (some of which incorporate thermal insulation) and which are described later in this chapter.

Post-production surface colouring is available for many sheet coverings on medium-span pitched roofs. BS 4904[5] selects 38 colours from the head standard BS 5252[6] which are appropriate for roofing.

Figure 5.4: Stainless steel on the roof of a church

TYPICAL ASSEMBLIES

Profiled steel or aluminium sheets are used in a number of assemblies as described in Box 5.1.

DESIGN CONSIDERATIONS

A comprehensive and flexible approach to roof design is essential to reduce the risks of moisture that may be associated with increased levels of thermal insulation.

External moisture

Wind-driven rain and snow are less likely to penetrate joints if the roof pitch is greater than 15° or the joints are sealed.

Penetrations through the external skin of the roof should be avoided; if this is not possible, the roof around the penetration should be sealed and insulation provided to reduce thermal bridging and subsequent condensation at the penetration/roof interface.

Detail around rooflights needs care to prevent moisture entering around the rooflight and to prevent condensation within sealed units.

Detailing to perimeter, internal or valley gutters should be simple, and should reduce the risk of blocking and subsequent overloading or flooding of weirs and overflows. Internal metal gutters can become extremely cold when exposed to flowing rainwater or melting snow, and internal gutters and associated penetrating pipework should be insulated with closed-cell, moisture-resistant insulation. Joints between sections of the insulation should be sealed with a vapour-resistant tape.

Internal moisture

An effective internal vapour control layer is imperative to the success of systems incorporating a cavity above the insulation. This vapour control layer must extend over the full area of the roof, and joints should be sealed with vapour-resisting tape. The vapour control layer may be a separate (polyethylene) layer or may be formed from an impermeable liner system.

In addition to this vapour control layer below the insulation, a permeable membrane may be needed above the insulation. This permeable membrane should direct any dripping condensate away to guttering. Effective detailing and subsequent installation of this membrane may be difficult around purlins and rooflights.

To reduce further the risk of moisture build-up within voids in the roof construction, any voids directly below the outer sheet should be ventilated, with inlets at eaves and ridge. The filler strips at the eaves and ridge should be perforated to provide the recommended ventilation.

Site-assembled, double-skin systems should be considered only where there is a low risk of moisture from internal processes or where extract ventilation is provided close to this moist process. Where the risk from internal processes is high (eg in swimming pools, steam or wet manufacturing), double-skin construction (with voids) needs careful consideration; if composite systems are used, the joints between panels must be sealed.

Ventilating air moisture

There is a small risk of condensation, arising from external ventilating air, the moisture coalescing and subsequently running or dripping onto the underlying insulation. A permeable membrane can be installed to direct this moisture towards gutters. If this risk is considered high and the detailing of a membrane impracticable, a site-assembled or prefabricated system should be considered that incorporates low permeability, closed-cell insulation which fully fills the void between the liner and outer sheet.

CHARACTERISTIC DETAILS

Basic structure

Roofs in the medium-span category are normally symmetrically pitched and are often now of trussed rafter design. Portal framed roofs, that are used more in industrial situations, are dealt with in Chapter 6.1.

The symmetrical pitched truss, used frequently in medium-span roofs, consists of sloping rafters over purlins spanning between the trusses, with some form of horizontal tie connecting the trusses together.

Box 5.1: Profile metal roofs — typical assemblies

Single skin

An uninsulated profiled sheet is fixed directly to the purlins. This form of construction is generally used only over storage buildings, warehouses, animal housing and uninhabited buildings or as canopies over workspaces.

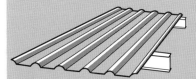

Single skin

Double skin

The most common type of profiled metal roof construction consists of a shallow profiled metal liner, a structural spacer system and an outer sheet that is normally of a deeper profile. Typically, the void between the two skins accommodates thermal insulation, usually mineral wool, and some form of vapour control on the warm, liner side of the insulation. A vapour permeable membrane can be placed above the insulation to direct any dripping condensate towards external guttering.

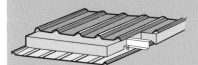

Double skin with zed spacer

Composite systems

These combine a liner sheet, rigid insulation and an external profiled weather sheet forming an integrated composite panel. Most composite panels are factory-assembled and incorporate a rigid plastics insulant that is foamed to fully fill the void between the inner and outer sheet.

Site-assembled composites consist of profiled inner and outer sheets and a separate pre-formed rigid insulant. This insulant may be plastics or mineral and, as it is rigid in nature and fully fills the void between the sheets, no structural spacer system or ventilation is required.

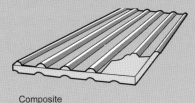

Composite

Boarded systems

These systems do not incorporate a metal liner but need a rigid insulation board supported below the external profiled sheet. The internal finish is not usually considered to provide vapour control so it may be simply the exposed face of the insulation board. These systems are uncommon and are suitable only where dry processes are enclosed and there is little risk from generated moisture penetrating the roof structure.

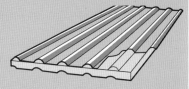

Boarded

Spacers

Used only in double-skin roofs, spacers create the cavity between the liner and external sheeting which accommodates insulation and appropriate membranes. They are structural elements and must be positioned correctly over the purlins or other structural elements and must be fixed securely to these elements.

Traditional zed spacers (see drawing of Double skin with zed spacer) are normally made from 1.5 mm-thick galvanised steel and are supported on a plastics ferrule. They are generally suitable for insulation up to 100 mm thick. If the insulation is thicker than this, the zed spacer may become unstable so it is advisable to use bracket and rail systems. These are more stable over a greater range of depths and are designed to reduce the effects of thermal bridging. The rails are formed from structural grade steel and the brackets incorporate a thermal break and vapour seal.

Bracket and rail

Not every bay will carry effective wind bracing. The older the roof, the more likely it is to be in timber; and the older the timber roof, the more likely it is to be jointed, tenoned or pegged rather than strapped and bolted together. Victorian times saw the widespread introduction of iron (and, later, steel) straps, stirrups and tie rods.

In roofs built in the 20th and 21st centuries, the basic material is usually either steel or timber with, since about 1950, a few roofs in aluminium.

Steel

Over medium spans, the rafters are normally rolled sections, though increasingly, sections such as welded tubes can be found. Tubular construction has at least one potential advantage over rolled sections in that, span for span, external surface area is less than for rolled, thus saving on protective coatings. However, care has to be taken that the tube ends are closed. Trusses fabricated from tubes may also be useful in dusty atmospheres where they may collect less dirt than rolled sections.

Timber

Trussed rafters for medium spans can be similar to those used over shorter domestic spans, though they may be scissor, Howe or Pratt types rather than fink or fan. Heavier bracing than for domestic span roofs is normal, particularly chevron bracing for various internal members in spans above 8–11 m (depending on design), though the general principles for bracing are more or less the same as for shorter spans (Figure 5.5).

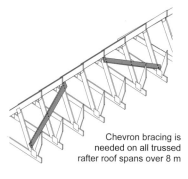

Chevron bracing is needed on all trussed rafter roof spans over 8 m

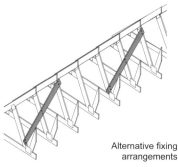

Alternative fixing arrangements

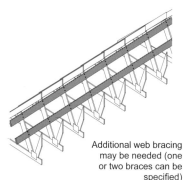

Additional web bracing may be needed (one or two braces can be specified)

Figure 5.5: Trussed rafters used over medium spans require bracing additional to that specified in domestic spans

Figure 5.6: A lead-covered base to a clock tower

of streaking from horizontal welts, is to set the sheets and welts at an inclination to the horizontal. Figure 5.6 shows four sheets intersecting at an arris in the clock tower. However, weathering details are easier to form where three sheets only intersect at arrises, as in Figure 5.7. Practically all the run-off occurs in streams at these arrises, and the catchment areas at the foot of each slope need to be adequate to avoid splashing or overshooting the gutters.

Stainless steel

Generally speaking, details appropriate to the use of zinc described in Chapter 3.3 may be used for fully supported stainless steel roof coverings. These coverings may be laid either as standing seam or roll cap, though some details may prove difficult to work because of the temper. Standing seams are usually 25 mm high.

Eaves and verges

Radiused eaves for both profiled and unprofiled metal sheet have been popular since the mid-1980s for medium-span metal roofs. Such eaves can be formed, in practice, only with warm roofs since there is no possibility for condensation formed in cold roofs to be drained out at the eaves. Crimping the sheet may be needed to form the radius but there is no evidence that this affects the performance of the material. Some designs of eaves and verges, however, have overlapping trim. Specifiers may need to pay particular attention to colour matching of trim.

With non-radiused eaves it is usual to employ filler pieces to close

Figure 5.7: A lead-covered broach spire

off gaps under the profiles. Some pieces may fall out after a period of time. In some existing roofs the lining may project to the gutter beneath any closure.

Abutments

Upstands at abutments to the roof in deeply troughed sheet metal cannot follow practice usual with other forms of covering as the depth of the trough prevents the use of standard methods of flashing. This means that specially shaped filler blocks have to be glued to the head of the troughs, and a generously sized flashing dressed over.

Hips

Hips can be formed in both profiled and unprofiled sheeted materials. They are usually covered with a generously sized, shaped ridge piece of the same material. In the absence of the ridge piece being profiled to match the sloping covering, it is important to ensure that filler blocks are in place to help in forming a weathertight joint that will prevent rain being blown up the slope and over the covered edge of the sheet (Figure 5.8).

Roof lighting

Extruded rigid PVC corrugated sheeting is dealt with in BS 4203-1[7]; corrugated translucent plastics sheets

Main roof areas

Tile, slate, etc.

Readers should see Chapter 2 for descriptions of basic roof coverings of tile and slate, and metal and other forms of sheeting.

Lead

Lead has been used over comparatively large pitched roof areas for centuries. It can be found on all pitches of roof up to the vertical. Fixings must be appropriate, for, on the steeper pitches, exposure to wind is usually severe. Creep may also be a problem; the traditional way of coping with this, and at the same time helping to ensure a reasonable degree of control of rainwater run-off and the minimising

Figure 5.8: A hipped, sheet metal covered roof

Figure 5.10: The roof space in King's College Chapel, Cambridge: a BRE researcher prepares to carry out an experimental smoke insecticidal treatment

made from thermosetting polyester resin (glass-fibre-reinforced) are in BS 4154-1[8]. Profiles are available matched to those in fibre-reinforced, steel and aluminium sheeting.

Readers should also see the section on Roof lighting in Chapter 3.1.

Service perforations

Readers should see the same section in Chapter 3.1.

Rainwater disposal

Readers should see the same section in Chapter 3.1 and also Figure 5.9.

MAIN PERFORMANCE REQUIREMENTS AND DEFECTS

Strength and stability

Many pitched roofs in the medium-span category are of trussed rafter construction. Readers should also see the same section in Chapter 3.1.

Although the collapse of trussed rafter roofs is rare, since most contractors are familiar with them, Case studies 6 and 7 provide reminders of the need to comply fully with advice for bracing and binding.

Where the structure of an old wooden roof has been affected by insect attack, and there is limited possibility of fixing scaffolding for access, it is possible to treat the timbers with insecticidal smoke. To be effective, however, the space needs to be more or less sealed and the treatment needs repeating regularly (Figure 5.10). Further advice is given in *Remedial treatment of wood rot and insect attack in buildings*[9].

So far as security against break-in is concerned, roofs of commercial premises storing high-value merchandise are only as secure as their weakest point. Therefore, they should be designed and constructed to provide a barrier that is as uniform as possible to deter and delay an intruder. In particular, roof lights may provide weak points in the barrier and therefore require special consideration.

Prescriptive requirements for the security of roofs and their component parts are contained in BS 8220: Parts 2 and 3[10], although many organisations concerned with high-security buildings, such as banks, have their own in-house standards. Because of the nature of the subject and particular applications, details of these requirements are not generally available.

Methods of testing and assessment for specific components, particularly glazing and locks, are covered by British and other national standards. British Standards are also available for bandit-proof (BS 5544[11]) and safety glazing (BS 6206[12]) which may be appropriate for roof glazing at risk.

Methods of assessment for other components, usually for use in low-risk areas, are being developed by

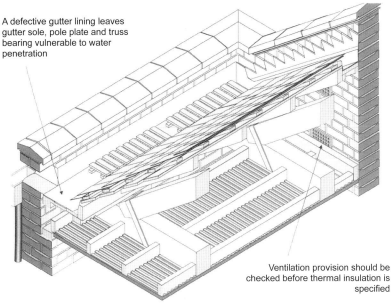

A defective gutter lining leaves gutter sole, pole plate and truss bearing vulnerable to water penetration

Ventilation provision should be checked before thermal insulation is specified

Figure 5.9: A rainwater disposal system on a large roof showing a secret gutter on the gable abutment and a large capacity, lead-lined gutter discharging to a hopper head on the external wall

BRE Global and the Loss Prevention Council. Some European national standards exist and there is a possibility therefore of a European standard eventually being produced. All roofs, even those incorporating a concrete slab, can be penetrated by determined intruders. Even opportunist thieves have been known to break through a concrete roof rather than try and penetrate a security door.

In general, therefore:

- roofs of low-level buildings should be designed and built to the degree of security required and to withstand the likely level of attack,
- openings should be kept to a minimum,
- architectural features which provide concealment or access to the roof should be avoided,
- fixings, particularly of roof sheeting, should be concealed or, if exposed externally, be tamper-proof.

Readers should see BS 5427-1[13] for information on the use of this category of covering.

Modifications for climate change

Mechanically fastened flat roofing systems

A typical mechanically fastened flat roofing system is:

- a structural deck, usually metal or timber-based, supporting a vapour control layer (reinforced bitumen membrane or polyethylene sheet or other sheet products),
- thermal insulation mechanically fastened through the vapour control layer into the structural deck,
- a waterproofing membrane supplied as a single layer and mechanically fixed at the seams. The fixings pass through the insulation and vapour control layer into the structural deck.

The number of fixings used to secure the insulation are determined by the wind-loading requirements (Figure 5.11). It would be a simple matter to double the number of insulation

Figure 5.11: Typical fixings on a metal roof

board fixings to withstand current wind loads and take account of the effects of climate change.

It may also be possible to apply this to mechanical fixing of the waterproof membrane. This would be determined by the number of fixings per metre length and it would be possible to increase this by two, for example.

The number of fixings per insulation board should be increased by a factor of two or two added to each number of fixings per metre run or per m². Rather than increasing the number of fixings, the pullout resistance of each fixing can be increased by a suitable factor such as 20%. The pullout strength of fasteners depends on the properties of the substrate into which they are fixed and the size of the fixing itself. Changing the size of fixing is a relatively inexpensive way of coping with future increased wind loading.

Dimensional stability

Cases have been brought to the attention of BRE where laminated glass has been used in sloping roof lights in spans greater than those recommended by the manufacturers and the glass has sagged. Deflections of up to 25 mm have been observed and, besides being unsightly, have potential effects on weathertightness. If the glass is inverted, it will return in time to more or less its original shape, though of course the defect may not be resolved for some time and the original cause needs to be corrected if the problem is not to recur.

Metal decks are subject to expansion and contraction which

need to be accommodated in the design.

Exclusion and disposal of rain and snow

Construction of pitched roofs over larger spans encountered in commercial and public buildings are carried out, in general, using materials whose basic characteristics have already been described in Chapter 2. It is arguably even more important that correct construction detailing is carried out on these roofs, which are more difficult to access for repairs than those of shorter span (Case studies 8 and 9). However, there are a few points to consider with these large areas which need correspondingly large drainage facilities.

Frequently, BRE investigators have seen domestic-sized rainwater gutters used on non-domestic construction. The gutters may only cope well with the increased water load from greater spans if there is more frequent provision of rainwater downpipes to limit the risk of overtopping the gutter. The only alternative, if downpipe frequency has to be restricted, is to provide gutters of significantly increased capacity designed fully in accordance with BS EN 12056-3[14] (Figure 5.12).

Energy conservation and ventilation

Thermal insulation is perhaps equally important in larger roofs as in domestic because of the similar scale of contribution that can be made to energy conservation and the need to comply with building regulations. But there are significant

Figure 5.12: A large gutter on a fibre-reinforced, corrugated sheet pitched roof

differences in both approach and in detailing.

Condensation as a result of cooling of the outside surfaces by radiation can occur in sheeted roofs as mentioned in Chapter 2.5. This risk is not only theoretical but also very real, and can have very unfortunate effects where the roofs are large. It is exemplified by the account in the National Building Studies Special Report 39[15] of the interruption of a public performance in the De la Warr Pavilion, Bexhill-on-Sea, in December 1935 by the noise of condensation falling on the ceiling of the auditorium from the sheeted roof above: the result of radiation to the clear night sky.

Many medium-span pitched roofs have very large glazed areas. Developments in multiple glazing and the availability of low-emissivity glass have led to designs where solar gain can balance the losses through the roof glazing.

Glazed areas are composed of two distinct parts: glass and framing. The contribution of the frame to the overall insulation value of the unit may be positive (timber or plastics frames) or negative (metal). Metal frames can incorporate thermal breaks to improve their insulation properties.

Cold roofs in this range of spans should have additional ventilation at the ridge equivalent to a 5 mm

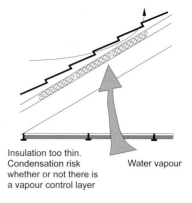

Insulation too thin. Condensation risk whether or not there is a vapour control layer   Water vapour

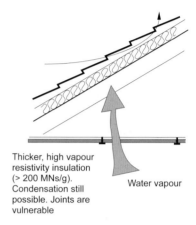

Thicker, high vapour resistivity insulation (> 200 MNs/g). Condensation still possible. Joints are vulnerable   Water vapour

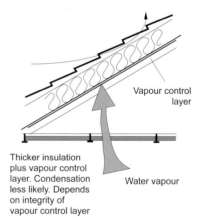

Vapour control layer

Thicker insulation plus vapour control layer. Condensation less likely. Depends on integrity of vapour control layer   Water vapour

Figure 5.14: Insulation in the slope of a roof to control condensation. If the sarking is permeable, ventilation through the tile laps will remove some water vapour

Figure 5.13: Evidence of condensation and 'white rust' (the corrosion product of zinc) forming on the underside of an uninsulated profiled steel sheeted roof. In a cold roof like this, adequate ventilation is crucial to performance

continuous gap (Figure 5.13). In contrast to cold roofs it is vitally important that site staff understand the different design concept inherent in warm roof design, and in particular the radical differences in ventilation provision (Case study 10).

Vapour control layers

Although vapour control layers* will be needed in some circumstances in small (usually domestic) roofs, they are more likely to be needed in medium-size and large roofs.

Interstitial condensation can sometimes be controlled by limiting the indoor humidity. However, there are many buildings, especially swimming pools and factories with wet processes, in which the humidity will inevitably be high. There are also some constructions that are so vulnerable to interstitial condensation that the internal humidity cannot realistically be lowered enough. If it is possible, placing thermal insulation on the cold side of the roof will keep all layers above the dewpoint. In many cases, though, the best solution will be the installation of a high vapour resistance layer on the warm side to prevent water vapour entering the roof construction (Figure 5.14).

There have been many and frequent instances of severe condensation occurring, more especially from construction moisture drying out, but also from other causes where the vapour control layer was ineffective. Examples of roof timbers with moisture contents in excess of 28% have been recorded, even in summer conditions.

As the achieved vapour resistance will depend at least as much on workmanship as on the design and integrity of the materials used, it is not realistic to specify a minimum vapour resistance to be achieved for the roof as a whole, though for the material to qualify as a vapour control layer it should have a vapour resistance greater than 200 MNs/g. There will be least risk of interstitial condensation where the vapour resistance of layers on the warm side of the insulation is at least five times greater than the vapour resistance of layers on the insulation's cold side.

Plastics films or metal sheets are the most satisfactory materials for forming a vapour control layer, either as separately installed layers or as finishes bonded to insulation board. Metallised polymer backing to plasterboard is intended to improve the thermal properties by reducing surface emissivities, but is rarely very effective as a vapour control layer if the joints are not positively sealed.

Joints in a flexible sheet vapour control layer should be kept to a

minimum. Where they occur, they should either be overlapped by a minimum of 100 mm and taped, or sealed with an appropriate sealant and should be made over a solid backing.

The same jointing methods should be used for tears and splits. If the control layer is bonded to a rigid board, joints between adjacent boards should be filled with an appropriate sealant and, preferably, taped as well. Penetrations by

* Of the various terms that are used, 'vapour control layer' is preferred to either 'vapour check' or 'vapour barrier' to emphasise that the function of the layer is to control the amount of water vapour entering the roof. The achievement of a complete barrier to the passage of water vapour is virtually impossible.

Case study 10: A draughty and energy-inefficient roof on a large public building

Numerous complaints by occupants that the building was draughty, and observations by maintenance staff that it was not as energy efficient as expected, led to BRE being commissioned to report on its condition and to recommend remedial measures.

The roof had been built in concrete tiles over a spun-bonded polyolefin membrane over thermal insulation boards laid on the slope of the roof over the rafters. (Although building elements other than the roof were involved, this case study is concerned only with the problems in the roof.)

The attic spaces of the main building were separated from the rooms below by a double ceiling. Many of the joints in the insulation boards over the rafters had been taped, but the taping was random and open joints up to 3 mm wide were noted. Insulation boards were found to be missing at valleys and wall heads, creating thermal bridges. Service pipes entered the roof space through crudely cut holes. Tile ventilators had been installed. Air movement in the attic spaces was measured and found to be in the range 0.2–0.3 m/s. Ventilation had also been provided in the eaves. Air movement at these locations ranged from 1.2–6.5 m/s.

The concept of warm roof design was relatively new at the time of construction and the builders had mistakenly provided ventilation to the roof voids and at the eaves, contrary to the recommendation not to ventilate warm roofs. This error was compounded by poor cutting and fitting of insulation boards, which provided through-routes for external air.

The BRE investigator recommended that:
- all tile ventilators should be blocked,
- thermal insulation should be installed where it was missing,
- eaves cavities should be filled with blown fibre,
- all pipe entry points should be sealed.

services should be kept to a minimum and gaps carefully sealed where they are inevitable. Draughts of moisture-laden air through gaps in vapour control layers are more significant than normal still air diffusion through materials, even if there are splits in the vapour control layer; it is therefore much more important to provide an air seal than to take elaborate precautions for making a total seal of the vapour control layer.

This is now even more important with the need to reduce air leakage in line with the Building Regulations (England & Wales) AD L[17]. The vapour control layer is likely to be the air barrier as well.

Air-handling devices

These include intakes and exhausts through the roof coverings and also possibly include trunking within the roof void. They may need to handle hot exhausts from local heating units. Ducts carrying air at different temperatures from ambient may need insulating to prevent condensation. Solutions will need to show how such ducts and terminals are to be incorporated in such a way that they do not prejudice other performance characteristics; for example:
- avoiding providing routes for flanking sound transmission,
- accommodating differential movements within the roof itself or between the roof and the remainder of the structure from which the duct is supported,
- preventing the spread of fire,
- airtightness,
- thermal bridging.

Fire precautions

For each purpose group, the requirements differ. Fire in trussed rafter roofs in which toothed connectors fail can lead to rapid failure of the trusses themselves. This is not normally a safety consideration in domestic buildings, but in more highly serviced buildings where there may be an ignition within the roof space, and especially where longer spans are involved, the extent of damage will be significant. In some cases there could be a personal safety problem too.

Figure 5.15: A test of the external fire performance of a pitched plastics roof covering to determine use in accordance with the building regulations in force at the time

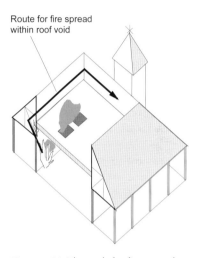

Route for fire spread within roof void

Figure 5.16: The path for fire spread in a roof perimeter void without cavity barriers

In addition to the risk from external sources of fire (Figure 5.15), a number of fires have occurred in the trussed rafter roofs of shopping precincts and stores where fire has travelled long distances through the uncompartmented triangular void of the perimeter leading to loss of the whole property (Figure 5.16). Timber soffits, poor fire stopping, and cavity barriers held with plastics fixings have all contributed to rapid spread of fire. Although not required for life safety, properly fitted cavity barriers might have reduced the extent of loss in many of these cases.

Readers should also see the section on Fire precautions in Chapters 3.1–3.5.

Lightning protection

Having decided that lightning protection to the structure is necessary (BS EN 62305-2[18]), it will be necessary to provide it according to the characteristics of the construction. In the case of steeply pitched roofs, just a few terminals sited on the highest parts of ridges should offer sufficient protection. With flatter pitches, the cone of protection afforded by terminals means that more of them are required than on steeper pitches.

Where metals are used on a significant scale in the roof construction and covering, they may be incorporated into the provision of lightning protection. Structural steel work, concrete-reinforcing bars and cleaning equipment anchor rails need to be bonded to the earthing system. Any metal above or on the structure which is close to the lightning protection system (eg metal handrails protecting means of escape routes) also may need to be bonded to the lightning protection system to prevent side flashing.

Provision must be made for connecting the earth strap from the roof down an external wall to the main earthing terminal. Conductors should be of aluminium or copper, the former being preferred as the latter tends to stain surfaces and accelerate corrosion of other metal components. Care is needed in specifying conductors to prevent electrolytic corrosion. It is possible to provide protection to conductors (eg by coating with PVC or other forms of sheathing).

The specification of lightning protection should be delegated to specialist consultants.

Daylighting

BRE Digest 309[19] describes a simple procedure for evaluating daylight factors in interiors.

Readers should also see Chapter 2.7.

Roof light glazing area

Average daylight factor (DF) is used as the main criterion of good daylighting. It is the ratio of indoor to outdoor daylight illuminance under the standard overcast sky. A daylight factor of 5% or more means that the space is well lit. A daylight factor of less than 2% means that electric light will be used all the time.

Daylight factor is given by the formula:

$$DF = \frac{WT\theta}{A(1-R^2)} \%$$

where:
DF = daylight factor
W = total glazed area of roof light aperture (m²)
T = transmittance of the glazing (normally 0.85 but can be as low as 0.6 if the glazing is dirty)
θ = the angle subtended in the vertical plane normal to the roof light by sky visible from the centre of the roof light (Figure 5.17).
A = floor area in m²
R = average reflectance of internal surfaces

To obtain a given daylight factor, the equation can be inverted:

$$W = \frac{DF\ A(1-R^2)}{T\theta} \%$$

The result of this prediction gives an initial glazing area, W, which can be used as a starting point in design, and in heat loss and solar-gain calculations. Later on, the exact shape and position of roof lights can be decided and, where necessary, the detailed daylight distribution checked, possibly using models or the detailed calculation methods given in BRE Digest 309[19]. Appropriate lighting controls can then be chosen to maximise the energy effectiveness of the daylight. BRE Digest 272[20] is also relevant.

Glare can occur from excessive contrast between the bright sky and a dark interior. For this reason, the internal surfaces of all roof light reveals should be light coloured.

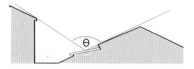

Figure 5.17: Sloping roof light showing θ, the angle subtended in the vertical plane normal to the roof light by sky visible from the centre of the roof light

Glare from the sky can also be reduced by:
• providing additional light to the ceiling (eg from windows),
• using adjustable blinds or curtains,
• splaying roof light reveals,
• using tinted glazing.

This last solution is not recommended in general because of the reduction in indoor daylight levels (average daylight factor being proportional to glass transmission as indicated above).

Usually the most important source of glare is direct sunlight. This should be controlled with shading devices or by having north-facing roof lights.

Solar heat gains occur throughout the year. During the summer they may cause overheating and discomfort for occupants; but in the winter heating season they can be beneficial, enabling a reduction in space heating load. Solar heat gains occurring through roof lights can be considerable and depend on:
• glazing area, heat gains being roughly proportional to the area of glazing,
• roof orientation,
• external obstructions such as other buildings and vegetation,
• type of glazing and shading device,
• time of day and year,
• weather,
• geographical location.

Section A2 of the CIBSE Guide, Volume A[21], contains methods for calculating solar heat gains. Section A2 gives solar radiation availability throughout the year; Section A5 shows tables of solar gain factors for a range of glass/blind combinations; Section A8 deals with the prediction of summertime temperatures in buildings (these are also to be found using the graphs in the BRE *Environmental design manual*[22]).

This type of approach can indicate whether measures such as increased solar shading or reductions in glazing areas would help in reducing unacceptably high interior temperatures.

Solar gains can also occur through opaque roofing. Their effects can be reduced by using reflective materials, substituting heavyweight for lightweight constructions, and by adding extra layers of thermal insulation on or near the outsides of the roofs.

Shading devices

Shading may be required to reduce solar heat gain, to prevent glare from the sun and localised thermal discomfort due to the sun shining on occupants, and to reduce or eliminate sky glare. Where such problems are likely to arise, all roof lights should be fitted with appropriate shading devices.

It is important to realise that some shading devices may not combine all of these roles effectively. For example, low transmission glazing will reduce solar heat gain and sky glare but has little impact on direct glare from the sun because of the high intensity of the sun's rays. Conversely, internal shading devices such as venetian blinds can eliminate glare from sky and sun but are less effective at reducing solar heat gain.

For some interiors, it may be acceptable to restrict summer sunlight using fixed elements like louvres or screens. However, retractable and adjustable shading such as may be considered for the vertical parts of roof monitors is often better suited to the low solar altitudes of the UK, especially on east- or west-facing surfaces. These systems should be easy to maintain and easy to operate. Adjustable external systems should be robust, or retract when necessary if vulnerable to wind damage.

Shading, particularly fixed shading, should not be designed so that it reduces interior daylighting in an unacceptable way. Some shading devices may also restrict natural ventilation.

Sound insulation

As mentioned in Chapter 3.1, sound insulation of low-pitched roofs depends largely on the type of material used for the deck. Approximate values are as follows:

- timber structure supporting timber boards (no ceiling): 20–25 dB
- timber structure supporting timber boards (with ceiling): 30–35 dB
- concrete deck not less than 200 kg/m²: 45 dB
- concrete deck not less than 200 kg/m² with roof lights approximately 10% of the roof area: 25 dB

There may well be circumstances where the roof is required to give enhanced insulation from airborne sound (eg where the building is near an airport). Airborne sound insulation of flat concrete roofs ranges from 45 dB to 50 dB, mainly depending on the weight of the roof. Some improvement may be achieved by adding, for example, a screed; but only if the existing structure can be shown by calculation to be strong enough to support it. Pitched roofs of tiles with suspended ceilings give insulation over the range 30–40 dB; if the structure permits, additional layers of plasterboard at ceiling level will help to improve the insulation value. Thin sheet roofs are unlikely to give more than 25 dB and can be as low as 15 dB. Roof lights reduce insulation values considerably, particularly if vented. Specialist advice will be needed.

Durability and ease of maintenance

Fully supported stainless steel sheeting

Type 304 austenitic stainless steel has been used for medium-span roofing on commercial and public buildings. Such material can be expected to last for 100 years, assuming the roof is self-cleaning and there is no build-up of contaminants such as chlorides. Type 316 stainless steel is used in more aggressive industrial and coastal conditions.

Rigid metal sheets

Sheets of rigid profiled (troughed) steel or aluminium (carried between purlins), whether they are single- or double-skinned, are more likely to be found on industrial buildings and are therefore dealt with in Chapter 6.1.

Other fully supported metal roofs

There is no essential difference between fully supported roofs in practice when installed over larger spans (Figure 5.18), though where long-strip coverings are used the movement will be correspondingly greater than with domestic spans.

Readers should also see the section on Fully supported metal roofs in Chapter 3.3.

Figure 5.18: A steeply pitched lead-covered roof. The overhang, which shelters the top of the slope from rainfall, gives a different appearance to this roof. It has no effect on durability

Slates and tiles

Assessments have been made in Chapters 3.1 and 3.2 about the durability of tiles and slates on short-span domestic roofs. There is no essential difference with respect to such materials over larger spans (Figure 5.19; Case studies 11 and 12). However, it is more important with non-domestic scale roofs not to let maintenance cycles slip (Figure 5.20).

Commercial buildings dating from the 19th century could well have decorative tiles that have long since become unavailable and cannot be

Figure 5.19: A non-domestic medium-span, plain-tiled, bonneted hip roof in good condition

Figure 5.20: In contrast with the roof in Figure 5.18, a long-forgotten valley in a heritage building about to be refurbished. Some tiles are delaminating and others have slipped. The gutter lining is completely unprotected

replaced (Figure 5.21). It therefore places responsibility on contractors to take great care to avoid damage to such items during refurbishment work.

WORK ON SITE

Access, safety, etc.

Readers should see this section in Chapter 3.1. Further advice applicable to larger spans is available from HSE (www.hse.gov.uk) and in their publication *Health and safety in roof work*[23].

Storage and handling of materials

In some circumstances, fixed-in-place building elements and components have to be protected for long periods before the building can be handed over (eg anodised aluminium roof lights).

Vulnerable items that might be damaged or stolen during work should be designed to be fixed at as late as possible in the construction process. The need for replacement has caused contract delays on a number of sites monitored by BRE. Timely deliveries can help, too, in preventing damage, preferably without jeopardising the construction timetable.

Figure 5.21: A hotel converted to offices. The irreplaceable Stanley Brothers glazed decorative tiles on the mansard have been disturbed and some have been broken during replacement of windows and rainwater gutters

Delivery, storage and stacking

Matters to which consideration should be given include:

- types of crating, palleting, etc.,
- sizes of units delivered,
- vulnerability to frost,
- vulnerability to dampness (whether packs are well drained),
- vulnerability to distortion (ground is not level or bearers are not packed up to level),
- vulnerability to accidental damage (particularly from passing vehicles or fork lift trucks within the stacking area),
- damage from transportation packing (materials and components are distorted because tight banding has not been removed once they are in store),
- shelf-life of products.

Glass and frames can suffer both moisture and mechanical damage following incorrect site storage.

Requirements for the correct storage of glass are given in BS 6262-7[24]. Primer-painted joinery and factory-coated metals are particularly vulnerable, as are PVC-U frames. Factory-sealed glazing units can have seals damaged by faulty handling and site storage. Units should be stored in original packing, well protected from mechanical damage and the weather.

Particular care which needs to be given to 'specials' should be noted. (Specials may take a long time to replace if lost or damaged.)

Plastics components should be protected from ultraviolet radiation and from contamination by sand, aggregates, oils and solvents.

Handling of roof trusses or long-span trussed rafters will need special consideration. The use of spreader bars for crane handling will be appropriate in most cases.

Restrictions due to weather conditions

When considering movement in large roofing components, particular attention may need to be given to the time of year when the installation is likely to be completed. This will enable alternative courses of action to be prepared whatever time of year the construction actually takes place. This need arises because sizes of components, which have been assumed at a maximum in summer conditions, may have a critical influence on the available space within a joint to accommodate thermal movement (and therefore subsequent performance) if assembly is delayed until cold weather.

Workmanship

Codes of practice for carpentry, joinery and general fixings are dealt with in BS 8000-5[25] and for slating and tiling of roofs and claddings in BS 8000-6[25].

Box 5.2: Inspection of medium-span commercial and public pitched roofs

The items listed in Boxes 3.11 and 3.13–3.16 apply here. Some items from these lists are repeated below because they are crucial to the performance of non-domestic roofs.

The particular problem features to look for are listed below.

- Inadequate arrangements for rainwater drainage and melting snow
- Unsafe access to roof glazing
- Rainwater penetration of complex roof shapes
- Sheeted roofs lapped incorrectly
- Creep of fully supported metal roofs
- Condensation under sheeted roofs
- Corrosion in metal truss members
- Wood rot and insect attack in timber truss members
- Absence of chevron and web bracing in timber trussed rafters

- Colour mismatching and poor colour fastness of sheets
- Absence of filler blocks at eaves, ridges
- Vulnerability of roofs to forced entry
- Sagging of laminated glazing in roof lights
- Lack of integrity of vapour control layers
- Sound transmission through services in roof voids
- Poor condition of lightning protection
- Inadequate daylighting through roof lights
- Poor condition of sun shading to roof lights
- Plastics fixings holding fire curtains

5.2 FLAT ROOFS

The traditional material for covering flat roofs on medium-span non-domestic buildings has been built-up felt or mastic asphalt. (The basic characteristics of these two types have been described in Chapters 4.1 and 4.2.) However, polymeric single-layer* materials are coming into general use. They are likely to be covered by CE marking in the future according to the requirements of BS EN 13956[26] (see Chapter 4.5).

The major differences between domestic and non-domestic buildings are the increased thermal and moisture movements in decks, and increased risk of deflection and subsequent ponding, due to the longer spans. In the case of roofs that are used for car parking, the roofs must be treated as floors and will therefore need fire resistance. Hospitals are also a special case from the point of view of fire: this topic is dealt with in Health Technical Memorandum 05-02[27].

Most polymeric membranes are either elastomeric or thermoplastic in character, though CSM (chlorosulfonylpolyethylene) is both. Furthermore:
- elastomeric materials tend to retain their elastic properties at very low temperatures and are relatively insensitive to temperature variations in service,
- thermoplastic materials become less flexible at lower temperatures, and exposure to high temperature and ultraviolet radiation degrades the polymer with a consequent increase in risk of cracking in service.

Elastomeric materials include EPDM (ethylene-propylene-diene monomer), normally in sheets 1 mm thick, IIR (isobutene-isoprene) and PIB (polyisobutylene), normally in sheets 1.5 mm thick. All polymeric materials exposed to UV will degrade in time. The time taken to degrade depends on the product formulation.

Thermoplastic materials include PVC-P (ie plasticised), CPE (chlorinated polyethylene), normally in sheets 1.2 mm thick, and VET (vinyl ethylene terpolymer). They are usually joined by solvent-welding, though some can be joined by hot-air welding. Also within this category come the various liquid waterproofing compounds useful for the more awkward shapes of roofs.

Although CE-marked products will become available in the future many of those currently on the UK market have obtained third-party certification.

The European Union of Agrément (UEAtc) has issued guidance for its members carrying out assessments of new thermoplastic materials[28]. The guidance covers such matters as:
- resistance to wind uplift,
- watertightness,
- resistance to thermal shock,
- dimensional stability,
- slip resistance,
- resistance to cyclic movement,
- resistance to static indentation,
- resistance to dynamic indentation,
- tear strength.

Methods of Assessment and Test have also been issued by UEAtc for:
- unreinforced PVC[29],
- SBS (styrene-butadiene-styrene) elastomer bitumen[30],
- EPDM[31].

The test methods referred to in BS EN 13956[26] are similar to those previously adopted by the UEAtc.

CHARACTERISTIC DETAILS

Basic structure
Many flat roofs in non-domestic buildings are of reinforced concrete. These roofs have become a common feature of multi-storey buildings and some of the illustrations included in this chapter are of this form of construction.

Roofs of timber and steel are also to be found (Figure 5.22). In some cases, the construction can be quite complicated, particularly among the building systems used in the school building programmes of the 1950s and 1960s (Figure 5.23). Since the changes in the early 1990s in responsibilities for the maintenance of many school buildings, surveyors invited to carry out inspections will need to make every effort to obtain the original drawings and

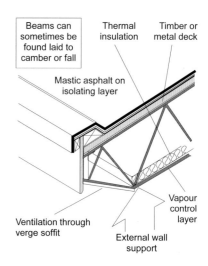

Beams can sometimes be found laid to camber or fall

Thermal insulation

Timber or metal deck

Mastic asphalt on isolating layer

Ventilation through verge soffit

External wall support

Vapour control layer

Figure 5.22: A mastic asphalt verge detail of a cold roof on a lightweight steel beam and column construction

---

* The term single-ply is often used.

MK. I.

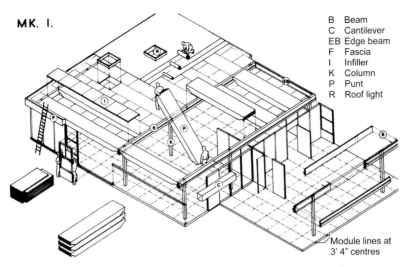

B   Beam
C   Cantilever
EB Edge beam
F   Fascia
I   Infiller
K   Column
P   Punt
R   Roof light

Module lines at
3' 4" centres

Figure 5.23: The 'Punt' system used in the Hertfordshire County Council school building programme of the early 1950s. Plywood roof units are shaped, as the name implies, like punts. Every component in the system was capable of being lifted by two men

Figure 5.26: An eaves kerb on a roof covered with BS 747 class 5 solar-protected felt

Figure 5.24: Lightweight galvanised steel lattice beams were used for the classroom spans (approximately 7 m) typical of the system-built schools of the 1950s. This particular example is of a Hills 8' 3" system

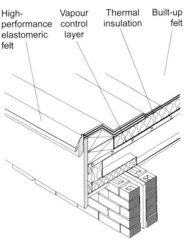

High-performance elastomeric felt   Vapour control layer   Thermal insulation   Built-up felt

Figure 5.25: Verge detail on a flat roof converted to a warm deck roof

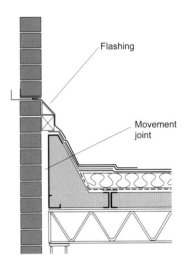

Flashing

Movement joint

Figure 5.27: A movement joint at an abutment in a warm roof. Flashings are vulnerable

other necessary information before carrying out any inspection work.

Timber flat roofs in medium-span construction are quite often constructed in ply-web beams of box or I section. These beams can be doubled, and also the webs stiffened with uprights, effectively making a composite box girder roof. The roofs normally have timber decks.

Steel lattices were also widely used in school building consortia systems in the years following the 1939–45 war (Figure 5.24).

Eaves and verges

Where there is little or no overhang at the eaves, it is necessary to provide a substantial upstand or parapet to give protection to the wall below (Figures 5.25 and 5.26).

Abutments

Some larger roofs that include beams or girders having relatively large movements need specially detailed joints at abutments (Figure 5.27).

Roof lighting and service perforations

Differences between service perforations are largely of scale. Flues may require special detailing (Figure 5.28).

Readers should also see the sections on service perforations in Chapters 4.1 and 4.2.

Rainwater disposal

Readers should see this section in Chapters 4.1–4.4 and Figure 5.29.

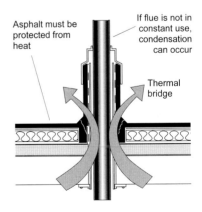

Asphalt must be protected from heat

If flue is not in constant use, condensation can occur

Thermal bridge

Figure 5.28: Flue projections through roofs need close attention to detail. Cyclical thermal movements may open cracks in the weatherproof layers at the bases of stacks where they project through roofs

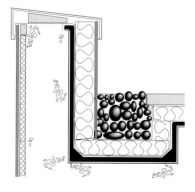

Figure 5.29: A parapet gutter in a mastic asphalt roof which has been converted to a warm deck inverted roof. Gutters need lining with insulation to avoid thermal bridging. They also need to be checked for adequacy

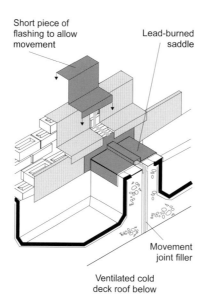

Figure 5.30: A movement joint through an asphalt covering

Movement joints

Built-up roofing will need to be provided with movement joints that reflect movement joints in the deck beneath (Figure 5.30). Although asphalt itself does not require the provision of joints to accommodate minor movement in the deck (any movement normally being taken up by the material), movement joints in the structure will need to be replicated through the asphalt.

MAIN PERFORMANCE REQUIREMENTS AND DEFECTS

Choice of appropriate decks

Many flat roofs on buildings built before the 1939–45 war had decks of reinforced concrete with the covering more than likely to be of asphalt. For medium-span buildings of more recent construction, the decks could well be lightweight; roofs of timber construction will normally, but not invariably, have timber decks.

Strength and stability

The structural design of roofs for buildings in the medium-span category is covered by the main British Standard codes of practice for the various materials:

- steel to BS 5950[32] (various Parts are appropriate to roofs),
- concrete to BS 8110-1[33],

- timber to BS 5268[34] (various Parts are appropriate to roofs).

Adaptation of medium-span flat roofs (eg to carry additional services) should not be attempted without first checking whether the existing structure is sufficiently strong to cope with any new loads. In particular, holes should not be cut in the webs of plywood box beams without calculating the effects.

The method of attaching the waterproof layer to the deck is an important first consideration

in the process of designing the weatherproofing system (Figure 5.31). There are three basic methods.

- Adhesive bonding to the deck or insulant, generally used for cold deck designs
- Loose laid and ballasted to prevent uplift, generally used for inverted designs
- Mechanically fixed through the sheet and insulation to the deck, generally used for warm deck designs

In the case of mechanical fixings, the numbers of these fixings must relate to the design wind speeds, remembering that suction is greatest on the edges of flat roofs (Figure 5.32). The fixings may be of strip or spot type (eg PVC material can be stuck to a PVC-coated metal plate spiked or screwed into the substrate).

Wind scour of ballast may occur, even on those flat roofs protected by parapets. Experience has shown that loose gravel will move under the scouring action of wind in storms, and may be blown off the roof causing damage to glazing or injury to passers-by. This problem and its solution are dealt with comprehensively in BRE Digest 311[35]. Paving slabs are less likely to be disturbed than gravel and may be

Figure 5.31: A roof light in a steel deck being prepared for asphalting. The deck spanning the troughs has not yet been completed. Expanded metal has been placed over the underlay on the side of the roof light and a chamfer partly fitted

Figure 5.32: Mechanical fixing of a single-layer roof membrane by drilling through the membrane into the substrate, screwing down a metal strip and welding on a cap sheet over the strip. Care needs to be taken to avoid rucking of the sheet during fixing (The detritus has still to be cleared away.)

preferable where the predicted risk of movement is high (see Chapter 2.1).

The strength properties of thermal insulation boards are important where pedestrian traffic is to be allowed on the roof. It is recommended that boards that carry paving and pedestrian loads should have a compressive strength of not less than 175 kPa. Where roofs are to carry heavier traffic, increased strength will be needed. The impact resistance of some insulation materials should be checked, particularly that of, for example, foamed glass. See also BS 6399-1[36], Clause 5.

Dimensional stability

The dimensional stability of reinforced concrete roof decks may sometimes be a cause for concern. Dimensional changes in the decks can cause damage to the supporting structure and, also, in some circumstances, give rise to distortion in the deck itself. The most important causes of excessive deformation are:
- thermal expansion and contraction,
- drying shrinkage and moisture movement,
- elastic deformation due to self weight and imposed loading,
- creep due to prolonged loading.

The exposure of flat roofs to solar radiation was referred to in Chapter 4.1 on domestic flat roofs. In the case of buildings larger in scale than domestic, the movements will be correspondingly greater and there may be other factors to take into account. Broadly speaking, as is the case with shorter-span roofs, roof decks in warm deck or inverted designs will experience lower temperature swings than those in cold deck designs. Whether or not temperature swings are converted into movements depends on the degree of restraint afforded to the deck by the supporting structure.

Cracking

If design provision is insufficient to accommodate the expansive size changes of the structure, cracking (eg of reinforced concrete decks) will result. In the case of fully bonded roof coverings, the cracks will be transmitted directly to the covering which may also crack, depending on its ability to accommodate movement, and lead to rain penetration. A fuller description of the risks of cracking in flat roofs, together with the diagnosis of their causes, is contained in *Cracking in buildings*[37].

In most cases, the defects caused by thermal movements can be readily recognised. They invariably show on plaster finishes. On a summer's day, changes in temperature will cause a roof to move, and cracks to open and close.

In unframed and framed buildings the cracking takes slightly different forms.

Unframed buildings
The whole roof slab tends to bow slightly upwards at the centre owing to the higher temperature of the upper surface and, at the same time, the slab spreads outwards. The two movements together tend to cause local cracking at the tops of walls and partitions, rather jagged when the walls are in line with the expansion movement, but otherwise with a simple type of horizontal break.

Framed buildings
The effects are generally more disfiguring. Beams tend to bow upwards and may take parts of the walls with them. The distribution of the cracking will depend on how the roof as a whole is restrained. In a simple symmetrical roof, movements will be outward from the centre. If the building abuts another so that the roof can only move in one direction, cracking will be localised at the unrestrained end and will be more severe there. Expansion takes place from fixed points such as stair and lift wells.

The remedy to reduce the amount of thermal movement is to place the thermal insulation above the deck creating a warm deck solution. The procedure given in Chapter 4.1 can be used.

Prevention of cracking in cold deck structures would require the provision of movement joints in spans exceeding about 10 m, or 5 m if the spans are restrained absolutely at one end.

Creep in concrete slabs or beams may be a problem in circumstances where it leads to excessive deflections. Excessive deflections should always be investigated. The ponding of rainwater, which is usually the first sign of deflection, may not always be due to local deformations but could be symptomatic of a more serious problem with the structure. BRE has seen more than one case where the cause of local deformations has been misdiagnosed and the hollows have been screeded over to assist rainwater drainage. This simply exacerbates the root problem and the roof again deflects, this time worse than before.

In warm deck designs it is crucial that the vapour control layer is not disrupted by size changes and therefore the vapour control layer should not be fully bonded to the deck.

Asphalt membranes on cold deck roofs are liable to low temperature cracking. Heat loss by radiation to the clear night sky in winter can reduce the temperature of dark surfaces to about −25 °C. The resulting shrinkage of the asphalt in its relatively brittle state can result in cracking, sometimes accompanied by loud bangs.

In decks used for car parking it is not usually feasible to create a warm deck roof because of the need for sufficient strength to resist the loads imposed by vehicular traffic.

Movement joints in roof decks used for car parking cannot be provided with upstands for obvious reasons, and any armouring of movement joints will need to be made flush with the deck. The bridging medium could well operate on the bellows principle using a suitable gasket; in most cases, there will also need to be some kind of sliding mechanical protection to the bridging medium.

Exclusion and disposal of rain and snow

Where a parapet bounds a metal-covered flat roof, melting snow may be a cause of water overflow. The problem arises where ice or snow blocks the gutter, creating a dam and the resultant ponded water overtops the upstands. The problem can in part be solved by using snow boards to keep the gutter and outflow clear (Figure 5.33).

In some cases, it may be thought worthwhile to install an automatic leak detection system. One such system works on the principle of a change in electrical conductance in a grid of metal tapes buried beneath, and insulated from, the outer surface of the covering. Water penetration is monitored by means of a microprocessor, and the source can be pinpointed and dealt with.

Where access is provided to roofs, the weathertightness of the roof covering is often not easy to maintain around handrail standards that perforate the waterproof layer (Figure 5.34). In general, the top of the upstands should be at least 150 mm above the surrounding roof, and the standard should be sleeved with the capping overlapping the upstand. Where holding-down bolts are needed to fix, for example, cleaning cradle rails, they should be seated on washers of neoprene or a similar material and covered with a suitable non-hardening mastic.

Rainwater entrapped during construction has proved to be a problem frequently diagnosed by BRE Advisory Service staff. It is evident that both construction water and trapped rainwater should be allowed to evaporate before roofs are enclosed by the weatherproof membrane, or be vented to the external air through hooded ventilators.

The design and detailing of kerbs, particularly those protecting movement joints (Figure 5.35), are crucial to the exclusion and disposal of rainwater; they should also be of sufficient height, not to be overtopped in storm conditions (Figure 5.36).

Figure 5.34: Perforation of a waterproof layer (eg for services, machinery and cleaning cradle rails) needs particular attention and careful detailing. Indeed there can sometimes be more perforation than roof!

Figure 5.35: A movement joint on a concrete slab roof. The jointing material has completely failed to accommodate the movements in the slab. This type of failure in specification has often been seen by BRE investigators

Figure 5.33: Drifting snow on a lead roof behind a parapet

Figure 5.36: The kerb at this hopper head proved too small to prevent overtopping in heavy rain

Energy conservation and ventilation

Thermal insulation products for the three basic kinds of flat roof (cold deck, warm deck sandwich and warm deck inverted) need to be selected carefully. Thermal insulation for flat roofs is dealt with comprehensively in BRE Digest 324[38].

For cold deck designs where the insulation is below the deck, most thermal insulation materials are suitable including loose fill. Where a span is greater than 10 m, extra ventilation is needed at the eaves and over the top of the thermal insulation equivalent to 0.6% of the total roof area.

For warm deck sandwich designs, either rigid insulation should be used, preferably with a low coefficient of thermal movement, or an overlay placed between the insulation and the membrane above. Extruded polystyrene boards are not recommended for this type of application.

In inverted designs of warm deck, rigid insulation is needed to support ballasting and roof traffic (this topic has been covered in Chapter 4.3). Rigid, closed-cell plastics or compressed mineral fibreboards are best, but organic materials are unsuitable. In inverted roof designs, the insulation panels rather than the deck may be subjected to thermal movements, allowing rainwater to reach the waterproof membrane below. The temperature of the deck is then lowered with the consequent risk of local condensation forming within the construction beneath where the rainwater is flowing. This phenomenon and its avoidance have also been described in Chapter 4.3.

Vapour control layers, where they are needed to control the passage of vapour (eg through joints), should not be bonded directly to corrugated metal decks since more than half the layer is unsupported. These vapour control layers should be laid only over a board on top of the metal deck.

Control of solar heat

Reflective treatments on the weather surfaces are an essential means of reducing solar gains for many

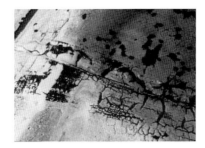

Figure 5.37: A site-applied solar-reflective surface on a BS 747[14] class 5 felt roof. The extent of this damage may look spectacular but it has had little degenerative effect

materials. A traditional coating was described in the section on prediction of temperatures in Chapter 2.5, though other more suitable coatings are now available (Figure 5.37).

Fire precautions

The performance of various kinds of covering with respect to fires from external sources has already been given in appropriate earlier chapters.

In cold deck roofs, combustible insulation laid beneath the deck between the joists can increase the risk of ignition by preventing the dissipation of heat. The addition of flame retardants can help, though there may be a problem associated with ventilation. BRE Digest 324[38] discusses this problem.

For warm deck inverted designs in which the insulation is covered by a ballasting layer of stones or slabs, fire risks from external sources are comparatively low. In warm deck sandwich designs, spread of flame over the covering will determine the extent of fire involvement of the roof. The breakthrough of the fire into the building will be determined largely by the material and the design of the roof deck. Interaction may, however, occur; for example, where bitumen and molten polystyrene may flow readily along troughs in a steel roof deck and, still burning, fall into the building below.

In commercial premises in the past, roofs have been extensively involved in the rapid spread of internally generated, self-supporting fires, often leading to substantial damage. The introduction of flammable insulants increases the risk. Very large insulated, cold deck,

Figure 5.38: Perimeter lightning conductors, correctly installed, on a flat-roofed office building

built-up roofs should have barriers to fire spread in the deck itself to limit the potential area of damage. Localising the effects of any fire, by allowing heat to escape, may be possible by introducing some form of ventilation.

For those cases where roofs are supported by steel construction and also serve the function of floors, provision of fire protection will be necessary in most situations. Reinforced concrete construction carries its own protection. For timber, BS 5268-4[34], Section 4.1 is relevant.

For most timbers exceeding 400 kg/m³ (which includes most commonly used softwoods and hardwoods) the surface spread of flame to BS 476-7[39] is Class 3. This may be uprated to Class 1 or Class 0 by appropriate methods of treatment. Section 4.1 of BS 5268-4[34] gives information on the rate of charring of various species of timber.

Lightning protection

Lightning protection is normally provided in the form of a grid of conductors placed at appropriate intervals over the whole of the roof, especially the perimeters. BS EN 62305-2[18] is relevant (Figure 5.38).

Sound insulation

Readers should refer to this section in Chapter 5.1.

Roof lighting

The calculation of daylight factor has already been referred to in Chapters 2.7 and 5.1. When adding roof lights to an existing flat roof, consideration needs to be given to their proximity to existing roof lights (Figure 5.39).

Figure 5.39: Plastics roof lights on a school building. From a fire standpoint, the two roof lights in the top left corner of the picture are too close together (see Figure 2.28 for permitted distances)

Durability and ease of maintenance

In a study carried out by BRE in about 1980, it was found that lightweight roofs with decks of timber or metal cost some six times as much to maintain as heavyweight roofs with concrete decks during the first 30 years of their lives. Although this situation no longer applies, there are certain points that surveyors need to bear in mind during inspections (see Box 5.3).

Decks

For concrete decks, deterioration could be due to a number of causes such as rusting of reinforcement through lack of sufficient cover to the steel. The rusting could be made worse by carbonation of the concrete. There is also the possibility of the use of high alumina cement concrete (HACC) in buildings built between about 1930 and the mid-1970s when the risks associated with this material were brought to the attention of the industry following several collapses of roofs. In essence, the mechanism of deterioration is as follows. Hydrated high alumina cement (HAC) undergoes a change in its mineralogical composition with time through a process known as conversion; associated with this is a loss in strength. In time, the strength could reduce to half the original value. Cover to reinforcement in most HACC beams is less than 20 mm and cracking due to

corrosion can take place. Chemical attack may produce further loss in strength where the components are subject to persistent dampness and, in some cases, the effects on strength have been very severe. Where this possibility exists, the question of durability should be addressed by a specialist[40]. Inspection for HAC deterioration is covered in Box 5.3 at the end of this chapter.

Since the 1970s there has been the remote possibility that deterioration may be caused by alkali silica reaction (ASR). Further information on ASR can be found in BRE Digest 330[41].

Short-span concrete deck roofs usually give visual warning of deterioration before danger of collapse is imminent, but longer spans may be more problematical, especially with roofs of box construction where the corrosion can proceed unseen. Problems may also arise where:

- the structure is post-tensioned,
- ducts may not be fully grouted, and
- tendons may be at risk of failure.

Identification of the kind of construction is therefore vital to proper monitoring of the performance of longer-span concrete roofs.

Coverings

The durability of built-up roofing has been covered in Chapter 4.1, asphalt in Chapter 4.2 and metals in Chapter 4.4.

Single-layer polymeric sheetings are estimated in third-party certificates to be capable of lives of 20 to 25 years, although experience of their use in other countries suggests that these estimates are conservative.

BRE has carried out tests of fatigue resistance, which has a direct relationship with durability, of single polymeric sheets. High standards of workmanship in the installation of polymeric sheets are crucially important to ensuring longevity.

A common fault seen in movement joints is where capping is attached to both sides of the upstands, leading to unacceptable strain on the capping.

Figure 5.40: Temporary patching on a flat felt roof pending a long-term solution. Questions of appearance could arise if such roofs are visible from other buildings

So far as liquid-applied materials are concerned, chemically cured systems seem to be more durable than emulsions; up to 30 years can be anticipated for the former.

Durability of balustrading needs careful consideration, particularly where ungalvanised mild steel is used.

Patching

Felted flat roofs in non-domestic buildings are vulnerable to damage caused by maintenance traffic, especially the feet of ladders that are not supported on boards. As in the roofs of domestic buildings, leaks may also arise from other causes; for example, walking on blisters will break them.

In domestic and non-domestic roofing, patching may provide a temporary solution until replacement of the whole roof covering becomes due. Where such roofs are seen from above, visual acceptability may become the governing criterion (Figure 5.40).

Where the damage is caused by movements in the deck, the bottom sheet of the repair must be partially bonded, as in Figure 4.20. Patching also depends on the quality of workmanship, especially cleaning off any stone chippings from the surface.

WORK ON SITE

Access, safety, etc.
Readers should refer to this section in Chapter 4.1.

Storage and handling of materials
Readers should refer to this section in Chapters 4.1 and 4.2.

Restrictions due to weather conditions
Any repairs that may be necessary to a reinforced concrete structure should be carried out in conditions no less strict than are appropriate for new work; for example, that the temperature of concrete is not allowed to fall below 5 °C (preferably not below 10 °C) and not to rise higher than 30 °C.

Liquid-applied systems should be laid in dry conditions and some need a dry spell to cure. Sprayed polyurethane foam systems need to be laid in completely windless weather conditions to achieve good performance, and such conditions are infrequent in the UK.

Readers should also see this section in Chapters 4.1 and 4.2.

Workmanship
As with many other roofing materials and methods, asphalting depends on the quality of workmanship. The vulnerability of some newly laid mastic asphalts to frost can lead to cracking (as noted earlier in the chapter) and protection is required. As for other materials, satisfactory long-term performance also depends considerably on the quality of site work.

Box 5.3: Inspection and testing of medium-span commercial and public flat roofs

When examining reinforced concrete structures for performance over time, and, in particular, finding apparent evidence of deflections and bow, it is as well to remember that the structure may have been manufactured and erected with these deviations. Permissible dimensional deviations given in BS 8110-1[33] for new construction allowed:

- 6 mm in members up to 3 m long,
- 9 mm in 3–6 m,
- 12 mm in 6–12 m, and
- a further 6 mm for every 6 m above 12 m.

BS 8110-1[33] has since been replaced by BS EN 1992-1-1[42].

Existing reinforced concrete roofs may show defects on routine visual inspections which will call for further testing. Visual evidence will include:

- cracking,
- corrosion and spalling,
- disruption of anchors in post-tensioned structures, and
- any evidence at all of water penetration.

Simple tests (eg checking the position of reinforcing rods using an electromagnetic covermeter in the manner prescribed in BS 1881-204[43]) may be appropriate for relatively unskilled people to use. Other, more sophisticated tests to BS 1881 should normally be left for experienced structural engineers to specify and interpret; for instance, those involving cutting cores for testing compressive strengths and the use of gamma radiography for detecting the presence of voids or for examining the filling of tendon ducts in pre-stressed concrete.

Radar has been used for detecting the presence of water in roof construction but it will not yield any useful information on the presence of corrosion of reinforcement; there are, though, other test methods that may be appropriate and further advice should be sought.

Load tests on completed structures or parts of structures may also be called for by the structural engineer in certain circumstances, and assessments carried out on cracking and recovery of the structure from deflection.

Roof coverings

Readers should refer to Boxes 4.3, 4.6, 4.7 and 4.8 at the ends of Chapters 4.1–4.4.

Mastic asphalt coverings used over lightweight decks, in conjunction with considerable thicknesses of thermal insulation, will need close inspection at abutments and verges.

High alumina cement (HAC)

Inspections of older roofs where HAC has been used should follow the recommendations of BRE Special Digest 3[44]. In the absence of chemical attack, the strength assessment guidance issued in 1975 and 1976 by the then Department of the Environment (DOE)[45] has been shown to be safe, but the risk of corrosion to reinforcement is an increasingly important consideration in the assessment of HAC components. The assessment should be carried out by specialists with experience of HAC structures. It is likely to be both disruptive and expensive but an outline gives an idea of what is involved.

With inspection of roofs containing HAC, the tasks are:

- confirming the use of HAC,
- establishing the section profile and the reinforcement,
- comparing with the allowable capacities of sections in the DOE guidance[45],
- if sections differ from the DOE guidance, calling in a specialist petrographer to examine lump samples by special test methods,
- assessing the shortfall in capacity and deciding whether to replace or strengthen affected structures.

Internal fracture tests and ultrasonic pulse velocity tests may be useful in identifying areas of very low strength, but are not likely to give the degree of assurance for a basis of appraisal.

Visual inspection of the soffits of components offers a viable method of monitoring the general structural condition of components where the protection to the reinforcement has been lost, but webs should also be inspected where possible. Where cracking near bearings is found, the beams should be propped and the circumstances assessed by a structural engineer.

5.3 VAULTS AND OTHER SPECIAL SHAPED ROOFS SUCH AS GLAZED ATRIA

This Chapter deals briefly with medium-span roofs of special shapes, including:
- barrel vaults and domes in reinforced concrete,
- lamella roofs in timber, and
- plastics or other glazed vaults such as are used over shopping malls.

A general description of the basic characteristics of shell roofs will be found in *Principles of modern building*[46].

Although (concrete) shells are an efficient means of spanning large areas, they cannot be used to hang loads of any significance. They are not much used, therefore, for commercial buildings. It is suspected that many of the shell roofs built in the years immediately following the 1939–45 war have been demolished by now.

In general, shells and vaults will have a waterproofed external surface, though not always. For an exposed stone vault, the selection and detailing of a suitable stone for repair is vital to longevity. The mechanisms of deterioration of natural stone are described in *The weathering of natural building stones*[47] (Figure 5.41).

The masonry hovels which encased the bottle ovens of north Staffordshire in the 18th and early 19th centuries also provide examples of vaults with no separate weatherproof covering. Since the surviving kilns are no longer in use to keep the structures relatively dry, deterioration ensues (*Potworks. The industrial architecture of the Staffordshire Potteries*[48] and Figure 5.42).

There are a few brick barrel vaults in heritage buildings though they were not always used on the top storeys. The massive masonry woollen mills of the Victorian era are examples of this form of construction (the so-called fireproof, shallow vaults held in cast iron column and beam construction with wrought iron ties). Apart from local crushing or spalling of units, there are no simple visual assessment criteria for masonry vaults and domes*.

CHARACTERISTIC DETAILS

Basic structure

Most examples of shell roofs are of reinforced concrete, usually cylindrical in form rather than domical, sometimes pre-stressed and/or post-tensioned. The shell may be very thin (as little as 50 mm) in order to reduce the dead load over comparatively large spans. Since the 1980s, ribbed glazed construction in various configurations has become popular.

Surveyors may occasionally encounter a lamella roof in timber, normally in the shape of a pointed

---

* Advice on the serviceability, maintenance and repair of masonry vaults and domes can be sought from BRE, the Heritage Departments or specialist consultants.

Figure 5.42: Brick masonry vaulted hovel. Very few of these once plentiful structures remain, though other roofs now benefit from the resulting absence of the pollution they once emitted. Plant roots will quickly damage mortar joints

Figure 5.41: This ribbed stone vault does not have a separate weatherproof covering. Durability of the roof depends on many factors including correct selection of the stone and its bedding planes, suitable mortar mixes, and air pollution levels

Figure 5.43: A lamella roof in timber

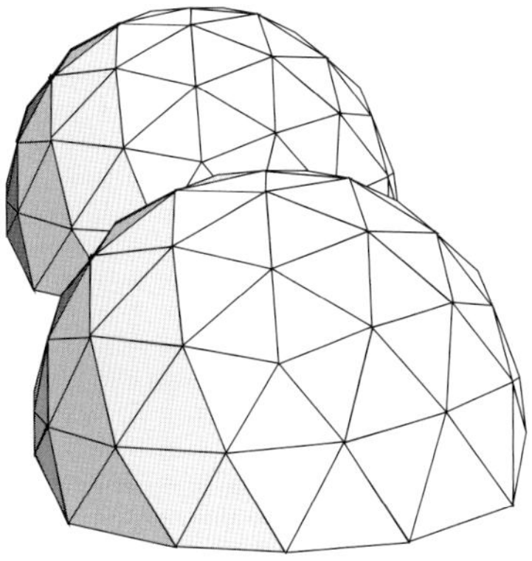

Figure 5.45: Geodesic domes

arch vault. Such roofs depend for their integrity on the soundness of the connections between the ribs (Figure 5.43), which are usually in the form of metal cleats, and on the connections between the sheathing and the ribs which may be nailed and glued. The actual lengths of timber used in these roofs are very short, spanning only two coffers.

Another form of roof which is still in existence, largely surviving from the 1914–19 war (when they were built in large numbers for aircraft hangars) is the Belfast Truss. Although the spans are large, the struts and ties were all made from comparatively short lengths of timber (Figure 5.44).

Geodesic domes are essentially of stressed skin construction characterised, as lamellas, by the lack of separate frames (Figure 5.45).

They have great inherent strength. They more often than not form the outer protective shells to structures having special functions (eg former radar installations). There has even been one at the South Pole! Construction will probably be fibre-reinforced plastics flanged plates bolted together through the ribs with some form of gasket sandwiched between the flanges, although double-skinned pillows have also been used. Geodesic domes are rarely encountered, though some houses roofed in this form have been seen in the west Midlands.

Roof lighting

Glass

Laminated glass can be made in curved shapes suitable for use in framed domes and vaults. Cast

glass is used for roof lights in small domes where the breakage risk is acceptable.

Plastics

Plastics roof light materials have been described in Chapter 2.7. Normally, if fully glazed barrel vaults cannot be glazed with cold-formed sheets, they are glazed with thermoformed plastics; acrylics are not used in the UK following the Summerland fire disaster in the Isle of Man, although they are still used in continental Europe. In the past, most roof lights have been made of PVC which is prone to discoloration (see the section on Durability later in this Chapter).

Other materials include:
- PMMA (polymethyl methacrylate),
- GRP (glass-fibre reinforced polyester),
- PVC,
- PC (polycarbonate),
- ETFE (ethylenetetrafluoro-ethylene) (Figure 5.46).

In concrete barrel shells, north lights were sometimes introduced in association with deep beams. Circular cast glass roof lights or lenses were sometimes used inset into concrete shells, often leading to rain penetration.

Since the 1960s most roof lights over larger spans have been framed up in steel or aluminium, and some very elegant structures are possible irrespective of the material forming the glazing (Figure 5.47).

Various polymer films are available that are designed to be applied to the inner or, in some cases, the outer surface of the glass. They are used to control light transmission or impact resistance. Also available are films to control ultraviolet light transmission to areas having light-sensitive contents.

Rainwater disposal

Predictions of water load are made according to BS EN 12056-3[14]. However, the removal of water from the eaves or ring beam of any dome demands a special design of gutters to follow the circular-on-plan shape which rather limits the choice of materials to form the gutter lining.

Figure 5.44: Belfast trusses at the Royal Air Force Museum, Hendon

Figure 5.46: ETFE geodesic biomes at the Eden Project, St Austell

Figure 5.47: A framed glazed dome illustrating a good understanding of the principles of natural lighting and the control of glare

The drainage of large barrel vaults is normally straightforward.

MAIN PERFORMANCE REQUIREMENTS AND DEFECTS

Strength and stability

The shell configuration is an efficient means of using materials over large uninterrupted areas. Certain points about design can be made that may prove of some assistance to the surveyor faced with producing a preliminary report on the condition of a shell roof, and having to decide

whether or not to refer matters to engineering consultants.

The first point to note is that reinforced concrete shells tend to have a large reserve of strength so that if one of the load-carrying mechanisms fails, another may be brought into play.

With very large spans, the bulk of steel reinforcement has been reduced by the use of high-tensile steel. This is usually post-tensioned. Post-tensioning also may be widely found in stiffening beams and trusses associated with shell roofs. Cracking in post-tensioned structures should always be thoroughly investigated.

Snow loads for curved roofs are difficult to estimate. It is assumed that these roofs are in the 'no access' category. The curved roof should be divided into at least five separate segments and the minimum loads appropriate to the mean slope of each segment determined.

Strength and stability considerations apply particularly to roof lights that may be subject to burglary, and any opening parts of the lights should be lockable.

Dimensional stability

Shrinkage of the concrete in shells is normally restrained by the reinforcement so cracks resulting from this phenomenon should be small and regular. However, where reinforcement is concentrated (eg in deep edge beams), the existence of larger cracks should be given particular attention.

Movement joints should be present in medium spans of sizes typically found in commercial and public buildings. These joints are necessary between shell and structural support as well as between shell and any large roof lights, and should be of at least 5 mm for each 10 m of span. They can occasionally be found within a single shell but more normally between individual shells at edge supports.

Exclusion and disposal of rain and snow

A number of different outer coverings may be used with a shell configuration although some form of continuous sheeting is favoured provided it can be restrained to the surface. The traditional solution for covering masonry domes has been sheet copper or lead.

Where expense was the first consideration, the concrete shell would probably have been covered with a three-layer built-up covering. It is now more practical and economical to cover with one of the single-layer coverings that have become available. In the past, even slates have been used although this demands a high degree of skill on the part of the craftsmen carrying out the work.

Canopies covering walkways are often glazed. They may take many different forms including barrels, dual-pitched, and even down to nearly flat (Figure 5.48). Wherever glazing is near the horizontal plane, there is always a premium on workmanship involved in placing gaskets or mastics if weathertightness is to be assured.

Figure 5.48: A canopy over a pedestrian walkway with nearly horizontal glazing. BRE experience suggests that shallow slopes cannot be expected to be self-cleaning

Roof lights are particularly vulnerable to penetration by fine, wind-blown snow, and the efficient closing of any louvres or other opening areas should be examined with care.

Where frameless horizontal roof glazing directly abuts vertical glazing in a clipped eaves detail, disfiguring staining caused by preferential run-off paths will invariably be encountered in time.

Energy conservation and ventilation

Some structures (eg public transport stations) may be open on one or more sides and this may have considerable effect on wind uplift acting on the roof as well as the more obvious effects on ventilation. There is no point in thermally insulating such structures.

Where reinforced concrete shells needed to be insulated thermally, sprayed coatings have been used to some extent. Surveyors should be aware of the possibility of the presence of asbestos. Other solutions used have included a thin coating of lightweight (vermiculite) screed over the concrete.

Fully glazed vaults are not always insulated with multiple glazing. Condensation, if it forms, will run down the sloping surfaces and may form pools at the foot of walls if not collected into channels.

Roof lights in general may be available in single, double or even triple skins with the latter category giving an acceptable thermal performance. Where smoke pressurisation systems are

in use, attention needs to be paid particularly to the airtightness characteristics of the glazing.

The sections on thermal insulation, condensation and ventilation in Chapters 5.1 and 5.2 also apply to this section.

Control of solar heat

Glazed domes and barrel vaults normally will require some form of sun-shading device to prevent heat build-up in summer conditions. Motorised systems are available though it would be a wise precaution to seek reassurance on the long-term performance of electromechanical devices.

Unheated conservatory areas can be used as pre-heat zones in buildings where consideration has been given to energy conservation measures. In a number of cases, though, economy in the use of energy has been negated where non-solar heating has been installed.

Fire precautions

The design of auditoria to comply with fire legislation is an issue not dealt with in this book. However, surveyors will need to consider such matters as means of escape, particularly of crowds escaping over roofs and the attendant requirements for smoke control (Figure 5.49).

In large and complex buildings the total fire safety engineering approach referred to in the Building Regulations 2000 (England & Wales) Approved Document B2[17] may be appropriate. Larger shopping centres were dealt with in BS 5588-10[49].

Sound insulation

The sound insulation of shell concrete roofs is similar to that of flat concrete roofs except that shells are usually thinner. Sound insulation in the range 40–45 dB can be expected from these roofs, depending on the amount of roof lighting.

If sound insulation is important in fully glazed vaulted roofs, the provision of laylights at the foot of the vault may help though there are maintenance implications. The most that can be expected, even with sealed outer skins, is insulation of around 40 dB.

Durability and ease of maintenance

Large glazed roof structures, especially those dating from Victorian times, will require very close inspection if potentially catastrophic decay and corrosion is to be identified and eliminated (Figure 5.50).

The opaque coverings of roofs within this category will suffer from the same kind of problems as domestic roofs already dealt with in earlier chapters (Figures 5.51 and 5.52).

Of the various kinds of vaulted roof structures, some of the more unfamiliar problems occur with concrete shell roofs. Because of the thinness of reinforced concrete shell roofs there is not much room for error in placing reinforcement, and very little concrete cover to hold back carbonation before the reinforcement becomes susceptible to corrosion. Particular attention needs to be given to checking the carbonation problem.

Roof lighting

There have been problems with maintaining the appearance and transparency of plastics domes and vaults of various configurations. The worst problems seem to occur with double-skin construction where, under exposure to the sun, the outer skin can get much hotter than the inner; this can lead to breakdown of the material, particularly where the glazing is tinted. The European Union of Agrément has issued guidelines for assessing new designs of roof lights[50]. The main points to

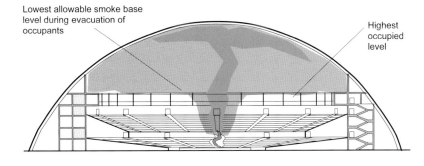

Figure 5.49: In buildings with large open areas (such as auditoria), the base level for smoke must never fall below head height at the highest occupied part of the building

Lowest allowable smoke base level during evacuation of occupants

Highest occupied level

Figure 5.53: A single-skin PVC roof light (seen from below) deteriorating from carbon formation caused by dehydrochlorination over a period of approximately 20 years. The black tape masks splits

Figure 5.50: Major restoration on the Palm House at the Royal Botanic Gardens, Kew. Parts of the structure, as well as the glazing, needed replacing

Figure 5.51: A copper-covered domed roof where the full patina has yet to form. The exact shade of the patina depends on the local pollutants

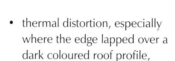

Figure 5.52: A copper-covered barrel vault approximately 30 years old. Although crinkling it appears still to be sound

note about the various materials are as follows.

PVC

PVC roof lights have been found to give a number of problems; those that have been affected were suspected of non-compliance with the relevant standard, BS 4203[51]. The problems have included:

- thermal distortion, especially where the edge lapped over a dark coloured roof profile,
- embrittlement and cracking at fixing points (embrittlement has occurred on prolonged exposure to weathering and may have been associated with loss of light transmission),
- loss of light transmission.

PVC roof lights weathered at low temperature (ie with free ventilation to both sides) tend to whiten on the surface. Sheets so affected may still remain essentially serviceable. At higher temperatures (eg where the structure traps hot air beneath the light), another mechanism

called dehydrochlorination can take over. This can be progressive, leading to the formation of a brown, ultimately black, carbonised surface layer (Figure 5.53). The darker the discoloration, the higher the temperatures reached in sunlight so that substantial heat distortion as well as loss of light transmission can result. Structures tending to trap hot air are most at risk. It is important to ensure that barrel vaults and pyramidal structures are adequately ventilated. The use of tinted sheets also tends to raise operating temperatures in sunlight. Double-skin roof lights have been known to turn completely black within seven years. Their durability hinges on the quality of the stabilisers used in manufacture (the best stabilisers being the most expensive) so the specifier has to decide whether to buy the less expensive short-life product (say of five years) or more expensive longer life one (say of 30 years). Surface-coated grades of PVC are also becoming available that offer substantially improved durability to sunlight.

Distortion of double-glazed roof lights made of PVC, often leading to rain penetration, has also occurred on an unacceptable scale. The problem has occurred mainly with low melting point PVC serving to provide low-cost venting in the case of fire: the PVC melts and falls away, and the fire is vented. However, it could lead to a health and safety issue. Summer temperatures are enough to distort the sheets.

Polycarbonate

Polycarbonate is not prone to discoloration so much as surface erosion due to weathering. This tends to lead to loss of surface gloss and reduction of optical clarity. Polycarbonate has relatively good retention of impact resistance and dimensional stability on weathering. Surface-coated varieties are available with significantly enhanced weatherability.

Acrylic

Acrylic sheet (polymethyl methacrylate) has very good retention of optical clarity on exposure to weathering but can be prone to crazing and embrittlement in the long term, especially when subjected to undue stress.

GRP

Glass-fibre-reinforced translucent sheets are dealt with in Chapter 6.4.

Glass

Many types of glass are available that are suitable for use in overhead glazing. However, a number of instances of breakages of glass used over shopping malls have been brought to the attention of BRE. Injuries to shoppers are not unknown. Unwired annealed and heat-strengthened glass should not, of course, be specified in single glazing or in either leaf of double glazing where there is a risk that breakage of the lower layer will not hold any dangerous shards. Ordinary laminated glass is, thickness for thickness, no stronger than annealed glass; its virtue is that broken shards tend to be held by the interlayer and, if recommended glazing techniques have been followed, the glass should be retained until safe removal is possible. Wired glass is weaker than unwired annealed glass of corresponding thickness but, as with laminated, the shards tend to

be held. Thermally toughened glass is stronger than annealed, both in terms of impact resistance and of thermal stress. When broken, the glass will not hold together though the pieces are relatively small. The best of both worlds can be obtained with sheets of laminated toughened glass.

In the case of roof lights glazed with thermally toughened glass, there have been some instances of spontaneous breakage which may constitute a safety risk in roof glazing, although these have been comparatively rare (Glass and Glazing Federation Data Sheet No 7.1[52]). The problem occurs because of contamination of the glass by nickel sulfide inclusions. There is nothing that the surveyor can do to guard against the risk in existing construction since it is one that occurs in manufacture. When specifying replacements, however, only glass should be selected that has been heat-soaked subsequent to toughening to allow the phase change which causes fracture to take place before installation.

Other causes of glass breakage include:
* inadequate attention to clearances on installation (at least 3 mm for single glazing, rising with sheet size, and more for double glazing),
* difficulties of access for routine maintenance and cleaning, and
* hail (Chapter 2.3).

There should be no direct contact between the glass and any rebated supporting sections which are made of hard tempered metal unless the supports are sprung.

Overhead glazing (usually thermally toughened) is sometimes fixed with clips instead of rebates and it is very important that any replacement is carried out competently.

WORK ON SITE

Access, safety, etc.

Readers should refer to this section in Chapter 3.1.

Box 5.4: Inspection of vaults and other special shaped roofs

Since concrete shells will undergo considerable movements over the relatively large spans for which they have been used, movement joints will have been necessary to accommodate changes in size. These movement joints will have usually been provided, of necessity, at the edges of the shell and therefore in the most vulnerable position, at the foot of the slope, for collecting grit and detritus off the roof. Surveyors should check that such joints are not bridged with debris or stone from the solar covering and are still capable of carrying out their original function.

The very largest spans may well have roller bearings similar to those used in bridge construction.

When carrying out an inspection of a large shell, there may be the appearance of sag in edge beams which is an optical illusion caused by their proximity to the arch of the shell. Taking measurements, rather than visual assessments, is therefore advised to confirm whether deflection has indeed taken place. Differential settlement of foundations of supporting structures is particularly of concern in the case of shell roofs.

The buckling of very large spans may occur also and may take the form of wave-like corrugations of fairly large dimensions. Buckling, though, is only likely to be a problem where the shells are very large or very thin.

Cover to reinforcement in concrete shells will be found to be as little as 10–12 mm so that any defect in the waterproof covering will have serious repercussions for the durability of the reinforcement.

The particular problems to look for are listed below.

Concrete shells or vaults

- Cracking of concrete
- Spalling due to carbonation of the concrete
- Creep, giving rise to excessive deflection
- Differential settlement
- Movement joints filled with detritus
- Presence of sprayed asbestos on soffits

Glazed shells or vaults

- Movement joints filled with detritus
- Lack of suitability and poor condition of apparatus providing access for cleaning
- Poor condition of glazing compounds or gaskets
- Corrosion in framing and clips
- Edge clearances to glazing material are inadequate
- Closing mechanisms of louvred openings in roof lights are inadequate

5.4 ROOF GARDENS (INTENSIVE GREEN ROOFS)

The design of earth-covered roofs is best based on the 'warm roof' principle, though it is probably misleading to speak of warm or cold roofs in this context where the temperature of the deck remains fairly constant whatever the provision of additional insulation.

Waterproof membranes in traditional roof gardens were often covered with a concrete screed to provide protection from mechanical damage or penetration by roots (Figure 5.54). Some plants have roots that can penetrate virtually any waterproof material, especially the softer organic materials like bitumen. A barrier membrane coated with copper or other biocide can protect the waterproof membrane from roots.

Robust drainage boards are available for intensive green roofs, able to withstand the weight of large quantities of soil and large vegetation without deformation.

A layer of insulating material such as glass-fibre or urethane foam is sometimes placed above the waterproof membrane, thereby creating an inverted roof. When rehabilitating such roofs, it may be possible to remove the paving slabs or other material used for weighing down the insulation to prevent wind erosion, and substitute a growing medium and vegetation in its place: a much more eco-friendly solution.

Roof gardens are sometimes proposed as a partial substitute for the original habitat lost in the re-development. In such a case, provided it has not been intermixed with sub-soil, there may be an argument for using the stripped topsoil from the actual excavation as a growing medium rather than a substitute artificial medium.

CHARACTERISTIC DETAILS

Basic structure
Figure 5.55 shows the section through the horizontal plane for the design and construction of roof gardens that have proved successful.

In practice, the only basic form of construction that has the necessary characteristics to take the large loads imposed by the growing medium is one made of reinforced concrete.

For roof gardens that are so large as to involve the use of garden machinery for maintenance, adequate access is considered at an early stage in design. If small trees are to be grown, then the drainage layers, as well as the soil depth, will be important (Figure 5.56).

Parapets
Parapets or changes of level will require the same standard of waterproofing as the basic structure, and consideration should be given to using a similar quality of construction to that used in tanking basements. Although the earth, once in place, will support the tanking, it is nevertheless necessary to make it self-supporting during installation. Upstand height will usually be beyond the self-supporting capability of the asphalt and will need to be

Figure 5.54: A roof garden on two levels on an office building (Photograph by permission of J Thomson)

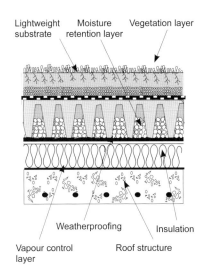

Figure 5.55: The essential components of a warm deck roof garden

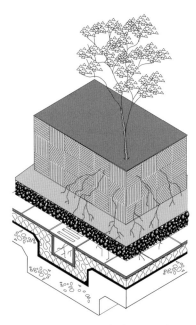

Figure 5.56: A warm deck inverted design should be used only where there is confidence that root systems will not penetrate the insulation layer

reinforced (eg with expanded metal lath).

MAIN PERFORMANCE REQUIREMENTS AND DEFECTS

Strength and stability

Deflections caused by the very large mass of soil over a roof should have been taken care of in the original design, but the owners or their representatives should be warned about introducing material and loads not foreseen by the designer.

Dimensional stability

The concrete deck should have, of course, a low drying shrinkage rate to avoid cracking. Once dry, the slab will be protected from extremes of temperature.

Exclusion and disposal of rain and snow, and artificial watering

Moisture retention is a desirable characteristic of a roof garden. In fact, earth covering the structure will range from being waterlogged to being slightly damp for long periods at a time; the water can be introduced by natural means (eg rain or snow) or artificial (eg sprinkler). The waterproof layer normally

should consist of asphalt laid in three coats to total thicknesses of 30 mm on horizontal surfaces and 20 mm on vertical. The relevant British Standard is BS 8218[53]. It may be possible in some cases to substitute an alternative, such as a polymer-modified felt, provided that mechanical damage during the installation of the roof will be avoided. Above the waterproof layer is a drainage zone that should consist of a no-fines material, for example of fired clay or sintered pulverised fuel ash (PFA).

Above the drainage zone, a geotextile sheet of non-woven plastics fleece material that allows water to percolate but which retains the fines from the earth, will be needed. Readers should also see Chapter 4.5.

Drainage

The design of the rainwater disposal system should take into account the potential for surcharging; while surcharging can occasionally be tolerated in ground-level disposal systems, it cannot be tolerated at roof level for obvious reasons. Additional allowances for periods of very heavy rainfall therefore may be needed. Various proprietary products of varying characteristics and thicknesses are available with which to form drainage layers, some combining compartmented cavities filled with drainage media of various kinds.

Energy conservation and ventilation

Thermal insulation used in the roof should not be affected by water, or take up water, to any significant degree. Extruded polystyrene has been successfully used provided it is of sufficient compressive strength not to deform under the soil loads.

Control of solar heat gain

Earth-covered roofs do not contribute to solar overheating and can help to reduce solar gain.

Sound insulation

Because of the very large mass of the roof, insulation against airborne sound (eg from aircraft) will be

excellent. Overall sound resistance depends on the performance of the weakest link in the chain and roof lights will reduce the sound-insulating qualities of earth-covered roofs considerably.

Durability and ease of maintenance

The majority of roof gardens have enjoyed a good record for watertightness but there have been some failures due to accidental perforation of the waterproof layer. Maintenance of a roof garden as a whole is a continuing commitment depending on the planting schemes adopted. In particular, tap rooting varieties of plants will need to be avoided. If the roof has thermal insulation above the waterproof membrane, there is a possible risk that root systems will travel underneath the insulation.

So far as the structure is concerned, the main point to be aware of is that the aggregate used in the construction of the reinforced concrete deck should not cause alkali silica reaction.

WORK ON SITE

Access, safety, etc.

Readers should see this section in Chapter 3.1.

Supervision of critical features

Since the waterproof layer, irrigation and drainage will be buried, it will be necessary to give close supervision to the installation of these components and systems, (especially where single-layer systems are vulnerable to damage during subsequent building operations)

Box 5.5: Inspection of roof gardens

The particular problems to look for are listed below.

- Water penetration
- Soil heaped over upstands and flashings
- Deterioration of the underside of the concrete slab

5.5 SWIMMING POOL ROOFS

During the late 1980s, deterioration of austenitic stainless steels was found in a small number of the 1400 or so public swimming pool and leisure pool buildings known to exist in the UK (*Corrosion of metals in swimming pool buildings*[54]). This fact, together with the higher temperatures and more aggressive chemicals such as chlorides and chloramines, and the increasing number of site investigations carried out by the BRE Advisory Service, merits special consideration of the roofs of these buildings (Figure 5.57).

MAIN PERFORMANCE REQUIREMENTS AND DEFECTS

Strength and stability
Since the late 1980s BRE has been advised of further instances of the failure of stainless steel suspension wires, and more bulging and collapsing of ceilings. The causes are examined in the section of this chapter on durability.

Energy conservation and ventilation
The internal atmosphere in a heated indoor swimming pool can be extremely humid. BRE has encountered many cases where condensation has occurred within the roof. Condensation, and its damaging consequences, are often hidden until well advanced, and can be particularly difficult and costly to rectify (Case study 13). Successful design of roofs for swimming pools demands a much more rigorous consideration of condensation risks than is needed for roofs of other buildings. A warm deck design is the preferred option for swimming pool roofs.

Figure 5.57: The potentially aggressive internal atmosphere in buildings housing swimming pools can cause corrosion of structural members and deterioration in other parts of building fabric

A possible alternative is to pressurise the roof void so that air moves only downward through gaps in the vapour control layer at ceiling level on a scale sufficient to prevent air carrying water vapour upwards. The pressurised air can be heated to prevent down-draughts. Condensation prevention depends on the fans running continuously and it is for this reason that small pools are often built with warm deck roofs instead.

BRE Digest 336[55] gives the basis of a satisfactory warm deck design.

Durability and ease of maintenance
Particular care must be taken about separating the metals used in the roofs of swimming pools since the corrosive atmosphere will accelerate corrosion of the less noble metals in contact with other metals.

BRE investigations have, however, revealed comparatively little corrosion of the metal indoor roof lining in swimming pools. The main risk is where profiled sheeting is manufactured with uncoated cut edges; in humid conditions these edges will corrode leading, initially, to staining of the sheet and, if neglected, underfilm corrosion causing blistering and detachment of the protective organic coating.

Under normal conditions in a pool environment, ferrous metals with an organic coating (paint or bonded plastics) should perform satisfactorily but bare metal must not be exposed. Under more severe conditions, as, for example, over flume pools, organic coatings alone are not good enough and the continually humid atmosphere prevents repainting from achieving a lasting solution.

Stress corrosion cracking of certain austenitic stainless steel suspension wires and other components (including brackets, bolts and fasteners, and wire ropes) has occurred in the aggressive contaminant-laden environment in swimming pool buildings (Figure 5.58). In laboratory tests, types 304 and 316 were found to fail in highly aggressive chloride-laden environments; these grades can be affected also by stress corrosion cracking. Stainless steels that have an improved resistance over standard austenitic grades should be used. Types 904L, 6% Mo, and 317LMN should be specified to replace types 201, 304, 316 and 321 in components that are not, or cannot, regularly be cleaned or washed. Plastics straps (eg nylon) are also known to fail; painted

Case study 13: Swimming pool slate roof

An outdoor swimming pool had been enclosed within a conventional slated cold roof and a suspended ceiling. During the first year the ceiling tiles became stained. They were painted, but the staining returned.

BRE was called in to advise the owner on the cause of the staining and possible remedial measures. From the site inspection it was discovered that the mineral fibre thermal insulation, which followed the profile of the roof, was damp. A ventilator grille was positioned along the entire length of one of the eaves but little or no ventilation provision was made at the other eaves.

There was no doubt that the cause of the staining was condensation forming on the sarking felt below the slated outer covering of the roof, running down the underside of the slope of the roof and dripping onto the insulation which, in turn, allowed water to drip through onto the ceiling tiles.

The roof had been designed on the cold deck principle with any condensation forming within the roof void being removed by ventilation. But the cross-ventilation actually achieved was totally inadequate to remove the sometimes large amounts of moisture formed above the pool.

The situation was almost certainly being made worse on cold nights by radiation where external temperatures fell below zero and clear night skies led to the temperature of the roof coverings falling below the ambient air temperature. The atmospheric moisture condensing on the underside of the outer surface of the roof then froze into a rime which, upon thawing, released considerable quantities of water onto the ceiling below.

BRE Digest 336[55] recommends converting a roof with this problem to a roof of warm deck design. This entails stripping the outer covering, laying a new deck, bonding to it a high-performance felt membrane to act as a vapour control layer, and overlaying with a thermal insulant unaffected by moisture (eg sheet extruded polystyrene) and a breather membrane before replacing the covering.

A possible measure that might have alleviated the problem, though not cured it, would have been to lay a breather membrane to act as a drainage layer, terminating at the eaves and so channelling the condensate into the gutter. Because of potential problems with fixing such a sheet, however, the chances of achieving a watertight membrane were small.

The principal aim of the advice given to the client was that the solution should create a true warm deck roof in accordance with BRE Digest 336[55]. The existing slates and battens needed to be removed and a deck, say of 19 mm ply with a water and boil proof (WBP) bond, fastened over the rafters. Then a high-performance felt should be bonded to the deck with hot bitumen. On top of this a 90 mm thick sheet of extruded polystyrene or similar insulant should be fixed, covered with a breather membrane, counter-battens, battens and, finally, the replacement of the slates.

---

aluminium and galvanised steel are alternatives, *but they must be regularly monitored.*

Further information on the durability of metals in swimming pool roofs is given in *Corrosion of metals in swimming pool buildings*[54] and, in the case of stainless steels, in *Stainless steel in swimming pool buildings*[56].

Figure 5.58: A suspended ceiling (from above) under the roof of a swimming pool. A suspension wire which has failed by stress corrosion cracking lies across the tee sections in the foreground. Collapse of the ceiling would follow if no corrective action was taken

WORK ON SITE

Box 5.6: Inspection of swimming pool roofs

In regular six-monthly inspections of the roof voids of swimming pool buildings, the particular problems to look for are listed below.

- Disturbance or discoloration of ceilings
- Bowing of supporting framework
- Condensation
- Corrosion of the underside of metal decking
- Pitting corrosion of structural steel
- Crevice corrosion of zinc or cadmium-plated ferrous fixings
- Brown stains, especially on the undersides of stainless steel components or in the interstices of wire ropes
- Stress corrosion cracks in certain austenitic stainless steel

components. Inspection should be made using a hand lens of at least ×10 magnification and with penetrating dye where necessary. This examination may not be practical on all components (eg thin wires) so any failure in a single wire should lead to replacement of all wires

- Cracking. If cracking is suspected, samples should be removed for laboratory test; also a structural engineer or BRE should be called in
- Conditions in the roof void exceeding 29 °C air temperature and 70% relative humidity. Pool management should be advised if these conditions exist

5.6 REFERENCES

[1] **BRE.** Stainless steel as a building material. Digest 349. Bracknell, IHS BRE Press, 1990

[2] **British Standards Institution.** BS EN ISO 9445: 2006 Continuously cold-rolled stainless steel narrow strip, wide strip, plate/sheet and cut lengths. Tolerances on dimensions and form

[3] **British Standards Institution.** BS 4868: 1972 Specification for profiled aluminium sheet for building. London, BSI, 1972

[4] **British Standards Institution.** BS 3083: 1988 Specification for hot-dip zinc coated and hot-dip aluminium/zinc coated corrugated steel sheets for general purposes. London, BSI, 1988

[5] **British Standards Institution.** BS 4904: 1978 Specification for external cladding colours for building purposes. London, BSI, 1978

[6] **British Standards Institution.** BS 5252: 1976 Framework for colour co-ordination for building purposes. . London, BSI, 1976

[7] **British Standards Institution.** BS 4203-1: 1980 Extruded rigid PVC corrugated sheeting. Specification for performance requirements. London, BSI, 1980

[8] **British Standards Institution.** BS 4154-1: 1985 Corrugated plastics translucent sheets made from thermo-setting polyester resin (glass fibre reinforced). Specification for material and performance requirements. London, BSI, 1985

[9] **Berry RW.** Remedial treatment of wood rot and insect attack in buildings. BR 256. Bracknell, IHS BRE Press, 1994

[10] **British Standards Institution.** BS 8220: Guide for security of buildings against crime. London, BSI
Part 2: 1995 Offices and shops
Part 3: 2004 Storage, industrial and distribution premises

[11] **British Standards Institution.** BS 5544: 1978 Specification for anti-bandit glazing (glazing resistant to manual attack). London, BSI, 1978

[12] **British Standards Institution.** BS 6206: 1981 Specification for impact performance requirements for flat safety glass and safety plastics for use in buildings. London, BSI, 1981

[13] **British Standards Institution.** BS 5427-1: 1996 Code of practice for the use of profiled sheet for roof and wall cladding on buildings. Design. London, BSI, 1996

[14] **British Standards Institution.** BS EN 12056-3: 2000 Gravity drainage systems inside buildings. Part 3: Roof drainage, layout and calculation. London, BSI, 2000

[15] **White RB.** Qualitative studies of buildings. National Building Studies Special Report 39. London, HMSO, 1966

[16] **British Standards Institution.** BS 747: 2000 Reinforced bitumen sheets for roofing. Specification. London, BSI, 2000. *NB: This Standard has been withdrawn and replaced by BS EN 13707: 2004*

[17] **Communities and Local Government (CLG).** The Building Regulations 2000.
Approved Document:
  B: Fire safety, 2006
  Volume 1: Dwellinghouses
  Volume 2: Buildings other than dwellinghouses
  L: Conservation of fuel and power, 2006
Available from www.planningportal.gov.uk and www.thenbs.com/buildingregs

[18] **British Standards Institution.** BS EN 62305-2: 2006 Protection against lightning. Risk management

[19] **BRE.** Estimating daylight in buildings. Part 1: An aid to energy efficiency. Digest 309. Bracknell, IHS BRE Press, 1986

[20] **BRE.** Lighting controls and daylight use. Digest 272. Bracknell, IHS BRE Press, 1983

[21] **CIBSE.** Environmental design: CIBSE Guide A. 7th edition. London, CIBSE Publications, 2006

[22] **Petherbridge P, Milbank NO & Harrington-Lynn J.** Environmental design manual: summer conditions in naturally-ventilated offices. BR 86. Bracknell, IHS BRE Press, 1988

[23] **Health and Safety Executive (HSE).** Health and safety in roof work. 3rd edition. HSG 33. London, HSE, 2008

[24] **British Standards Institution.** BS 6262-7: 2005 Glazing for buildings. Code of practice for the provision of information. London, BSI, 2005

[25] **British Standards Institution.** BS 8000: Workmanship on building sites. London, BSI
  Part 5: 1990 Code of practice for carpentry, joinery and general fixings
  Part 6: 1990 Code of practice for slating and tiling of roofs and claddings

[26] **British Standards Institution.** BS EN 13956: 2005 Flexible sheet for waterproofing. Plastic and rubber sheets for roof waterproofing. Definitions and characteristics. London, BSI, 2005

[27] **Department of Health.** Guidance to support functional provisions in healthcare premises. Health Technical Memorandum 05-02. London, The Stationery Office, 2007

[28] **European Union of Agrément (UEAtc).** UEAtc special technical guide for the assessment of single layer roof waterproofing. Paris, UEAtc, 1991

[29] **European Union of Agrément (UEAtc).** Directives for the assessment of roofing systems using PVC sheets without reinforcement, loose laid under heavy protection and not compatible with bitumen. Method of Assessment and Test (MOAT) No 29. Paris, UEAtc, 1984

[30] **European Union of Agrément (UEAtc).** Special directives for the assessment of reinforced homogeneous waterproof coverings of styrene-butadiene-styrene (SBS) elastomer bitumen. Method of Assessment and Test (MOAT) No 31. Paris, UEAtc, 1984

[31] **European Union of Agrément (UEAtc).** Special directives for the assessment of roof waterproofing systems with non-reinforced vulcanized EPDM. Method of Assessment and Test (MOAT) No 46. Paris, UEAtc, 1988

[32] **British Standards Institution.** BS 5950: Structural use of steelwork in building (various Parts are appropriate to roofs). London, BSI

[33] **British Standards Institution.** BS 8110-1: 1997 Structural use of concrete. Code of practice for design and construction. London, BSI, 1997

[34] **British Standards Institution.** BS 5268: Structural use of timber. London, BSI (various Parts are appropriate to roofs, inc. Part 4: Fire resistance of timber structures)

[35] **BRE.** Wind scour of gravel ballast on roofs. Digest 311. Bracknell, IHS BRE Press, 1986

[36] **British Standards Institution.** BS 6399-1: 1996 Loading for buildings. Code of practice for dead and imposed loads. London, BSI, 1996

[37] **Bonshor RB & Bonshor LL.** Cracking in buildings. BR 292. Bracknell, IHS BRE Press, 1995

[38] **BRE.** Flat roof design: thermal insulation. Digest 324. Bracknell, IHS BRE Press, 1987

[39] **British Standards Institution.** BS 476-7: 1997 Fire tests on building materials and structures. Method of test to determine the classification of the surface spread of flame of products. London, BSI, 1997

[40] **Currie RJ & Crammond NJ.** Assessment of existing high alumina cement concrete construction in the UK. Proceedings of the Institution of Civil Engineers: Structures and buildings 1994: 104 (February): 83–92

[41] **BRE.** Alkali–silica reaction in concrete. Part 1: Background to the guidance notes. Part 2: Detailed guidance for new construction. Part 3: Worked examples. Part 4: Simplified guidance for new construction using normal reacitivity aggregates. Digest 330. Bracknell, IHS BRE Press, 2004

[42] **British Standards Institution.** BS EN 1992-1-1: 2004 Eurocode 2: Design of concrete structures. General rules and rules for buildings. London, BSI, 2004

[43] **British Standards Institution.** BS 1881-204: 1988 Testing concrete. Recommendations on the use of electromagnetic covermeters. London, BSI, 1988

[44] **Dunster A.** HAC concrete in the UK: assessment, durability management, maintenance and refurbishment. Special Digest 3. Bracknell, IHS BRE Press, 2002

[45] **Building Regulations Advisory Committee.** Report by sub-committee P (high alumina cement concrete) BRAC (75) p 40. London, The Stationery Office, 1975. Addendum No 1; BRAC (75) p 59, 1975. Addendum No 2; BRAC (76) p 3. London, Department of the Environment (DOE), 1976

[46] **Building Research Station.** Principles of modern building, Volume 2: Floors and roofs. London, The Stationery Office, 1961

[47] **Schaffer RJ.** The weathering of natural building stones. Shaftesbury, Dorset, Donhead Publishing, 2004 edition

[48] **Baker D.** Potworks. The industrial architecture of the Staffordshire Potteries. London, Royal Commission on the Historic Monuments of England, 1991

[49] **British Standards Institution.** BS 5588-10: 1991 Fire precautions in the design, construction and use of buildings. Code of practice for shopping complexes. London, BSI, 1991

[50] **European Union of Agrément (UEAtc)** UEAtc guide for the agrément of individual rooflights UEAtc guide for the agrément of continuous strip rooflights Paris, UEAtc, 1989

[51] **British Standards Institution.** BS 4203: 1980 Extruded rigid PVC corrugated sheeting. London, BSI, 1980

[52] **Glass and Glazing Federation.** Non-vertical overhead glazing: guide to the selection of glass from the point of view of safety. Glazing Manual Data Sheet No 7.1. London, Glass and Glazing Federation, 1994

[53] **British Standards Institution.** BS 8218: 1998 Code of practice for mastic asphalt roofing. London, BSI, 1998

[54] **Moore JFA & Cox RH.** Corrosion of metals in swimming pool buildings. BR 165. Bracknell, IHS BRE Press, 1989

[55] **BRE.** Swimming pool roofs: minimising the risk of condensation using warm-deck roofing. Digest 336. Bracknell, IHS BRE Press, 1988

[56] **Nickel Development Institute et al.** Stainless steel in swimming pool buildings. A guide to selection and use. Birmingham, Nickel Development Institute, 1995

6   MEDIUM-SPAN INDUSTRIAL ROOFS

This category includes the many types of medium-span roofs over industrial buildings. Structural forms vary considerably, including single-storey factory buildings, with a wide variety of trusses, north lights and monitors. Glazing includes both part and full, often in patent glazing which is dealt with separately in Chapter 6.2. Older roofs in the category are covered in slate, asbestos cement or steel sheeting.

Newer roofs are commonly in troughed steel sheeting with thermal insulation, either installed separately or incorporated in the sheeting. However, many factory roofs built before the 1960s are unlikely to be thermally insulated.

Although described here mainly in the context of industrial buildings, roofs of similar construction will be found in other buildings of a commercial nature (Figure 6.1).

Figure 6.1: Metal sheeted roofs at an Inverness-shire ski resort. The frequency of fixings on the slopes reflects the degree of protection required for buildings in an area where weather conditions can be extreme. Advice on weather conditions that may be encountered in any part of the UK can be obtained from the Meteorological Office

6.1 SHEETED PORTALS, NORTH LIGHTS, MONITORS AND SAW-TOOTHS

These types of roof are normally covered by sheets of:
- profiled coated steel,
- profiled coated or non-coated aluminium, or
- coated fibre-reinforced polymers.

The characteristics of the material for aluminium sheet have been dealt with in Chapter 3.5; galvanised steel without further coating is dealt with in Chapter 6.3.

Profiled sheeted roofs may be either warm or cold deck as with other forms of pitched roofs dealt with in Chapter 3. In the warm deck roof, the thermal insulation is normally sandwiched between, and in full contact with, the two skins of material. The insulation can either be provided in a composite sheet or installed separately. In the latter case, the thermal insulation is pre-formed to fill all the spaces between the sheet profiles or the void between the sheets may be foamed. Provided the sheets are sealed to each other, no separate vapour control layer is required.

In a cold deck roof, the thermal insulation is usually above the purlins with a ventilation space between the outer sheet and the insulation. A separate vapour control layer and breather membrane are required. Most existing examples of roofs in this category are cold roofs. There are substantial condensation risks with roofs of this design and, wherever possible, it is recommended that warm roofs are constructed instead.

There have been considerable advances since the early 1980s in the use of cold-rolled steel structural sections in conjunction with profiled sheeting. With such thin members forming the structure, corrosion will have potentially more damaging effects than on, for example, hot-rolled steelwork simply because a larger proportion of the cross-section area can be affected. There is a further trend for short on-site construction times which indicates an increasing demand for the use of cold-formed steel sections in the future.

CHARACTERISTIC DETAILS

Basic structure

Portal frames consist of interconnected roof and column sections. They are normally fabricated from steel sections, welded to form a continuous section, and can span to over 40 m. The webs of the frames can be either of solid or lattice construction (Figure 6.2), though the latter will be more expensive both to fabricate and to maintain. While it is possible to weld on site, final connections are usually bolted. It will normally be found to be easier to repair or alter welded structures than bolted structures under the cover of the existing roof.

Concrete portals are also used for shorter spans where their dead weight is not so critical.

Timber portals may also be encountered. There is a wide range of sections used depending on spans and heights. Designs include glulam portals as well as those fabricated from stiffened plywood boxes.

So far as non-portal frames are concerned, several different shapes are considered in this chapter: the trusses or girders supporting the roof may be flat or pitched, dual or single.

Since Edwardian times, steel has obviously been the favoured material in industrial buildings

Figure 6.2: A lined steel lattice roof over industrial premises under construction in 1959

fabricated into trussed girders of Warren, Pratt or even Vierendeel form (Figure 6.3). There are still many riveted steel roofs in service, but bolted and welded trusses and girders have largely replaced them, certainly since the 1939–45 war. Ties are often cambered in spans over about 15 m.

Main roof areas

Sheeted coverings may be of various materials, profiles and finishes. Nearly 100 different profiles in coated steel sheet are available in the UK and each is available in a bewildering variety of colours. Where the same material is carried from roof to wall via a rounded, rolled or curved eaves (unless the sheets are interrupted for the insertion of a gutter), the same profile is needed for both surfaces. See also the discussion of sheet materials in Chapter 4.1.

There has been an increasing tendency for metal sheets to run the whole length of the pitch of the roof to obviate the need for head laps. These sheets are, however, all laid to overlap at the side, though

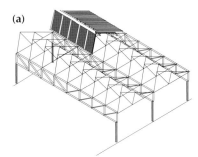

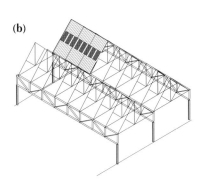

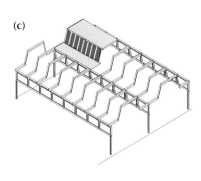

Figure 6.3: (a) Warren girders with north lights, (b) Pratt girders with saw-tooths, (c) Vierendeel girders with monitors

sometimes the lap is formed by a third member such as a cover strip rather than a lap of adjacent sheets. Where head laps are necessary, the lap is normally sealed unless:

- the pitch is greater than 15°, or
- the lap is greater than 200 mm, or
- the exposure is sheltered.

It is important, especially when Zed purlins are made from relatively thin sheet steel, that fixings are installed in the top flanges truly perpendicular to the top surface of the flange (Figure 6.4). Although some of the figures in this chapter show fixings into the ridges of sheets, fixings into the valleys normally give good performance unless those fixings are driven incorrectly. Rivets are

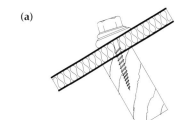

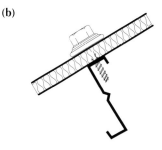

Figure 6.4: Sheets through which fixings have been driven at an angle into a purlin may distort and leak (a). Fixings should always be driven perpendicular to the top flange of the purlin (b)

commonly used to fix trim. Fibre-reinforced sheets are not made in such long lengths as metal sheets and therefore joints in the slope of the roof are more common.

Eaves

Overhanging

A drip created at the end of low-pitch sheeting is recommended to prevent water from being blown back under the eaves. In cold roof designs there should be provision for the condensate forming within the void to drain down to the eaves via the underlining or a suitable membrane (Figure 6.5). It is not possible to achieve this with a rolled eaves.

Clipped

Clipped eaves are normally constructed as a rolled eaves where the sheets are effectively profiled in three dimensions with crimping where necessary (Figure 6.6). Of course the wall and roof sections must be of identical profile. Rainwater will need to run away from the foot of the vertical cladding unless the section is interrupted higher up for the insertion of a gutter (Figure 6.7).

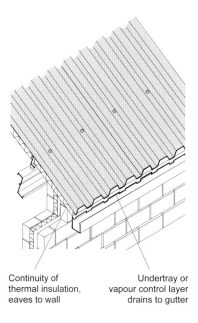

Continuity of thermal insulation, eaves to wall

Undertray or vapour control layer drains to gutter

Figure 6.5: Eaves detail providing drainage paths to the eaves for the removal of condensation forming on the underside of the top sheet

Figure 6.6: Deteriorating profiled fibre-reinforced sheeting being overclad with crimped profiled metal

Roof lighting

Three factors are important in decisions about roof lights.

- Daylighting must be adequate in both quantity and distribution.
- There should be freedom from excessive glare.
- Sunlight penetration of the building will need to be controlled in the roofs of industrial premises.

In practice, the design of roof lights will depend not only on lighting requirements but also on the structural design and the arrangements for overhead services (Figure 6.8). Detailed design of

(a)

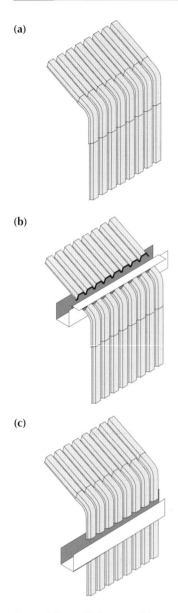

(b)

(c)

Figure 6.7: A rolled eaves without provision of guttering (a) and with two methods of incorporating a gutter (b) and (c). In both (b) and (c), to avoid condensation, it will be necessary to insulate the gutter to the same standard as the remainder of the wall and roof

Figure 6.8: Roof light and vent incorporated into the 30-year-old asbestos cement sheeted roof of an industrial building

(a)

Maximum 11%, minimum 5%

(b)

5% uniformity

(c)

Maximum 9%, minimum 5%

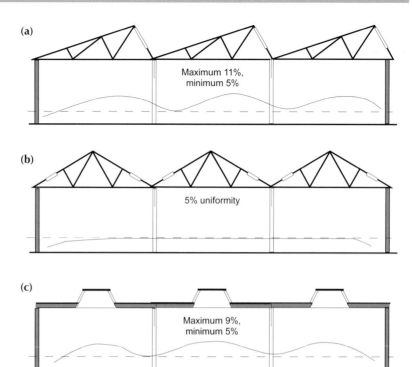

Figure 6.9: Daylight factors achieved with various types of roof and roof light profiles: (a) north lights, glass area 20% of floor area, (b) continuous strips, glass area 10% of floor area, (c) cranked monitors with sloping glazing, glass area 16% of floor area; see also Figure 6.10

lighting includes the points listed below.

• Main lighting from the roof can often be supplemented by reflected light from interior surfaces provided these are in a colour that has a suitably high reflection factor.

• In most working environments, the aim should be to provide as even a distribution of light as possible. Approximate daylight factors given by different roofing configurations are shown in Figure 6.9.

• Glare is usually at its maximum when the outside sky contrasts markedly with dark surfaces adjacent to the roof lights. To avoid glare the view to the outside is screened and the brightness of surrounding surfaces is raised.

• Sunlight penetration may often be sufficiently controlled by glazing that has a diffusing effect.

Figure 6.9 illustrates different examples of daylight factor achieved at the working plane for the various types of roofs shown in Figure 6.3. It is clear that:

Figure 6.10: A cranked monitor roof of similar configuration to that shown in Figure 6.10(c)

• the level of natural illumination depends on the type of glazing adopted as well as on the area of glass (Figure 6.10),

• horizontal or low-pitched glazing will generally be more efficient than vertical or steeply pitched glazing, provided it is kept clean.

Efficiency will also be reduced if there are obstructions, either external (eg adjacent roof surfaces) or internal (eg ventilation trunking), or if the colour of the internal surfaces is dark.

MAIN PERFORMANCE REQUIREMENTS AND DEFECTS

Strength and stability

The spanning capability of some of the newer composite sheets, where the thermal insulation is incorporated between two metal sheets, is greater than that of uninsulated single sheeting, profile for profile, and some economy in purlin spacing may be achieved. Alternatively, some roofs will be found that have sandwich construction with insulation installed separately (Figure 6.11).

In the case of cold-formed sections (eg Zed section purlins) the thin material which gives advantages of lightness and high strength-to-weight ratio has some disadvantages. Cold-formed sections are particularly susceptible to local buckling when subject to compression and local buckling will affect the overall behaviour of the member. This is why edge stiffening is employed on many cold-formed sections. Some roofs may be of stressed skin design, particularly in the form of folded-plate roofs. The use of light-gauge steel in this form of roof results in a very economical structure.

For connections in these thinner materials, resistance spot welding or mechanical fasteners are more suitable than bolting or arc welding.

In some forms of roofing (eg where spans are greater than around 4.5 m), it is essential to fit sag bars to prevent rotation of purlins. Sag bars are fitted differently in roofs of slopes over and under about 10° where direction of rotation can reverse. However, even where fitted, BRE investigators have noted that they do not always succeed in preventing rotation. Stronger purlins may be called for.

Impact damage risk
So far as metal sheeting is concerned, there may be concern over impact resistance both to a hard and to a soft body. Damage could occur during clumsy maintenance.

Tests are usually carried out at:
- 7.5 Nm and 15 Nm for a hard body, and

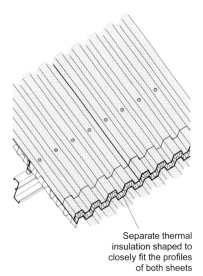

Figure 6.11: Composite sheeting using separate profiled insulation

Separate thermal insulation shaped to closely fit the profiles of both sheets

- 125 Nm and 250 Nm for a soft body.

There should be no indentation visible. These standards are achievable though slight indentation may be acceptable for some products at the higher soft body impact level. Steel sheet of 0.7 mm thickness in Z1 quality is usually specified to avoid the worst damage, but some profiles may be satisfactory in thinner sheets.

As far as fibre-reinforced corrugated and flat sheets are concerned, these are relatively fragile. Some of the properties are described in BRE Information Paper IP 1/91[1]. The characteristics of these sheets are different from the asbestos cement sheets that they are designed to replace. Cracking may occur, even in roofs that are as little as two or three years old, though its cause may be as much due to fixing deficiencies or inappropriate specification as to material deficiencies (Figure 6.12).

Exclusion and disposal of rain and snow

Sheeted roofs operate on the lap principle. The side lap is normally provided for by an overlap of one complete undulation of the sheet but it is wise to make sure that the lap is away from the direction of the prevailing wind. In areas subjected to high driving rain, sealing the lap may be advisable.

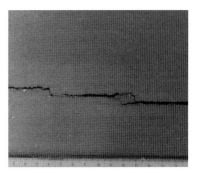

Figure 6.12: Cracking in PVAC (polyvinyl acetate) fibre-reinforced sheeting

There may also be a problem with rain penetration of side as well as head laps and, in some circumstances (eg when the pitch is very shallow, ie less than 10°) it might be better to seal the side laps as well as the head laps. The problem may be exacerbated by the build-up of detritus in the valleys where it can form a dam causing rainwater to overflow both head and side laps (Figure 6.13).

Head laps are more vulnerable to rain penetration and capillary action. On sheltered sites where the head lap is at least 200 mm and the pitch is greater than 15° it may not be necessary to seal the joint. In all other cases, especially on exposed sites, two seals in each lap are recommended: one above and one below the fixing to the purlin (Figure 6.14).

Where curved or rolled eaves are in areas having relatively high pollution levels, care needs to be exercised to avoid run-off from the roof from staining the wall surfaces.

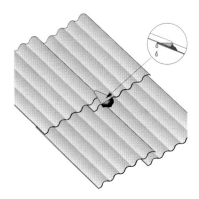

Figure 6.13: Detritus can build up in valleys of sheeting and lead to water penetrating to the interior of the building

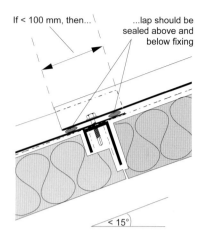

Figure 6.14: Seals above and below the fixing to the purlin to prevent water penetration

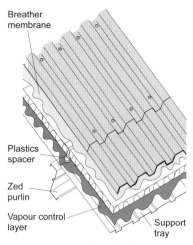

Figure 6.15: A traditional design of a profiled sheeted roof with Zed purlins and plastics spacers

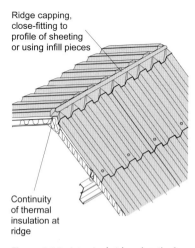

Figure 6.16: A typical ridge detail of a profiled sheeted roof

A gutter above the change of pitch is recommended to reduce this risk (see Figure 6.7b).

Energy conservation and ventilation

Factory-bonded composite panels are frequently used for the roofs of large-span buildings such as factories. The panels comprise a rigid foam insulation sandwiched between two steel or aluminium sheets spanning over pressed metal purlins to create a warm roof. Various profiles and finishes are available. Thermal insulation thicknesses range from 40–100 mm, giving U values in the range 0.4–0.2 W/m² °C.

But cold roof designs are also possible where the insulation is placed above the purlins with a ventilation gap over the insulation and under the outer sheet. A vapour control layer is essential (Figure 6.15). Condensation occurring within the construction is a substantial risk with cold roof designs. As described in earlier chapters, condensate very often forms on the inner face of the outer sheet due to radiation to the cold night sky, and it can freeze there and subsequently thaw. Condensate can also saturate the insulation and corrode fixings and the edges of sheet metal. Condensate will eventually drip from purlins or sheet fixings onto ceilings and into the interior of the building.

Where thermal insulation is applied separately to an in-situ

construction of cold roof design, the use of closed-cell insulants should be considered. Provision for draining condensation will need to be made down the slope of the roof, past purlins and lining tray upstands, and out to gutters at the eaves. Adequate ventilation is also needed above the insulation from eaves to ridge with vents at both ends of the slope; special eaves filler strips may need to be provided to ensure this ventilation and to keep birds out. Proprietary products are available. A vapour control layer of at least 500-gauge polyethylene is needed over the lower lining sheet with joints lapped and sealed with suitable tape.

Thermal bridges, especially at ridges, tend to be an unfortunate feature of these designs where sloping sheets are unlikely to butt closely and where extra insulation may be required under ridge cappings and valley gutters (Figure 6.16).

Adding thermal insulation to existing north lights is not easy. The aim is to achieve as even a coverage as possible, not forgetting the undersides of valley gutters and the tops of rainwater pipes which, if uninsulated, can create a thermal bridge with consequent risk of condensation (Figure 6.17). Internal rainwater pipes provide a source of condensation when moisture inside a building condenses on the surface of pipes cooled by the passage of rainwater. These pipes should be wrapped with a closed-

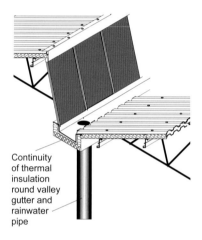

Figure 6.17: Continuity of thermal insulation round valley gutters and rainwater pipes in a north light roof is crucial to avoid condensation

cell insulation material or insulation covered in a vapour control layer. Where an existing roof is to be entirely replaced, then the latest version of AD L1A & L2B of the Building Regulations 2000 (England & Wales)[2] will apply. Similar requirements exist in Scotland[3] and Northern Ireland[4].

Control of solar heat gain

Delamination and undulation in sandwich sheets was thought, at one time, to be caused by solar heating and consequent expansion effects on the outer skin breaking the bond with the thermal insulation layer. However, recent opinion suggests that this risk is not high. A more likely cause of delamination is overtightening of fixings, particularly

Figure 6.18: Misalignment of purlins puts insulated sheets under strain causing delamination which may be made worse by solar heat

where the purlins spanned by the sheets are not quite in line (Figure 6.18). Solar heating may aggravate this problem.

Modifications for climate change

Profiled sheet roofing systems

Profiled sheet roofing systems (profiled fibre-cement, glass-reinforced plastics, polycarbonate, metal, bitumen fibre, tile-effect panels and sandwich panels) are often fixed with primary screws located to a pattern dictated by the profile and by the need to prevent leaks at the end laps. In some cases, the fixing pattern already specified is over-designed as far as anticipated wind loads are concerned so there is sufficient over-design to cope with future climate change. For some profiles, manufacturer's literature will recommend additional primary screws in zones of peak loading.

Secondary fasteners, including stitcher screws, used to fasten one sheet to another are usually only for waterproofing and would not need to be increased in number. Concealed fix and secret fix profiles may not have any space to accommodate additional screws.

How wind load is transmitted to the structural frame supporting the roof is known as the load path. This includes purlins, the inner liner sheet, spacers and the external weathersheet. It is usual for insulation to be placed between the inner liner and the external weather sheet. The spacer system can be a

weak link in some insulated roofs and some roofs have failed because of it.

Some metal roofs are supplied as a number of components. The transfer of loads from the metal roof sheet covering to the main structural frame is complex and may not be considered in total at the design stage. Increasing the number of primary fixings used to attach metal sheets to the roof structure may not result in an improvement of resistance to wind uplift of the complete roof construction. For this reason no modification is recommended. More fasteners in the external sheet would not add to the wind resistance if load paths to the structural frame remain the same.

Flashings and trims

These are often the first to suffer wind damage, in part because the standard of workmanship during fixing may not have been well supervised. Providing more fasteners per metre run is the only solution in some cases. Aluminium fasteners can break as a result of thermal movement and increased and more frequent high temperatures will cause significantly more problems.

Fire precautions

The performance of various kinds of roof covering in fire have already been described in earlier chapters (see also Figure 3.126). In double-skin sheeted roofs, however, cavity barriers may be required depending on the position of the thermal insulation and on other criteria (Figure 6.19).

Cavity barriers may be formed from:
- 0.5 mm steel,
- 38 mm timber,
- 12.5 mm calcium silicate cement or gypsum-based board,
- polyethylene-sleeved mineral wool under compression.

Other materials that have been tested may also be suitable.

Roofs, particularly in industrial buildings, often will need to be perforated for the fitting of smoke vents.

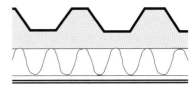

Cavity barriers not required. The insulation should make contact with both skins of the roofing

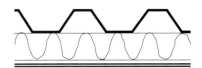

Cavity barriers necessary

Figure 6.19: Requirement for cavity barriers in double-skin sheeted roofs

Portal frames

Where any section of a portal frame forms part of or supports a wall which needs to be fire resisting, all of the frame must be fire resisting since it is impossible to distinguish between the 'column' and the 'rafter'. Therefore guidance is given (eg in AD B2 of the Building Regulations (England & Wales) 2000[2]) on an acceptable alternative to making the rafter sections fire resisting.

The sheet metal roofing widely used on this category of building needs to be considered from a fire precautions point of view. Profiled sheeting of galvanised steel or fibre-reinforced cement, or pre-painted steel or aluminium with a PVC or PVDF (polyvinylidene-fluoride) coating is expected to achieve AA designation in most cases.

Daylighting

Daylight factors achieved by various forms of roof light have been referred to earlier in this chapter. So far as structural and installation requirements are concerned, double-skinned roof lights are available in a range of profiles to match those of sheet materials. Siting needs careful consideration since spanning capabilities are less than for normal sheeting.

BRE has noted a number of cases where moisture, caused either by rain penetration or condensation, has been trapped under the plastics skirts of roof lights, thereby creating conditions for corrosion. Inspection fairly soon after work has been completed on the installation of roof lights should identify the problem before corrosion begins.

Sound insulation

A profiled sheet steel roof with minimal thermal insulation will give approximately 30 dB sound insulation.

If it is necessary to consider methods of reducing internally generated noise from passing through the roof, and if it is not possible to reduce the noise at source, consideration could be given to the construction of an inner lining. Filling large cavities with, say, mineral wool will give a greater improvement than will a close-fitting sandwich, though the condensation risk in the cavity should be assessed. Specialist advice will be necessary.

Readers should also see this section in Chapter 5.1.

Durability

The deterioration of coatings of profiled metal sheets has caused concern to BRE since the 1970s with a number of cases being investigated (Case study 14). Even disregarding the many cases where materials suitable only for internal use have found their way onto the outside of buildings, there still remains a number of examples of premature deterioration (Figure 6.20).

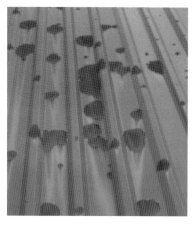

Figure 6.20: Random corrosion on profiled steel sheeting

A general problem (referred to the BRE Advisory Service on a number of occasions) is that of inconsistency in the appearance of surface coatings. There are two main considerations:
- variations in the shades of new coatings,
- surfaces weathering to quite different shades after a few years' exposure.

Some manufacturers also seem to have had problems in matching colours in sheets offered as replacements.

A survey was carried out in 1992–93 of the experiences of chartered surveyors with organic coated roof sheeting (reported in *Survey of performance of organic-coated metal roof sheeting*[5]). The survey was prompted by reports of deterioration occurring some four or five years after installation instead of the 15–25 years before first maintenance would be required, as claimed by manufacturers. Although there was no real indication of how representative the responses were in relation to the situation throughout the UK, this survey showed that there were grounds for some concern about the durability of coated sheets on the 2400 buildings covered in the responses. Four out of five buildings had PVC-coated sheets and 1 in 20 had PVDF-coated sheets. Two in five buildings had cut edge corrosion and about 1 in 7 random corrosion elsewhere on the sheets. Of those roofs treated by painting, about 1 in 5 were still unsatisfactory after treatment.

Further information on defects is available in the report *Durability of cladding*[6]. This study showed that delamination predominated as the cause of failure in a proprietary plastics coating; 66% of all cases reported this cause over an average life of eight years. Colour changes were also reported in 20% of the cases after about five years. With other types of PVC, cut edge corrosion occurred in 50% of the cases examined and delamination occurred in 25% of cases. With PVDF-coated sheets, cut edge corrosion occurred in between 50% and 66% of cases.

The British Board of Agrément published a Method of Assessment and Testing for metal sheet roofing[7], including steel and aluminium, that sets out the criteria against which products are assessed for certificates. The properties assessed include:
- loading characteristics,
- thicknesses,
- alloy,
- temper,
- fire characteristics and performance,
- durability,
- maintenance requirements,
- cleaning requirements.

The durability of composite sheets depends on whether there is any risk of delamination of the thermal insulation from the metal sheets or internally within the foam itself. The risk of delamination is undoubtedly higher with the darker colourings on the outer surface since higher temperatures are reached. Specifiers should enquire of the manufacturer whether there is such a risk and what guarantees are given.

There is considerable knowledge about the durability of profiled asbestos cement sheets. Asbestos cement gradually becomes brittle over the years making maintenance difficult. The roofs of some buildings at BRE cracked at the ridges about 25 years after installation; they have been successfully patched, temporarily, though access without breaking the existing sheets further was not easy (Figure 6.21).

Some of the fibre-reinforced sheets that first became available in the early 1980s have been known

Figure 6.21: Asbestos cement sheeted roofs temporarily repaired by patching with white mineral-surfaced torched-on felt rather than replacing whole sheets

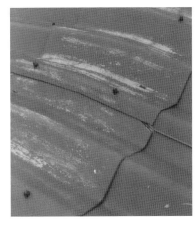

Figure 6.22: The surface coating on this fibre-reinforced sheet has been lost after just over five years of exposure. This usually happens on ridges and eaves rolls

Case study 15: Industrial roof in fibre-cement profiled sheets

The BRE Advisory Service had been asked to inspect a roof over some industrial premises that had suffered water ingress after a period of four years from installation. The specification for the sheeting included metal lining panels with insulation between the linings and the profiled top sheet. The sheets were laid to a 10° pitch over a slope of some 30 m.

Longitudinal cracks were evident in the full length of some sheets. Where the cracks had occurred in the troughs of the sheet, water had run down the liner until meeting the purlin and leaking into the building. It was noted that the cracks could have been produced during manufacture and not noticed or reported at installation.

The cracks did not always occur at or in the line of the fixings. The longitudinal edges of the sheets were observed to be bowed. Sealant, applied between the laps of the sheets, had been squeezed out and was visible in the tail laps. The sealant had not been consistently placed to the weatherside of the fixing bolts.

The conclusion of the BRE investigator was that, although it was unlikely that there had been overtightening of the fixing bolts, there was considerable lateral restraint to the sheets. With the known propensity for warping and bowing in some fibre-cement sheets, he decided that restraint in the lateral direction of the defective sheets was the most likely cause of the cracking.

to deteriorate rapidly, but there are good examples too. The more commonly used materials include:
• cellulose-reinforced calcium silicate,
• PVAL (polyvinyl alcohol) reinforced Portland cement,
• glass-fibre-reinforced Portland cement.

Among the defects seen to occur by BRE investigators (not necessarily on all the above materials) are:
• irregular crazing,
• splitting (eg from fixing holes),
• curling,
• pitting,
• efflorescence,
• loss of surface coating (Figure 6.22),
• warping leading to cracking (Case study 15).

The European Union of Agrément has issued a Method of Assessment and Test for corrugated and flat fibre-reinforced sheets[8]. Not all the mechanisms of failure are understood as yet, though, as with manmade slates, manufacturers are well aware of the problems that have occurred and hopefully their incidence should be reduced in future. This area is inadequately served by standards.

Specifiers should enquire from potential suppliers the extent of their warranties since most manufacturers cover only the cost of materials should replacement be necessary. As with manmade slates, specifiers may also wish to see installed examples of particular products before committing themselves.

Corrosion

The long-term performance of metallic roofing components will be dictated to a large extent by their resistance to corrosion (Case study 16). It is common practice for the protection to the underside of composite metal sandwich sheets to be of much lower specification than that to the upper surface. If there is any serious risk of condensation forming on the underside of profiled sheets, the underside will need suitable extra protection. The manufacturer will need to be consulted to identify a compatible coating. Where the complete roof is built up in situ from separate linings, thermal insulation and outer sheetings, care will need to be taken that no thermal bridges are formed within the construction (Figure 6.23). Also, the risk of condensation forming on the underside of the outer sheet depends on the integrity of the inner lining acting as a vapour control layer.

Bimetallic corrosion of metals should be guarded against, both in fixings and between fixings and claddings if the latter are of metal. Contact between two metals does not necessarily cause corrosion, but the wrong combination of metals under particular conditions (including the presence of moisture) will accelerate the corrosion of the

less noble. This can apply even to two different grades of stainless steel.

Atmospheric corrosivity

Most metallic cladding systems, as well as some non-metallic systems, are supported on a metallic frame. These frames are likely to be of plain carbon steel, aluminium alloy or stainless steel with the material being selected as much for its strength or its light weight as for its corrosion resistance.

The degree of corrosion resistance necessary for the roof-supporting structure will vary with

The roof of an industrial building was covered in profiled sheet steel. The surface protection had begun to break down after a life of about four years and was inspected by BRE investigators after six. The sheeting on the roof was covered with corrosion spots varying in size from a pinhead to palm size. The location of the corrosion spots was widespread: the edges of overhangs, the tops of overlaps, the middle of ridges, and around bolt fixings. Deformed areas of sheeting showed signs of corrosion staining. Sections of the roof sheeting were removed for examination in the laboratory.

The manufacturer alleged that the sheeting was galvanised to BS 2989[9] (since withdrawn) with a minimum of 10 μm alkyd coating on the underside, and the upper surface coated with a pre-treatment of zinc phosphate, a chromate sealing rinse, and a top coat comprising a thermosetting primer and a 25 μm acrylic silicone.

The samples were examined under a microscope and some tiny blisters were found indicating under-film corrosion. Also visible was cracking of the top coating formed when the sheet was processed. A pinhole detector was used to locate discontinuities in the coating, and numerous defects were found and recorded. The coating thickness was measured by mounting the sample in a cold-setting resin, and grinding and polishing to produce a finish suitable for examination under the microscope. The organic coating was weighed and the thickness of the galvanising was determined in accordance with BS 2989[9].

Neither the organic coating nor the galvanising was found to conform to specification and, consequently, the sheet was quite inadequate to meet its performance requirements over a reasonable life.

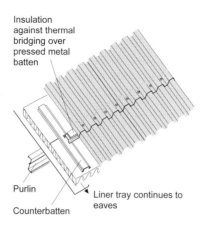

Insulation against thermal bridging over pressed metal batten

Purlin

Counterbatten

Liner tray continues to eaves

Figure 6.23: A liner tray under profiled metal roofing with separate thermal insulation

the conditions expected to occur behind the external surface.

Protection of steel
Plain carbon steel frames need some protection, the main options being hot dip galvanising or organic coatings or both. The required thickness of a zinc coating will be determined by the required life of the cladding. The size, chemical composition and method of manufacture of the component members of the frame will determine whether that thickness will be possible to achieve. The steelwork in most existing buildings will have been protected to BS 5493[10]. A new standard that partially replaces this is now available, BS EN ISO 12944[11]. Classification C1 of the new standard applies to interior steelwork.

Readers should also see this section in Chapter 7.1 for further reference to this standard as it affects steelwork exposed to the elements.

Protection of aluminium
Aluminium, normally used in construction in the form of an alloy, is relatively durable. An oxide skin forms when exposed to the atmosphere, providing a natural protective layer. To meet the wide range of service conditions it is important to select an appropriate alloy, but copper-bearing aluminium alloys should be rigorously avoided.

Mill-finished aluminium may be used without further protection provided the surface is regularly

washed to remove atmospheric deposits. However, depending on service conditions, aluminium can be protected in other ways (eg plastics coatings or anodising). In any case, the frame and the roof covering must be metallically and chemically compatible or be isolated from each other.

Thermal insulation
The existence of thermal insulation immediately below a roof covering can also complicate the situation since moisture can collect in some insulation materials and be retained, as it were, within a poultice in contact with the frame. While it may seem that the major risk is from rain penetration, the possibility of condensation increasing the risk of corrosion, both to the structure as well as to the internal face of the metal roofing, should always be considered; indeed, in some circumstances, condensation has been shown to constitute a significant risk (Case study 17). Ventilation of a cavity is the best way of reducing the risk though, of course, there should be no inadvertent reduction of thermal insulation value.

When detailing the installation of transparent or translucent corrugated roof sheeting with similar profiled sheets of other materials, it should be remembered that any sheet lapping over an under-sheet of a dark colour can cause the upper sheet to soften in high temperatures. PVC is particularly affected and the worst cases occur where the sheet has been specified with a low melting point in order to form a relatively cheap source of venting in case of fire.

Moves towards greater use of double-skin construction for reasons of thermal insulation, where the temperature of the outer skin will be further raised, makes matters worse.

Fixings
The risk of corrosion to fixings is high in roofs of the kind described in this chapter, so specifying stainless steel fixings may be appropriate. There is some evidence, however, of loss of fixity in conditions of condensation. From the numerous

The problem showed as severe dampness in the roof which was sometimes so bad that drips fell from the insulated suspended ceiling. The roof comprised a profiled metal deck carried on timber purlins. At the site inspection, although actual dampness was not present, there were signs of staining. Investigation suggested that penetration of the outer surface was not the cause and that it was more likely to be condensation. No attempt had been made to cross-ventilate what was essentially a roof of cold deck construction. In particular, the design of the roof prevented ventilation at the eaves. The problem would have been exacerbated by the phenomenon known as cold night radiation where the temperature of the deck falls below that of the outside air temperature.

The recommendation was to form a warm deck roof, using the existing outer covering of the roof as an effective vapour control layer on which the new thermal insulation could be placed.

types of proprietary fixings available for roof sheetings, whether they are rivets or screws, self-drilling and self-tapping, or just self-tapping, or whether they need to be fixed into pre-drilled holes, choosing the right fixing for the job is vital. In particular, sandwich sheets are probably fixed best with multi-diameter threaded shank devices so that outer and inner skins are not inadvertently forced apart by the threads of single-diameter devices during the fixing procedure.

Where bolted connections have been used in the structure, the bolts are usually one of two kinds:
- 'black' bolts which provide a loose fit in their holes,
- close-tolerance bolts which are machined to tight clearances.

High-strength friction grip bolts (sometimes termed 'pre-loaded') are normally used in standard clearance holes; they depend, though, on great care being taken in surface preparation and protection, as well

as precision in tightening, to achieve their design strength. They should not be interfered with at any time after installation without the express permission of a structural engineer. The surveyor or architect inspecting a structure should be aware, however, that the hardened steel washers normally used with these bolts can bite into the softer steels of structural members at the time of installation; this is not necessarily the result of deformations in service. Close-tolerance bolts are rarely used these days although some older structures have them (see Case study 15).

Ease of maintenance

The lives of most metal sheets can be prolonged with proper maintenance. Manufacturers nowadays are prepared to give indications of periods to first maintenance in particular environments. Typical periods to first maintenance in the case of coated metal roofs might range from 10 to 20 years, but actual performance as opposed to theoretical performance will vary considerably according to factors such as:
- nature of surface coating,
- sheet profile,
- colour of the surface and, hence, temperatures reached,
- pollution of the environment,
- pitch and orientation of roof, and, hence, the deposition of pollutants affecting the coating.

Although fibre-reinforced sheets can, with care, be pressure-cleaned of dirt and algae, BRE experience with the durability of surface-sealing coats after cleaning is not favourable, and some asbestos cement roofs treated at BRE have reverted to their original drab appearance after five years. However, at least one firm offers a 10-year warranty for its reconditioning services which may be a reflection of the care with which the work needs to be done.

WORK ON SITE

Access, safety, etc.
Readers should see this section in Chapter 2.1.

Storage and handling of materials
Mishandling packs of sheets has been known to cause cracking. The problem occurs where sheets being craned onto the roof in packs are not supported underneath (Figure 6.24). The catenary shape adopted by the sling will tend to bend the lower sheets; and the greater the number of sheets lifted together, the greater the risk of cracking.

Restrictions due to weather conditions
One of the most important considerations for the handling of large sheets on site is the strength of the wind. Large sheets should not be handled in wind speeds (3-second gust speeds) greater than 12 m/s.

Arc welding, particularly of structural sections, should not take place on site when rain or snow is falling since the welding electrodes need to be kept dry. Wind can affect the shielding needed by gas welders.

Workmanship
Where coated steel sheets are specified, it will be necessary to pay particular attention to:
- preventing overtightening of fixing bolts which could cause deformation of the sheets (see Figure 6.18),
- eliminating sheets where the coating has been scratched or abraded during handling and fixing.

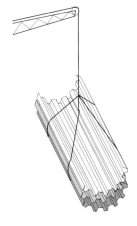

Figure 6.24: The sling tends to break the sheets in the pack if a spreader bar is not used

Care is also needed if roof lights are to be assembled on site since badly fitted lights can lead to water ingress. For this reason, factory-assembled roof lights may offer a lower risk of failure.

So far as the basic roof frame is concerned, if any welding is to be done on site the question of quality of the welds will need to be established. Unless there is an obvious failure, it will be unusual to suspect a failure in a weld in the steel structure of an existing roof since welds are subjected to close scrutiny on fabrication. If a failure is suspected during a routine inspection, testing may be possible (eg for surface cracks, by magnetic particle inspection or penetration by fluorescent dye). Less appropriate to existing structures are radiographic and ultrasonic methods of inspection since these methods may not be easily available to surveying and architectural practices. In any case, the results of testing by such techniques should be interpreted by engineers.

Coatings of precoated sheets can be removed before repainting, but the work may be expensive and requires a great deal of care. Various sections of BS 7079[12] may be helpful in this respect.

Supervision of critical features

It is most important with profiled steel sheeting that any cut ends are coated at the time that the roof covering is assembled. It is also considered to be good practice to further coat the underside of any exposed eaves since the protection on the underside is normally less effective than that on the upper surfaces of sheets.

Accurate measurement of the positions of purlin centres for fixing of sheets is essential.

When sheets are replaced it will be necessary to drill out all the old rivet fixings of trim so that distortion can be avoided in the sheets which remain.

Box 6.1: Inspection of sheeted portals, north lights, monitors and saw-tooths

The particular problems to look for are:

- Corrosion of structural members
- Distortion of structural members, especially buckling of cold-formed sections
- Distortion of sheets adjacent to any which have been replaced
- Corrosion of sheets, both externally and internally
- Delamination of insulated sheets
- Breakdown of seals between sheets
- Loosening of sheet fixings due to wind effect
- Corrosion of fixings

6.2 PATENT GLAZING

Patent glazing is dealt with here, in the context of medium-span industrial buildings, since, more often than not, it is associated with roof lighting for buildings of this type.

The term *patent glazing* covers sloping or vertical surfaces formed from specially shaped bars of metal carrying sheets of glazing approximately 600 mm wide and set at a pitch that is normally greater than 15° (BS 5516[13]).

Glazing is of sheet glass or plastics sheets of various kinds:
- wired annealed glass to BS 952-1[14])
- toughened glass,
- laminated glass,
- polycarbonate sheeting,
- PVC sheeting,
- PMMA (polymethyl methacrylate) sheeting.

The glazing is usually framed in steel or aluminium bars which are protected in various ways, eg:

- lead sheathing (for steel),
- PVC sheathing (for steel),
- powder coating (for aluminium) (Figure 6.25),
- mill finish (for aluminium),
- composite metal and plastics.

Patent glazing is normally planar in shape though it is possible to provide segmentally curved shapes.

CHARACTERISTIC DETAILS

Basic structure
The patent glazing bars, whether of traditional or inverted design (Figure 6.26), are always supported off the surrounding structure directly, and not via separate sub-frames.

Glazing may be supported either from the two longitudinal sides of the sheet or by all four edges using intermediate cross-transoms carried from the longitudinal bars. Glazing which is supported on two edges may be subdivided in its length by non-loadbearing cames which allow

a weatherproof seal to be formed across the slope.

Eaves
A short section of sloping roof will need to be inserted under the glazing to master the wall thickness. This arrangement though, carries the risk of it becoming a gathering place for detritus with consequent unsightliness. Stops to prevent the glazing sliding down the slope are normally bolted to the ends of the bars.

Abutments
Flashings are required to cover the junctions in the same way as those needed for other kinds of abutments (see also this section in Chapter 2.3 and Figure 6.27).

An abutment is usually provided in the form of a gap between the patent glazing and the adjoining structure, flashed with a metal flashing dressed down over the special capping on the end bar and

Figure 6.25: Cast wired glass in coated aluminium patent glazing. Note: the flashings have not been dressed down correctly

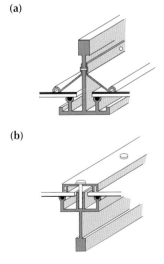

(a)

(b)

Figure 6.26: Traditional (a) and inverted (b) patent glazing bar shapes

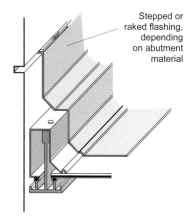

Stepped or raked flashing, depending on abutment material

Figure 6.27: Abutment of patent glazing against a wall

the glazing or over the edge of the glazing in the case of top abutments. Flashings at the head of the slope should overlap the patent glazing by at least 75 mm and incorporate an anti-capillary gap formed under the flashing. For shallow slopes of less than 20° the flashing should be sealed to the glazing.

If the patent glazing is used as an inset roof light at approximately the same slope as an adjoining non-glazed roof covering, consideration should be given to installing a kerb or downstand over which the flashing can be taken. Otherwise, fixing the flashing so that it is effective may be difficult. In lower abutments the patent glazing must adequately overlap the adjoining material.

MAIN PERFORMANCE REQUIREMENTS AND DEFECTS

Strength and stability
Patent glazing must take loads just as any other kind of roofing. Among the points to which particular attention needs to be paid are designing for 3-second gust speeds for the design of infilling and using BS 6399-3[15] for estimating snow loads, especially for shallow pitches where snow could adhere to the glazing.

Nomograms are given in BS 5516[13] for minimum thicknesses of glass and plastics sheeting used as infilling.

An assessment of safety aspects should be carried out. There is an obvious risk of injury to persons

underneath glazing if it breaks and a discussion of the factors involved appears in Chapter 5.3. There are basically two kinds of scenario that can be accepted for patent glazing, either:
• no break whatsoever, or
• break safely.

No break is probably limited to polycarbonate sheet, though even this can fall out of the bars if the edge cover is insufficient or the catastrophe severe. Wired annealed glass has been acceptable in the past though it may not be particularly safe, especially if the wires are severed in the break. Unwired annealed glass will not break safely. Adhesive plastics sheet applied over annealed glass can offer some reduction of risk. Toughened and laminated glass both have reasonably safe breakage characteristics.

Low-level patent glazing can be vulnerable to forced entry, even if fixings are accessible only from the inside, and the risk of this happening should be assessed. Risks were discussed in general terms in Chapter 5.1.

Dimensional stability
Although patent glazing, from its basic design, can accept a small amount of movement in surrounding construction by reason of the side clearances in the glazing, any movement joints in the main structure must be reflected in the patent glazing.

Movement joints in the glazing can be formed by setting two bars adjacent to each other with a cap flashing over.

The movements to be allowed for in plastics glazing include:
• for polycarbonate and PVC sheeting, 3 mm in every metre,
• for PMMA sheeting, 5 mm in every metre.

In all cases, a minimum edge cover of 15 mm is required, though this may need to be increased where the bearing is on two sides only.

It is possible to lap two edge-supported glazing sheets to cut down the total length of material required and to provide a movement

accommodation joint (as, indeed, is almost invariably done in glazing horticultural greenhouses). In the case of patent glazing, it is usual to provide a blade seal of neoprene, EDPM or other suitable plastics material to provide a weathertight seal. (It should be noted that lapping of glazing provides a potential lodging place for detritus and algae which can be unsightly, and it is almost impossible to clean without dismantling the assembly.)

Exclusion and disposal of rain and snow
The original intention of patent glazing was that it did not involve putty, and the pre-formed strip or cord underneath the glass was not the primary weather seal but merely the inner air seal on a two-stage drained system of jointing. Over the years, this simple principle has to some extent been disregarded and now the weathertightness of some so-called patent glazing seems to depend on adhesion of the seal provided between bar and glazing. A wide variety of seals has been used such as pre-formed tapes, cords and strips in both solid and cellular form, and gun-applied mastics. British Standards are available for gun-grade materials of certain polymer types though not for tape sealants (see BS 6213[16], BS 6093[17], BS 8000-16[18]).

There is, however, a second line of defence with patent glazing bars and that is in the shape of the channel which is formed in the section at each side of the bar underneath the seating for the glazing. The route to the outside must be kept open for any rain penetrating the outer cover and running down the channel in the bar.

If the glazing is to be replaced at any time, care must be taken to ensure that the deflection characteristics of the bars are appropriate to the kind of glazing to be specified or damage could result.

Snow guards should be fitted above patent glazing (see BS EN 12056-3[19]).

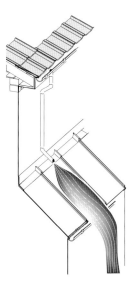

Figure 6.28: Rainwater discharging from a rainwater pipe over patent glazing will leave deposits on the glass and overshoot the gutter

Rainwater pipes should not discharge over patent glazing (Figure 6.28). There are two reasons for this:
- the whole of the water load will be concentrated on a narrow section of gutter at the foot of the patent glazing with consequential risk of overshooting in all but the lightest of rainfall,
- the glazing will quickly become streaked with dirt.

Energy conservation and ventilation

Much patent glazing comprises single sheets of glass giving minimal U values. However, it is possible to use double glazing of which the simplest form is twin-walled plastics sheeting. Unless the bars are thermally broken, the advantages of double sheeting generally will be negated.

Condensation will form on the underside of single glazing in many circumstances. At a slope of around 30° or shallower, condensate will tend to drip off the glazing sheets depending on their surface characteristics and the length of the slope. The risk of condensation forming should be assessed and provision made at the foot of the glazed slope to drain it to the outside.

Control of solar heat

If solar-control films are used, the resulting temperature gains of the underlying materials will need to be monitored. The thermal safety of glass will be reduced by the application of film. The BRE Advisory Service has seen cracking of south-facing patent glazing to which solar-control film has been applied after installation of the glazing.

Fire precautions

Unwired glass at least 4 mm thick in patent glazing roof lights can be regarded as likely to achieve an AA rating[2] for surface spread of flame, and polycarbonate and PVC roof lights which achieve a Class 1 rating by test may also be regarded as having an AA designation.

Fire resistance is another matter. Provided the edge cover is sufficient, wired glass can normally be used where fire resistance is necessary; for example, where the roof in which roof lights are incorporated is used as a floor.

Durability

The durability of the glazing bars depends on the degree of protection afforded to them. (See Chapter 3.3 where the durability of metals has been discussed in general terms.)

Glass used as infilling is normally durable, though laminated glass may deflect under its own weight to form a concave shape (see Chapter

5.1). This is reversible, though it will occur in the opposite direction when inverted unless further preventive measures are taken. The wires in wired glass at cut edges have been known to corrode with unsightly consequences. (See also Chapter 5.3 where the durability of plastics sheets has been discussed in general terms.)

Where lead flashings are used above roof glazing of any kind, it is necessary to consider the effect of run-off onto the glass and the consequent formation of lead carbonate and lead sulfate which is nearly impossible to remove. The lead flashings should be treated with a patination oil to encourage an insoluble patina to form. Construction below the patent glazing may also suffer discoloration (Figure 6.29).

Ease of maintenance

Access for maintenance may need to be by mobile platforms unless the installation has been designed to provide support for crawl boards (Figure 6.30); there are many patent glazing systems that have been designed to accept crawl boards in accordance with BS 5516[13].

When glazing is being replaced, care should be taken to ensure that glass edges are cleanly cut and not nibbled. Edge irregularities are a source of subsequent fracture when the glass expands from solar heating.

Figure 6.29: Water flowing in streams from the patent glazing bars on this roof has deposited dirt on the slates below in unsightly streaks

(a)

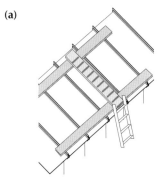

(b)

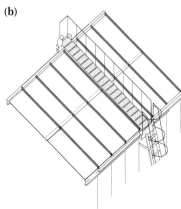

Figure 6.30: Short-span, shallow-pitched patent glazing should be designed to take crawl boards for maintenance traffic (a) but this provision must always be checked. Longer spans should have permanent handrailed walkways, preferably designed to be movable so that all parts of the glazed area can be reached (b)

WORK ON SITE

> ### *Box 6.2:* **Inspection of patent glazing**
>
> The particular problems to look for are:
>
> - Corrosion of bars, particularly at bearings
> - Lifting of cames to provide weather seals between adjacent panes
> - Lifting of flashings
> - Anti-capillary gaps not present under flashings
> - Flashings not sealed at slopes less than 20°
> - Glazing type is unsafe for the use of the building
> - Channels at bottoms of slopes not free to drain
> - Snow guards not in good condition
> - Wires in glass corroding at the edges
> - New lead flashings not treated with patination oil

6.3 TEMPORARY AND SHORT-LIFE ROOFS

Galvanised steel sheet to BS 3083[20], laid to the requirements of CP 143-10[21], is quite often used for the coverings of buildings without further protection. Also to be found are roofs covered with corrugated fibre-reinforced saturated bitumen sheets.

This chapter includes roofing for agricultural and horticultural buildings since different standards normally operate for these types of building. With certain exceptions for buildings within defined distances from habitable accommodation, the building regulations for England and Wales, Scotland and Northern Ireland[2–4] do not apply.

Also included in this chapter are brief references to air-supported or inflated structures used as temporary sports halls and similar structures, but tents and marquees are outside the scope of the chapter. Large-span fabric-covered framed roofs are dealt with in Chapter 7.1.

CHARACTERISTIC DETAILS

Main roof areas

Rigid sheet coverings

Corrugated (sinusoidal) steel sheets are normally available in widths of around 600 mm and upwards, and are self-supporting over purlin spacings of 1–2 m depending on gauge and profile. If pitch lengths are short (of the order of 3 m or so) and laps can be eliminated, pitches of 10° will be satisfactory. Pitch needs to be increased to 15° if laps are needed with the laps being about 150 mm. Laps can be sealed with a non-hardening mastic in exposed areas. Fabrication to large radius curvature has been possible for many years as evidenced by the many (ancient) Dutch barns and Nissen huts to be seen.

The galvanising of corrugated sheets is by dipping after manufacture, so cut end corrosion ought to be eliminated. Fixing is achieved by means of hook bolts placed through the ridges to steel purlins or galvanised drive screws to timber purlins. Plastics washers should be used to provide a weatherproof seal.

Fabric coverings

There are some short-span, fabric-covered roofs which are supported on light tubular steel frameworks. The example shown in Figure 6.31 has performed well for the 25 years since erection.

Eaves

Rigid sheet coverings

Galvanised steel eaves normally project 50 mm to gutters but are rarely found with underlays at this point. They do, however, need shaped fillers made, for example, from expanded plastics to prevent rain being blown underneath. The shallower the pitch, the more likely rain is to be blown up the underside of the sheet.

Verges and ridges

Rigid sheet coverings

Verges can either be left uncapped, with a cover mould fitting closely under the corrugations at the wall head, or be capped with a pressed steel angle flashing covering the last three corrugations and lying flush with the wall surface in a simulated bargeboard profile. Many ridges of roofs of galvanised steel have been constructed with ridge cappings that do not follow the profile of the corrugations. These ridges will leak as a result of driving rain being blown up the slope unless fitted with shaped filler pieces.

Figure 6.31: Twin-walled, PVC-coated, nylon-sheeted roof on a light steel framework. After 12 years, deterioration was confined to a small split which was easily repaired

Roof lighting

Rigid sheet coverings

Plastics sheet coverings, profiled to match the corrugations of the rest of the roof, are the usual form of roof light in a temporary building.

Early examples of glass-fibre-reinforced polyester (GRP) sheets have not given satisfactory performance, becoming weathered and obscured in a relatively short space of time and well inside the lifespan of the surrounding sheeting.

GRP sheets vary widely in thickness and strength. Poor quality types can weather extensively leading to exposure of fibres, dirt retention and loss of strength (Figure 6.32). Provision of better polyester resins or of surface protection in the form of adherent fluorinated polymer film can give substantially improved durability. Essentially, durability can be enhanced in manufacture by the incorporation of a fine tissue near the top surface which holds a resin-rich surface coating. Dimensional stability is generally good and the material is less prone to embrittlement than PVC. However, it could suffer loss of strength if allowed to deteriorate extensively.

The appropriate British Standard for corrugated plastics translucent sheets matching the profiles of the opaque parts of roof coverings is BS 4154[22].

MAIN PERFORMANCE REQUIREMENTS AND DEFECTS

Strength and stability

Rigid sheet coverings

Although the normal structural design codes govern most aspects of the design of these types of roof, there are special provisions for certain categories of building. For example, BS 5502-22[23] makes special provision for wind loading on canopy roof structures (see also BRE Digest 284[24]). There is also simplified provision for snow loading on glasshouses: 0.3 kN/m^2 measured over the whole area plus 0.9 kN/m^2 measured over the lower third of each valley slope or the lower 3 m of each slope, whichever is the smaller. According to the material forming the structure, there is also special provision for the sizing and spacing of structural members, especially purlins. Surveyors concerned with these structures will need to familiarise themselves with the many relevant parts of BS 5502[23].

Fabric coverings

The design and construction of inflated structures is a specialised field. Among factors to be considered are the likely numbers of people occupying the structure for each particular category of use, whether domestic, industrial or public assembly.

Figure 6.33: An air-supported structure in the 1970s. BRE has carried out investigations and experiments on a variety of these

Materials for inflated structures may be either coated fabrics or films in various combinations (Figure 6.33). The reinforcement may be, for example, glass-fibre, nylon, polyester or polypropylene; and the surface coating of PVC, chloroprene rubber, silicone or fluoropolymer compositions.

Exclusion and disposal of rain and snow

Rigid sheet coverings

BS 5502-22[23] states that the structure of a building used for agriculture should be weatherproof and constructed so that it will not transmit moisture due to rain or snow to any part of the structure or its contents which would be adversely affected by the moisture. So far as the roof is concerned, therefore, the requirement differs little from other buildings. The main consideration is whether the contents of the buildings would be adversely affected.

Fabric coverings

BRE tests on deflections in air-supported structures (a) in still air under snow load and (b) in moderate wind without snow load, have shown there was no danger of premature collapse due to instability. Some practitioners have indicated that there is no need to provide for snow loads where the interior of the structure is heated to at least 12 °C.

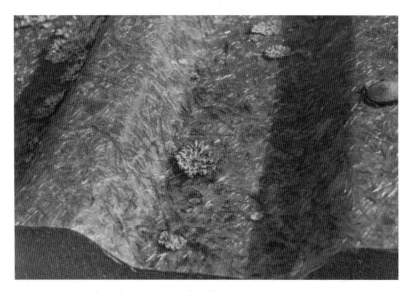

Figure 6.32: Weathered, corrugated, glass-fibre-reinforced polyester sheet after about 20 years exposure. Light transmission has been severely curtailed

Energy conservation and ventilation

Special buildings may be needed to provide special conditions for the storage and handling of agricultural and horticultural crops. The design of roofs for these structures falls outside the scope of this book, but the various parts of BS 5502[23] provide for the design of many such kinds of buildings.

Rigid sheet coverings

Condensation is often found on the undersides of single-skinned metal roofs, depending on circumstances, but particularly so in the case of animal housing. The condensation is more likely than not to drip from pitches of 10–15° rather than running down to the eaves[25].

Fabric coverings

There can be no thermal insulation in single-skin air-supported structures. Fresh air supply and the simultaneous removal of stale air, carrying with it excess water vapour, should provide adequate control of condensation but drips may occur from the relatively flat areas near the apex of a roof.

Fire precautions

Rigid sheet coverings

BS 5502-23[23] covers buildings exempt under building regulations, and which are used for agriculture and horticulture.

Fabric coverings

Adequate means of escape need to be provided for inflated structures in case of fire and the materials from which they are made should not readily support combustion. Although most products give off toxic fumes when burnt, reassurance may need to be sought on the toxicity of fire products from certain kinds of coated fabrics (eg PTFE).

Sound insulation

BS 5502-32[23] covers agricultural buildings.

Durability and ease of maintenance

Rigid sheet coverings

Ferrous metals, except stainless steels, are normally in themselves insufficiently durable for use externally and require additional corrosion protection. This additional protection can either be a metallic coating (eg zinc) or an organic coating (eg PVC) or a combination of both (a duplex coating). The life of ferrous metals used in temporary buildings is directly related to their protective coatings.

The most common metallic coating is zinc, although other hot dip coatings (eg of aluminium and aluminium–zinc alloys) are available. The life of a zinc coating is dependent on the environment to which it is exposed and to the thickness of the zinc. The thickness of the zinc coating required to give protection depends, in turn, on many factors, but, in particular, cladding sheeting is sometimes formed from pre-galvanised post-formed material; the methods of post-forming the sheet (after galvanising) generally restrict the total zinc coating weight, in this case to around 275 g/m^2 on both sides. With thicker coatings, the zinc may crack and spall on bending. This thickness is unlikely to provide adequate protection on its own and further protection (eg in the form of a coating of plastisol or PVDF) will be required.

One type of ferrous metal which can be used in certain circumstances without additional protection is a weathering steel. These steels have a low rate of corrosion and can weather to an attractive colour. However, there is a major drawback as the run-off from the material is rust-coloured and will cause staining

Figure 6.34: Bedmond Chapel, a corrugated steel building that has been well maintained since it was erected in 1889

to adjacent materials; consequently, they have not been used much in roof coverings.

BS 5502-22[23] defines various categories of lifespan for agricultural buildings: 10 years or less, ranging up to 50 years. Given a required life for a building, the life of the roof can also be decided, as can the lives of the component parts of the roof such as the cover to be provided to reinforcement in a concrete roof beam.

Part 21[23] of this same standard specifies precautions to be taken to prevent the corrosion of metals, particularly those in contact with cement or concrete. Galvanised steel sheet for use in normal conditions of exposure should be protected by a minimum of 350 g/m² zinc coating on both sides or an aluminium alloy coating of 135 g/m². In a marine environment, a minimum zinc coating of 275 g/m² each side with factory-applied organic coatings to BS 5427-1[26] is required, or 350 g/m² of zinc plus primer and two coats of paint. These should give lives of at least 10 years depending on the internal conditions of the building. For lives in excess of 10 years, a heavier coating of zinc will be needed (eg 610 g/m²) which can only be applied to a pre-formed sheet. As stated in *Principles of modern building*[25], the durability of corrugated steel sheets depends on their effective and continuing protection (Figure 6.34). Site cutting

of sheets is a particular point to watch since cut edges rarely receive the additional protection recommended by manufacturers.

If sheets are unpainted, they can be expected to begin to deteriorate from about the 20th year in relatively unpolluted areas (Figure 6.35) and rather earlier in more polluted areas. Regular painting with a paint system compatible with the previous coating will maintain their condition. Once rusting has begun, priming with one of the rust-inhibiting primers is effective, but whether this is economic depends on the areas to be treated. It is much better to take the decision on preventive maintenance before rusting begins. BS EN ISO 12944-5[10] describes suitable protective paint systems.

There are special requirements for materials subject to highly corrosive internal atmospheres, such as stock housing.

GRP (in, for instance, roof lights) that has deteriorated has been mentioned earlier in this chapter. There is no way in which the translucency of old sheets can be restored by site treatment and they would need to be replaced.

Fabric coverings

With air-supported structures, the coated fabrics used in their manufacture need to be carefully monitored to check for deterioration. In particular, long-

term performance relates very much to the loading history of the structure. Some fabrics may stretch if the structure is kept inflated over long periods of time. In one test carried out by BRE, a PVC-coated, flame-proof, light stabilised fabric had stretched about 7% in the weft direction after continuous inflation for three-and-a-half years. A nominated design life for each structure and strict monitoring of performance during use should be specified.

Experience in the USA with air-supported roofs made to inferior specifications suggests that when these structures fail (by, say, puncturing of the fabric or defective pumps), they have been difficult to repair. In a number of cases complete replacement has been necessary.

The durability of other tensioned fabrics, such as polyester coated with flexible PVC, is probably around 10 years, though glass-fibre or polyester fabrics coated with silicone or fluoropolymers are expected to last much longer.

WORK ON SITE

Access, safety, etc.

Most roofs covered in a rigid material will require the use of crawl boards, ladders or other forms of mechanical access equipment, as well as safety harness, when carrying out inspections. Inflated structures can be deflated for inspection.

Readers should also see this section in Chapter 3.1.

Figure 6.35: The roof of a farm building in 24-gauge corrugated galvanised steel erected around 1958 in a relatively unpolluted area. No corrosion was apparent for 30 years. Rusting is now confined to the exterior and to the centre of each of the corrugated steel sheets where the galvanising appears to have been thinner than elsewhere. There is some detritus from overhanging trees

---

Box 6.3: Inspection of temporary and short-life roofs

The particular problems to look for are:

- Roof pitches less than 15° if sheets are lapped
- Deterioration of plastics washers in roof fixings
- Missing ridge fillers
- Inadequate light transmission
- Condensation on single-skin roofs
- Corrosion at cut edges
- Fabric-covered structures nearing the end of their design lives

6.4 REFERENCES

[1] **West JM & Majumdar AJ.** Durability of non-asbestos fibre-reinforced cement. Information Paper IP 1/91. Bracknell, IHS BRE Press, 1991

[2] **Communities and Local Government (CLG).** The Building Regulations 2000.
Approved Document:
    L: Conservation of fuel and power, 2006:
        1A: New dwellings, 2A: New buildings other than dwellings, 2B: Existing buildings other than dwellings
London, The Stationery Office. Available from www.planningportal.gov.uk and www.thenbs.com/buildingregs

[3] **Scottish Building Standards Agency (SBSA).** Technical standards for compliance with the Building (Scotland) Regulations 2009.
Technical Handbooks, Domestic and Non-domestic:
    Section 6: Energy
Edinburgh, SBSA. Available from www.sbsa.gov.uk

[4] **Northern Ireland Office.** Building Regulations (Northern Ireland) 2000.
Technical Booklets:
    F: Conservation of fuel and power.
        F1: Dwellings, F2: Buildings other than dwellings. 1998
London, The Stationery Office. Available from www.tsoshop.co.uk

[5] **Cox RN, Kempster JA & Bassi R.** Survey of performance of organic-coated metal roof sheeting. BR 259. Bracknell, IHS BRE Press, 1993

[6] **Ryan PA, Wolstenholme RP & Howell DM.** Durability of cladding. A state of the art report. London, Thomas Telford, 1994

[7] **British Board of Agrément (BBA).** Precoated metal sheet roofing and cladding. Method of Assessment and Testing No 34. Garston, BBA, 1986

[8] **European Union of Agrément (UEAtc).** Technical guide for the assessment of the durability of thin fibre reinforced cement products (without asbestos) for external use. Method of Assessment and Test (MOAT) No 48. Paris, UEAtc, 1991

[9] **British Standards Institution (BSI).** BS 2989: 1992 Specification for continuously hot-dip zinc coated and iron-zinc alloy coated steel flat products: tolerances on dimensions and shape. London, BSI, 2004. *NB: This Standard has been withdrawn.*

[10] **British Standards Institution (BSI).** BS 5493: 1977 Code of practice for protective coating of iron and steel structures against corrosion. London, BSI, 1977

[11] **British Standards Institution (BSI).** BS EN ISO 12944 Paints and varnishes. Corrosion protection of steel structures by protective paint systems. Various Parts. London, BSI
    Part 5: 2007 Protective paint systems

[12] **British Standards Institution (BSI).** BS 7079: 2009 General introduction to standards for preparation of steel substrates before application of paints and related products. London, BSI, 2009

[13] **British Standards Institution (BSI).** BS 5516: 2004 Patent glazing and sloping glazing for buildings. London, BSI, 2004
    Part 1: Code of practice for design and installation of sloping and vertical patent glazing
    Part 2: Code of practice for sloping glazing

[14] **British Standards Institution (BSI).** BS 952-1: 1995 Glass for glazing. Classification. London, BSI, 1995

[15] **British Standards Institution (BSI).** BS 6399-3: 1988 Loading for buildings. Code of practice for imposed roof loads. London, BSI, 1988

[16] **British Standards Institution (BSI).** BS 6213: 2000 Selection of construction sealants. Guide. London, BSI, 2000

[17] **British Standards Institution (BSI).** BS 6093: 2006 Design of joints and jointing in building construction. Guide. London, BSI, 2006

[18] **British Standards Institution (BSI).** BS 8000-16: 1997 Workmanship on building sites. Code of practice for sealing joints in buildings using sealants. London, BSI, 1997

[19] **British Standards Institution (BSI).** BS EN 12056-3:2000 Gravity drainage systems inside buildings. Roof drainage, layout and calculation. London, BSI, 2000

[20] **British Standards Institution (BSI).** BS 3083: 1988 Specification for hot-dip zinc coated and hot-dip aluminium/zinc coated corrugated steel sheets for general purposes. London, BSI, 1988

[21] **British Standards Institution (BSI).** CP 143-10:1973 Code of practice for sheet roof and wall coverings. Code of practice for sheet roof and wall coverings. Galvanized corrugated steel. Metric units. London, BSI, 1973

[22] **British Standards Institution (BSI).** BS 4154: 1985 Corrugated plastics translucent sheets made from thermo-setting polyester resin (glass fibre reinforced). London, BSI, 1985

[23] **British Standards Institution (BSI).** BS 5502: Buildings and structures for agriculture. Various Parts including:
    Part 21: 1990 Code of practice for selection and use of construction materials
    Part 22: 2003 Code of practice for design, construction and loading
    Part 23: 2004 Fire precautions. Code of practice
    Part 32: 1990 Guide to noise attenuation
London, BSI, Various dates

[24] **BRE.** Wind loads on canopy roofs. Digest 284. Bracknell, IHS BRE Press, 1986

[25] **Building Research Station.**
Principles of modern building, Volume 2:
Floors and roofs. London, The Stationery
Office, 1961

[26] **British Standards Institution (BSI).**
BS 5427-1: 1996 Code of practice for
the use of profiled sheet for roof and wall
cladding on buildings. Design. London,
BSI, 1996

7 LONG-SPAN ROOFS

For the purposes of this book, long-span roofs are taken to be predominantly those of clear spans over 30 m. The variety of designs is legion, as are the materials from which they are made. However, they do tend to have one characteristic in common: most are found over single-storey buildings or grandstands where people congregate in large numbers. For this reason alone they demand special consideration, particularly with regard to their structural safety.

Figure 7.1 A long-span roof, Stansted Airport

7.1 ALL KINDS OF LONG-SPAN ROOFS

This chapter will deal only with the increased performance requirements and construction details that stem from longer spans together with, for example, possible greater deflections and more onerous rainwater disposal problems.

In roofs with spans of around 50 m, the conventional basic structure consists, more often than not, of lattice construction spanning effectively in two dimensions. Although it is possible to construct long-span roofs of concrete (eg by prestressing; Figure 7.2), the self weight of the structural material is considerable. In practice, these large roofs are nearly always of metal: either steel or occasionally aluminium (Figure 7.3).

A recently constructed shell roof at the Imperial War Museum, Duxford, forms a notable exception. This long-span roof is formed of two shells of precast concrete units tied together. Short vertical ribs on the lower panels provide a fixing for the upper panels. The roof forms a half ellipse on plan (Figure 7.4).

Laminated beams of wood (so-called glulam) have been used in some of the very longest spans but are relatively few (Figure 7.5). Typical sections can be 500 × 2000 mm spanning over 30 m in flat or arched forms. These units are mostly imported into the UK since it is uneconomic to maintain domestic production lines for relatively few roofs. Warren, Belfast and Bowstring timber trusses have also been used in long-span roofs.

Cold-formed sections in steel are widely used in construction (eg in space frame structures with welded or bolted joints). Protection from corrosion in such structures needs to be given particular attention (see Chapter 7.2).

Many long-span roofs are of unique design (eg Figure 7.6); unsurprisingly, they present obvious difficulties to those responsible for their maintenance and repair in the absence of particular maintenance instructions prepared by the original design team. Since the cost of replacement of these roofs is usually very large indeed, they

(a)

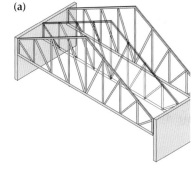

(b)

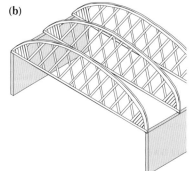

(c)

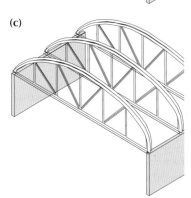

Figure 7.2: A long-span shell roof in reinforced concrete constructed in the 1950s

Figure 7.3: (a) Warren, (b) Belfast, (c) Bowstring girders and trusses

Figure 7.4: Shell roof at the Imperial War Museum, Duxford

BS EN ISO 12944[1] and BS EN ISO 14713[2].

Since movements of the roof surface of tensioned structures are often considerable, it is usual for the deck to consist either of tensioned flexible sheeting or for comparatively small stiff panels to be clipped to the cables so that movements can take place relative to each other. The latter solution demands flexible jointing between the panels.

Flexible sheeting is usually of woven material, weather-proofed with one of a variety of plastics. Those that have been used in the UK include:

- PTFE woven fibre,
- PTFE coating on woven glass-fibre,
- PVC-coated polyester.

The distances that fabrics can span between supports (whether of cable, rod or truss) vary considerably according to such criteria as material, thickness, method and evenness of tensioning, and direction of warp and weft of woven fabrics; no simple rules can be given. Generally speaking, the glass-fibre-reinforced fabrics will span the greater distances (see Figure 7.6).

Roof lighting

Most large roofs will require at least part of the area to be glazed (Figure 7.9). In the case of tensioned fabrics, the light transmission ranges widely (from approximately 10–15% for coated glass-fibres to some 60%

Figure 7.5: A long-span glulam purlin roof under construction

are normally subjected to close inspection on a regular basis and, indeed, may even be instrumented for regular measurement of structural behaviour. Very little guidance of a general nature can be given for the inspection of these roofs (Figure 7.7).

CHARACTERISTIC DETAILS

Basic structure

Structures that are nominally flat are often cambered in manufacture or assembly, irrespective of the material of construction, so that snow or access loads do not cause deflection below the horizontal.

Tensioned structures, in which suspended cables form the main structural members in tension, can be used where the spans to be covered are very large. This form of construction often lacks rigidity and, consequently, the deck and covering

fluctuate continuously when under load. The towers supporting the cables need frequent inspection and maintenance. Some roof designs, however, employ a combination of techniques that assist each other in providing a balanced structure (Figure 7.8).

An alternative to cable support is the use of stiff tension members which follows standard structural steel detailing more closely. This type of structure is frequently to be found in sports stadia and grandstands. Since the steel is outside the main weatherproof surface of the roof, special protective measures will be needed. There is more guidance in

Figure 7.6: A plastics fabric roof

Figure 7.7: A novel means of spanning an iconic structure

for woven PTFE) depending on how clean they are and whether or not internal liners are needed to improve thermal insulation or fire-performance characteristics.

Readers should also see this section in Chapters 2.6, 5.3 and 6.1 relating to non-fabric roofs.

Access for cleaning and maintenance

Most large-span roofs will have built-in provision for access for cleaning roof lights, either by means of travelling cradles or catwalks. With flat roofs designed for access there should be few problems provided indentation loads from access ladders or platforms are avoided.

MAIN PERFORMANCE REQUIREMENTS AND DEFECTS

Strength and stability

The majority of collapses of roofs coming to the attention of BRE have been with comparatively long spans (BRE Digest 282[3]). The bigger a building is, the more complex it is to determine its response to the loads imposed on it. In long-span roofs, the components that make up the roof can be more susceptible to permanent deflection due to dead load and more flexible in their response to live load than are the shorter spans. General guidance on procedures and reporting is

given in the Institution of Structural Engineers' report *Appraisal of existing structures*[4]. All structural appraisals of long-span roofs should be carried out by a competent engineer, though some guidance may be

given to surveyors to determine the circumstances when engineering appraisals become necessary (see Box 7.1 at the end of this chapter).

Long-span timber roofs may represent the greatest unknown to a surveyor who is perhaps more familiar with steel and concrete. Glulam roofs can present particular problems, being prone to creep on a significant scale. There have been a number of failures in plywood beams, many showing signs of creep. Most of the creep takes place in the first year following installation but long-term displacements of timber long-span roofs can be substantial. The glues normally used in glulam roofs will be satisfactory for at least 50 years where the timber is protected from the weather, and upwards of 25 years where not protected. BRE Digest 340[5] and TRADA Wood Information Sheet 1-6[6] give further guidance.

There is some reassuring evidence from Canada of the long-term durability of glulam construction obtained from tests on 30-year-old

Figure 7.8: An external skeleton suspended roof at the National Exhibition Centre, Birmingham

Figure 7.9: An example of a unique design of roof lighting at Waterloo International Railway Station

structures. Glulam beams have been shown to possess a factor of safety of 1.53 after a 30-year life, representing some three-quarters of the original strength when new and still perfectly adequate for their purpose.

Movement of the structure of a glulam roof may be considerable, and movement joints need to be inspected frequently to ensure that detritus does not build up in them. Care also must be taken not to hang loads from the structure (eg lighting equipment) that were not foreseen by the designer and taken account of in the initial design.

Dimensional stability

In most long-span roofs, where movements resulting from solar gains and wind loads may be considerable, the deck does not usually contribute materially to the stiffness of the structure, though of course it must be anchored securely to it.

The most important factors that cause instability or deformation in a structure are:

- thermal movements,
- reversible and irreversible movements due to drying shrinkage, moisture movement or creep in concrete or timber members,
- elastic deflection due to the self weight of the roof and applied loading,
- ageing which may lead to brittleness in certain types of material.

Exclusion and disposal of rain and snow

Rainwater disposal from all long-span structures (including, especially, tensioned structures) will have been given due consideration by the designer (BS EN 12056-3[7]).

Deflections are often considerable in roofs with spans of around 50 m or more and the need to provide for deflections must be accommodated in the falls to be provided for drainage purposes.

Siphonic-assisted storm water drainage

The use of siphonic-assisted storm water drainage was in its infancy when the first edition of this book was prepared. Subsequently, there has been increasing interest in the use of the technique and it has been used on both refurbished and new buildings. For the larger building, it offers a number of potential advantages over hitherto universally specified gravity drainage systems, including:

- suction-induced full-bore pipe flows and hence a reduction in the number of downpipes, which is particularly useful when duct space is limited (There may also be a concomitant reduction in below-ground drainage runs.)
- removal of the necessity for sloping horizontal discharge pipes, as a consequence of which this system can be more easily accommodated within floor voids.

Design guidance is contained in BS 8490[8].

The outlets each consist of:
- a conventional grille to exclude leaves,
- a large-diameter shallow bowl with anti-vortex plate surrounded by baffles, which have the effect of restricting the volume of air carried into the system, and
- a small-diameter outlet to carry the discharge to waste.

When rain falls, the bowl is filled, the air is excluded and a suction is induced, causing the outlet to flow at full bore.

On any roof, snow may drift and lodge against abutments or it may accumulate in long valleys. It is important that gutters and valleys should be covered with snow boards so that water from rain or melted snow is not prevented by accumulated snow from reaching the downpipes.

Energy conservation and ventilation

It is unlikely that all long-span roofs will need to have thermal insulation to the standards called for by building regulations. More often than not, they are used over stadia and sports facilities which may be, in any case, partially open to the outside air. Where fabric roofs cover atria, with a U value approximating to single glazing, special consideration needs to be given to heating, and, especially, avoidance or disposal of condensation forming on the underside.

Where, however, the building is fully enclosed and is heated, the practical limitations on ventilation of long-span roofs via the eaves will almost certainly require the warm deck concept to be used.

Control of solar heat gain

Cables will stretch considerably under the action of solar heating.

Light-coloured or white surfaces will promote durability since they reflect more of the sun's energy than conventional roofs.

Fire precautions

The performance of various kinds of covering in fire has already been given in Chapters 4.1 and 4.2. In very large buildings such as factories, means of escape often present problems to the designer, and a 'fire engineered solution' (involving, for example, sprinklers) will need to be carried out on the roof structure.

In the case of glulam construction where the roof also functions as a floor, the structural members must be checked for fire-resisting properties; the charring rate to be assumed for such material is the same as for solid timber for most of the glues in common use. Protective treatments are available, usually applied under pressure. The tendency of the wood to embrittle has been overcome. So far as fire is concerned, it is the metal fastenings that are the Achilles' heel of these beams since the metal conducts heat to the interior of the wood. The relevant standard is BS 5268-4.1[9].

Fabric roofs need special consideration. If any fabric is sufficiently near to the seat of a fire, there is a strong possibility that it will melt rather than burn, though this does depend on the materials. Where it does melt, the products of combustion will automatically be vented to the outside air to the benefit of any occupants, but, of course, there is the attendant risk to occupants from the molten material falling on them from above.

Sound insulation

Fabric roofs are virtually transparent to sound. Some long-span roofs have decks of lightweight fabricated metal because of weight demands. These roofs can be given improved sound reduction (over 40 dB(A)) by using cavity construction and sound-absorbent materials.

Durability and ease of maintenance

Steel

Structural steelwork exposed to the elements should be protected to criteria selected from BS EN ISO 12944[1]. Conditions in the interior of buildings have already been referred to in Chapters 3.1 and 6.1, but the Standard covers five further categories depending on the severity of the environment to which the steelwork is exposed. The full list of categories includes those given in Tables 7.1 and 7.2.

It is expected that most external steelwork will be in the High durability category, although in polluted areas the time-to-first-maintenance will probably be less than 15 years because of the difficulty in depositing enough zinc on the steel (about 1500 g/m² is the practical maximum). The zinc coating may be sealed or unsealed. Other coatings for shorter lives may include metal spraying and various paint systems such as zinc-rich, drying oil, silicone alkyd chemical resistant, and bitumens of various kinds; certain systems may also be used in combination to enhance durability.

All parts of the BS EN ISO 12944[1] are relevant. They are listed in the references.

Cable saddles and intermediate suspension points are likely to be the parts of the structure most vulnerable to deterioration and will need to be inspected even though access may not be straightforward (Figure 7.10).

Timber

Surveyors should be alert to the possibility that some species of imported timber used in long-span roofs show inactive wood-boring insect holes at the time of delivery. Inactive infestations should not reduce the utility or durability of the wood, though appearance may suffer.

Fabrics

The durability of PTFE-coated glass-fibre is expected to be at least 20 years, and will probably be 30 years. It is unaffected by ultraviolet light. PVC-coated fabrics will probably be less durable (perhaps up to 15 years).

Fabric coverings, generally, are vulnerable to mechanical damage and care needs to be taken to limit vandalism, particularly from the use of missiles. Patching on site is less satisfactory than that undertaken in the controlled conditions of a factory. Also, fabrics may stretch in service with glass-fibre-reinforced

Table 7.2: Durability requirements for structural steelwork[1]

Category	Description	Lifespan (years)
L	Low	2–5
M	Medium	5–15
H	High	> 15

Table 7.1: Level of environmental protection required for structural steelwork[1]

Category	Description	Type of environment
C1	Very low	Interiors
C2	Low	Rural
C3	Medium	Urban or industrial with low sulfur dioxide pollution or coastal with low salinity
C4	High	Industrial areas and coastal with moderate salinity
C5-I	Very high	Aggressive industrial atmospheres
C5-M	Marine	Coastal and offshore with high salinity

Figure 7.10: Access to the masts in this cable-stayed roof is for the most part from solid decking. However, part of the suspension system is over curved roof glazing where access for inspection is not so straightforward

fabrics. Bird soiling may be present as a result of roosting on the cables and struts forming the rigging, and will need hosing off periodically.

In the 2004 edition of Approved Document B of the Building Regulations (England & Wales) 2000[10], guidance is given on structures covered with flexible membranes and PTFE-based materials. Any flexible membrane covering a structure (other than an air-supported structure) should comply with the recommendations given in Appendix A of BS 7157[11]. Guidance on the use of PTFE-based materials is given in the BRE Report, *Fire safety of PTFE-based materials used in buildings*[12].

WORK ON SITE

Access, safety, etc.

With long-span roofs it is more likely than not that mobile working platforms will need to be used for access unless permanent access has been provided (Figure 7.11). HSE offers advice on the safe use of mobile working platforms[13].

Readers should also see this section in Chapters 3.1 and 4.3.

Box 7.1: Inspection of all kinds of long-span roofs

The particular problems to look for are:

• Rainwater penetration

• Wear on cable saddles

• Defective corrosion protection on tensioning devices

• Corrosion of metal members

• Wood rot and insect attack in timber members

• Inadequate light transmission

• Cleaning equipment in poor condition

• Movement in the structure

For the assessment of long-span roofs, the surveyor should look for:

• Signs of movement in the roof or its supports and if any of these movements are progressive

• Bowing of members and if any components carrying compressive loads show signs of bowing

• Completeness of connections and any signs of distress

• Deterioration of components, eg:

  • signs of corrosion in metals

  • signs of rot or insect attack in laminated timber beams

  • signs of rainwater penetration or condensation

• Signs of overloading of the roof (eg by repairs).

• Alterations to the original roof design and whether this was with the knowledge and agreement of a competent engineer

• Signs of accidental damage

fabric performing better than woven plastics.

Tensioned fabrics are largely self-cleaning under the action of rain, although rainwater flowing over the surface of the material will tend to concentrate at seams in the fabric and lead to uneven deposition of dirt and detritus. Surface deposits could well build up in sheltered areas. The best performance in this respect is given by PTFE-coated

Figure 7.11: Permanent provision for access on a large roof in London

7.2 SKELETAL STRUCTURES

This book does not deal with large-span structures built in Victorian times: for example, the magnificent spans of the major railway stations (see Figure 1.8). There is too much variety for the brief descriptions contained in this book, and every structure will be more or less unique.

CHARACTERISTIC DETAILS

Basic structure

The majority of large-span structures dating from the early years of the 20th century will be of trusses and purlins in lattice steel, conventionally bolted together to provide a rigid structure. Two of the largest such buildings are shown in Figure 7.12(a). A similar example built in more recent years is shown in Figure 7.13.

Although some examples of this type of roof in the past have been constructed in situ at high level, it is entirely practicable to carry out most of the construction at or near ground level with obvious benefits of improved safety and ease of access. A prime example of this is Terminal 5 at Heathrow Airport (Figure 7.14).

Another efficient method of spanning large clear spaces is the space frame. The most widely used form of space frame in the UK employs lightweight steel sections with the units being welded up in the factory before transport to site where the final connections are made with bolts. These steel units are normally between 1 and 2 m square in plan and in a similar range of depths. There have been a few examples in aluminium, though these seem largely to have been confined to exhibition buildings. Alternatively, the space frame can be built up on site from individual

Figure 7.12: (a) Airship hangars at Cardington, Bedfordshire, one of which was operated by BRE as a major fire research facility. (b) The interior of a hangar

Figure 7.13: A large steel truss roof at the Imperial War Museum, Duxford

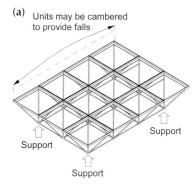

(a) Units may be cambered to provide falls

Support
Support
Support

(b)

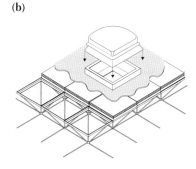

Figure 7.15: (a) A typical space frame construction. (b) Roof lights can normally be inserted into any bay of a frame

structural hollow section tubes or rectangular sections with welded threaded connectors, the members being screwed into standard nodes.

With either technique, the deck may be nominally flat, cambered to achieve roof drainage (Figure 7.15a), or arched, with the arch tending to be used over the longest spans. Support for the space frame is normally provided at positions inset from the perimeter of the building and less frequently at the perimeter. Comparatively large cantilevers are possible. Decks (eg of plywood) are usually fixed down directly to the top chords of the units. They are normally finished with one of the many single- or multi-layer finishes available. Roof lights will also be needed.

Eaves and verges

Space frame roofs may have vertical fascias or mansard and cornice section shapes following the profile formed by the diagonal members of the units. Anti-sag rods may be encountered. Eaves are normally fitted with proprietary gutters bolted to the downstands of the units and flashed into the gutters. Verges usually consist of a proprietary section clamped to the upper surface of the deck.

Abutments

Older large-span buildings are, in the main, tall and freestanding so far as the roofs are concerned, with ancillary buildings adjoining the walls at lower levels. Space frames, however, may well abut other

buildings which are higher. Where the edge upstand of a space frame abuts another building, there will need to be a movement joint. Often the joint is in the form of a gutter. However, the gap will need to be flashed in the usual way.

Roof lighting

The space frame can take conventional kerbed roof lights at any module position (Figure 7.15b). Alternatively, some structures may be directly glazed using toughened flat glass set on double EPDM (ethylene-propylene-diene monomer) lattice gaskets and clamped to the top chords with stainless steel clamps and bolts. Intermediate joints are sealed with a silicone sealant.

Rainwater disposal

It is a truism that large-span roofs collect large amounts of rainwater, but to experience a heavy rainstorm underneath a roof light inside one of the large-span buildings already described can be like standing underneath a waterfall, such is the ferocity of the run-off. Gutters cannot always cope, no matter how ludicrously large they appear to be.

Figure 7.14: Terminal 5, Heathrow Airport, under construction. The roof structure was completed at ground level and jacked into position (Photograph by permission of BAA)

The most that can be done in the circumstances is to direct the run-off over surfaces where ground-level drainage can cope.

The designers of many large buildings make provision for rainwater disposal internally, particularly where external drainage is not feasible, but the risk of overflow should be carefully considered.

Regular maintenance to ensure that blockages do not occur is obviously necessary.

MAIN PERFORMANCE REQUIREMENTS AND DEFECTS

Strength and stability

With very long spans, especially where large cantilevers are involved as in the roofs of grandstands, particular care has to be taken by the designer to limit deflection under snow load. Older roofs in this category will also need to be monitored, especially after heavy snowfalls.

Space frames are relatively robust for the purpose of hanging service loads from the structure (Figure 7.16), though any proposal to increase such loads during alterations to the building should be checked out with an engineer.

Dimensional stability

Long-span roofs of the type already described may be prone to dynamic problems when wind generates resonance in the structure. Structures can be checked after completion without the need for personal access. BRE has done considerable work on assessing the performance of these roofs using a laser interferometer to check that they conform to the assumptions used in the design. This technique may also be useful where suspicions are aroused in relation to older structures.

Exclusion and disposal of rain and snow

The cause of any ponding on the roofs of long-span structures should be correctly diagnosed. BRE has known cases where pondings were thought to be local to the decks and were brought up to level; they proved to be cosmetic solutions since the root causes lay with deformations in the underlying structures.

Fire precautions

The performance of various kinds of covering in fire has already been described in Chapters 2.6, 4.1–4.4 and 5.2.

Durability and ease of maintenance

In some space frames, the nodes for assembling the structures may consist of special multi-faceted castings into which the sections forming the struts and ties are welded or bolted. These nodes, being well underneath the deck, are the most likely parts of the structure to be accessible and can be inspected for signs of movement or fracture of welds. Some aluminium nodes may consist of tubes crimped into serrated slots and these can be inspected for signs of fatigue cracking.

WORK ON SITE

Access, safety, etc.

New space frames are frequently assembled on the ground, sometimes complete with services, tensioned for correct camber, and hoisted into position by cranes positioned at intervals round the perimeter (eg Figure 7.17). Access to

Figure 7.16: Space-frame units supported from conventional structural steel lattice beams and columns

Figure 7.17: A section of a Nenk system space-frame assembly being hoisted into position

existing space frames will need to be treated as any other roof, bearing in mind the relatively thin sections of the units.

The usual provision for access is by moveable cradle or by fixed catwalks. Tightening up of health and safety legislation has led to the need to shroud fixed access ladders and to provide permanent fixing points for temporary ladders. Where a space frame is fully glazed, access to the upper surface for cleaning can only be by mobile platform with appropriate height and reach above the eaves[13].

It may be feasible to carry out significant parts of the assembly at ground level, and to jack the semi-complete structure into place (see Figure 7.14).

Readers should also see this section in Chapter 4.1.

Workmanship

Welding is frequently employed in the jointing of space frames. Points of particular concern in relation to any necessary welding on site include the experience of the welders and their awareness of the particular details called for by the designer, especially with regard to tolerances.

Box 7.2: Inspection of skeletal structures

The particular problems to look for are:

- Rainwater penetration
- Reduction in camber
- Corrosion of metal members
- Wood rot and insect attack in timber decks
- Inadequate light transmission
- Cleaning equipment in poor condition
- Movement in the structure, particularly after heavy snowfalls

7.3 REFERENCES

[1] **British Standards Institution (BSI).**
BS EN ISO 12944 Paints and varnishes.
Corrosion protection of steel structures by
protective paint systems. London, BSI
Part 1:1998 General introduction
Part 2: 1998 Classification of
environments
Part 3: 1998 Design considerations
Part 4: 1998 Types of surface and
surface preparation
Part 5: 2007 Protective paint systems
Part 6: 1998 Laboratory performance
test methods
Part 7: 1998 Execution and supervision
of paintwork
Part 8: 1998 Development of
specifications for new work and
maintenance

[2] **British Standards Institution (BSI).**
BS EN ISO 14713: 1999 Protection against
corrosion of iron and steel in structures.
Zinc and aluminium coatings. Guidelines.
London, BSI, 1999

[3] **BRE.** Structural appraisal of
buildings with long-span roofs. Digest 282.
Bracknell, IHS BRE Press, 1984

[4] **Institution of Structural Engineers
(ISE).** Appraisal of existing structures. 2nd
edition. London, ISE, 1996. (NB: 3rd
edition is due for publication in 2009)

[5] **BRE.** Choosing wood adhesives.
Digest 340. Bracknell, IHS BRE Press,
1989

[6] **TRADA Technology.** Glued
laminated timber. Wood Information
Sheet (WIS) 1-6. High Wycombe, TRADA,
2003

[7] **British Standards Institution (BSI).**
BS EN 12056-3: 2000 Gravity drainage
systems inside buildings. Roof drainage,
layout and calculation. London, BSI, 2000

[8] **British Standards Institution (BSI).**
BS 8490: 2007 Guide to siphonic roof
drainage systems. London, BSI, 2007

[9] **British Standards Institution (BSI).**
BS 5268-4.1: 1978 Structural use of
timber. Fire resistance of timber structures.
Recommendations for calculating fire
resistance of timber members. London,
BSI, 1978

[10] **Communities and Local
Government (CLG).** The Building
Regulations 2000.
Approved Document:
  B: Fire safety, 2006
London, The Stationery Office. Available
from www.planningportal.gov.uk and
www.thenbs.com/buildingregs

[11] **British Standards Institution
(BSI).** BS 7157: 1989 Method of test
for ignitability of fabrics used in the
construction of large tented structures.
Appendix A. London, BSI, 1989

[12] **Purser DA, Fardell PJ & Scott GE.**
Fire safety of PTFE-based materials used
in buildings. BR 274. Bracknell, IHS BRE
Press, 1994

[13] **Health & Safety Executive (HSE).**
The selection and management of mobile
elevating working platforms. Construction
Information Sheet 88. London, HSE,
2008. Available as a pdf from www.hse.
gov.uk/pubns

8 INDEX

OTHER TITLES IN THE BRE BUILDING ELEMENTS SERIES

Building services

BR 404

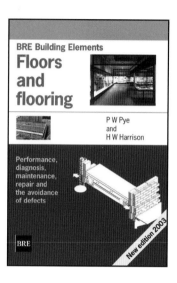

Floors and flooring

BR 460

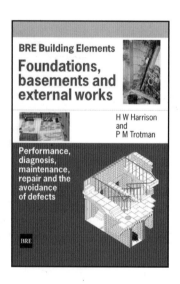

Foundations, basements and external works

BR 440

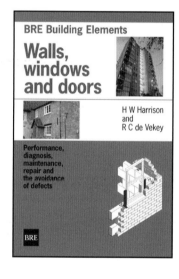

Walls, windows and doors

BR 352

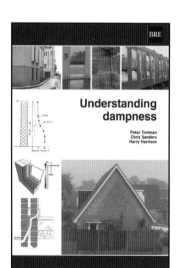

Understanding Dampness

BR 466

To order, please contact IHS BRE Press

Telephone – +44 (0) 1344 328038

Email – brepress@ihs.com

Web – www.brebookshop.com

WITHDRAWN

182209